THE
HISTORY
OF
THE
FUTURE

THE
HISTORY
OF
THE

OCULUS, FACEBOOK,

AND THE REVOLUTION

THAT SWEPT

VIRTUAL REALITY

FUTURE

BLAKE J. HARRIS

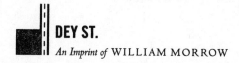

DEY ST.
An Imprint of WILLIAM MORROW

HarperCollins books may be purchased for educational, business, or sales promotional use. For information, please email the Special Markets Department at SP-sales@harpercollins.com.

FIRST EDITION

Designed by Renata De Oliveira

Library of Congress Cataloging-in-Publication Data has been applied for.

ISBN 978-0-06-245596-3

19 20 21 22 23   LSC   10 9 8 7 6 5 4 3 2 1

*For Katie, who
stood by my side
as a one-year
project became
three . . .*

CONTENTS

FOREWORD BY
ERNEST CLINE

I WAS BORN IN MARCH OF 1972. IT WAS A PERFECT TIME TO BE BORN A GEEK, because the geeks were about to inherit the earth.

I don't think I actually realized I was a geek until May of 1977, which was when the very first *Star Wars* film was released. I was five years old, and when I saw it for the first time I nearly lost my mind. I was suddenly unable to talk or think about anything other than *Star Wars*. It became my first obsession. *Star Wars* was the very first thing that I ever "geeked out" over. But it definitely wouldn't be the last.

The following year, I received an Atari 2600 for Christmas and yet another obsession was born. To me, those Atari cartridges in the old shoebox beside our TV were more than just games. They were *computer simulations*. They allowed me to simulate driving a tank or piloting a starship through an asteroid field. One of these games—*Adventure*—even allowed me to take control of an avatar inside a virtual world. I could navigate its labyrinths, pick up items, slay dragons, storm castles, and search for hidden treasure. Sure, my avatar was just a square, and the virtual world was rendered with blocky two-dimensional graphics on my TV screen. But to me, that computer-generated kingdom seemed like a real place. Playing the game was like being transported to another reality, without ever leaving my living room.

I didn't realize it then, but I'd just become part of the first generation of human beings to have the ability to play video games at home. And a few years later, when I received a TRS-80 Color Computer for my birthday, I became part of the first generation to have home computers, too. Around the same time, my family bought our first VCR, another newly affordable invention that would drastically alter the course of my life.

A few years later, in junior high, I discovered another early form of virtual reality when I began to play fantasy role-playing games like *Dungeons and Dragons* with my friends. These games allowed players to simulate other worlds using the most powerful computer in existence—the human brain. Using paper, pencils, a few rulebooks, and some polyhedral dice, you could create an entirely new reality, which existed only inside the collective imaginations of you and your fellow players. *D&D* allowed me to experience a crude (but still totally immersive) form of virtual reality long before there were computers capable of generating one.

I didn't get my first taste of "real" virtual reality until the early '90s, when I attended Gen Con, one of the world's largest gaming conventions. I waited in line for over an hour to play a new kind of game called *Dactyl Nightmare*, on a stand-up Virtuality 1000CS machine. The VR helmet was bulky, the controller was awkward, and the blocky polygon graphics did a poor job of simulating reality. But I still remember being blown away by that brief experience, because of the vast potential the technology showed. Even in this early primitive form, putting on a virtual reality headset felt like being transported into another world—a digital reality where you could interact with other real people, through their game avatar. It was amazing! Once the headsets got smaller and the computer graphics got better, virtual reality was going to change the world by creating a completely new one. It seemed inevitable.

In 1992, a film called *The Lawnmower Man* was released and it began to play on a seemingly endless loop on HBO and Cinemax. It had a ludicrous story line, but the movie's depiction of virtual reality's future potential stoked my anticipation for this evolving technology even further. So did Neal Stephenson's novel *Snow Crash*, which was published the same year. I was completely floored by Stephenson's stark vision of a sprawling virtual world called the Metaverse that millions of real people around the world were able to access with a pair of VR goggles. *Snow Crash* built upon the concept of cyberspace that William Gibson had introduced in his 1984 novel, *Neuromancer*, by extrapolating VR technology even further. With a programmer's eye for detail, Stephenson laid out how a persistent globally networked virtual reality like the Metaverse might function. He described its evolution and architecture, along with its culture, laws, and ecology. He also hinted at the potential social and economic side effects of living a dual life, split between the real world and a virtual one.

Reading *Snow Crash* left my mind reeling with the implications of our virtual future. It also convinced me that the immersive VR technology depicted in the novel was probably only a few years away from becoming a reality.

But it wasn't. Aside from a few false starts and failed prototypes, virtual reality would continue to languish in an unrealized limbo, just beyond our collective reach, while the final decade of the twentieth century gave rise to a different form of world-altering technology instead—the internet.

I got my first job in the computer industry in 1995, working as a technical support representative for CompuServe in Columbus, Ohio. I spent my days there helping people learn how to access and use the fledgling World Wide Web, and over the next five years, I watched as the internet rapidly evolved from something that most people had never even heard of into a service that nearly everyone was using, every single day, for just about everything. In the blink of an eye, the internet changed nearly every aspect of human civilization and how it functioned. It erased our borders and gave rise to a new digital country—one where the members of humanity all around the world were now connected to one another, every second of every day.

During this time, video games and computer graphics also continued to evolve and improve at an astonishing rate. In 1997, the very first massively multiplayer online role-playing game, *Ultima Online*, was launched. It was soon followed by other MMORPGs like *EverQuest* and *World of Warcraft.* These online games represented the very first persistent virtual reality simulations, which were populated by hundreds of thousands of real people who accessed them from different countries all over the globe. Players began to live alternate lives through their game avatars, even though they could only see the digital world they inhabited through the two-dimensional window of their computer monitor. For a lot of people, even this early crude form of virtual reality already seemed more compelling than the real world.

In several of the call centers and IT companies where I worked during this time, I recall seeing dozens of my coworkers bring a laptop with them to the office every day, so that they could play *World of Warcraft* from their cubicles while taking nonstop tech support calls. Even more astounding was when I learned that MMORPG players had started to sell swords, armor, and other virtual magic items on eBay for *real* money. These were objects that didn't actually exist—they were nothing but a collection of ones and zeroes on a game server somewhere. But these virtual items still had value in the real world, because real people spent real hours of their lives inside the game's virtual reality, and so everything that went on in there was still important to them. If a magic sword made it easier to defeat your virtual enemies or impress your virtual friends, then people were willing to pay real money for it. For the first time in history, people were suddenly able to earn a living in the real world by buying and selling goods in a virtual one.

I also remember how fascinating it was to watch as people began to develop real-world relationships inside these games. People would meet, become friends, and fall in love with each other—sometimes without ever setting foot on the same continent. For many players, the emotional bonds formed in a virtual world were just as strong as their friendships in reality—sometimes even more so, because interacting through avatars gave you total control over how you appeared to others.

I realized I was witnessing the emergence of a completely new kind of human relationship—one that had never existed before in our history.

But even as MMORPGs gave rise to the first virtual worlds, the early promise of virtual reality technology continued to remain unfulfilled and just out of reach. As we reached the dawn of the twenty-first century, VR seemed to have receded back into the realm of science fiction. Films like *Existenz* and *The Matrix* presented a fairly sinister vision of VR, placing the technology at the center of a Cronenberg body horror parable, or depicting it as a machine-manufactured prison for the human mind. As Morpheus put it: "The Matrix is the world that has been pulled over your eyes to blind you from the truth."

I loved those films, but that wasn't how I believed the future of virtual reality would actually play out, if and when the technology ever actually became available. When I surfed the internet, it already felt like I was entering another reality, one where all the world's pop culture was mashed up together. It seemed obvious to me that the internet would eventually evolve beyond its two-dimensional origins, into a sprawling virtual universe that was filled with planets instead of websites. Instead of a web page devoted to *The Lord of the Rings*, the virtual internet I was envisioning would contain a whole planet for Tolkien fanatics—a virtual re-creation of Middle Earth, where fans could experience the fictional world firsthand. Or they could jump into a teleporter and visit any of the planets featured in *Star Wars*, or *Star Trek*, or any other fictional world that had ever been concocted by the human imagination. It would be every geek's dream. The ultimate video game. A virtual utopia where you could go anywhere, do anything, and be anyone. I even had the perfect name for it—the OASIS.

When I began to think about what sort of person might create this virtual utopia, I immediately thought of Richard Garriott, the eccentric video game designer who created all the *Ultima* games, including *Ultima Online*. He was known for cosplaying as his video game avatar, *Lord British*, at press events and conventions. He was also famous for holding elaborate haunted house events at his mansion in Austin, Texas, which was rumored to be filled with hidden rooms and secret passages. His larger-than-life personality also reminded me of the character Willy Wonka, and when I made that connection, an idea suddenly occurred to me: What if Willy Wonka had been a video game designer instead of a candymaker? And what if he held his golden ticket contest inside his greatest video game—a sprawling virtual reality that had replaced the internet?

The moment that notion entered my head, I knew I was onto something. It seemed like a great premise for a screenplay or maybe even a novel. So I continued to flesh it out. I remembered that back when I was playing *Adventure* on my Atari 2600, I'd managed to discover a secret room inside the game, where its creator, Warren Robinett, had hidden his name. This was the very first video game Easter egg, and discovering it was one of the most thrilling childhood memories. I wondered, What if my Wonka-esque game designer hid his own Easter egg somewhere inside his virtual universe, and then after his death, he held a posthumous contest to find it? The first person to find the egg would win his fortune, along with ownership of his game company and control of his virtual kingdom.

That got me thinking about what sort of tests and challenges my eccentric game designer, James Halliday, would leave behind to find a worthy successor, and another idea occurred to me: all the riddles, puzzles, and clues leading to the hidden Easter egg could be linked to the dead billionaire's various pop-culture passions—his favorite books, movies, video games, cartoons, and TV shows from his youth. I loved this idea, because I knew that by making Halliday's passions mirror my own, it would allow me to pay tribute to all the things I loved, by weaving references to them into the story.

My mind was suddenly flooded with ideas for pop culture puzzles, and I began to fill notebook after notebook with them, creating a rough outline for what would eventually become my first novel, *Ready Player One.* It would take me nearly eight years to finish it, writing in my spare time in the evenings and on the weekends while continuing to work my day job doing tech support. Sometimes I would get frustrated and set the novel aside for a few months to write a screenplay. But I never gave up on *Ready Player One.* I was determined to finish the book, even though I wasn't sure if I would ever be able to get it published once I did.

I finally finished my first draft of *Ready Player One* near the tail end of 2009. My agent put the book up for auction a few months later, in April of 2010, and to my shock a bidding war erupted over the US publishing rights, with Random House emerging as the winner.

The very next day, another bidding war broke out in Hollywood over the film rights, which ended up selling to Warner Bros., with me attached to write the screenplay. In those forty hours, my whole life changed. But I wouldn't begin to get a sense of how profoundly it had changed until the following year, when the book finally came out.

When *Ready Player One* was published on August 11, 2011, it became an instant international bestseller. What I didn't know at the time was that my book would soon find its way into the hands of a brilliant young inventor named Palmer Luckey, who was already developing the virtual reality headset design that would finally bring affordable, usable VR to the masses.

I'm still amazed by the timing of it all. In 2009, at the same time I was finishing *Ready Player One*, Palmer was cobbling together his first prototype VR headset, which he called the PR1. He was only seventeen years old, and he built it in his parents' garage.

Looking back, I now realize that Palmer and I both had the same goal. I wrote a science fiction novel about virtual reality because I was fascinated by the concept and wanted to imagine its vast potential and limitless application. Where is this technology that I've been promised for decades? And what will it look like, if and when it ever actually becomes a reality?

The difference is that instead of just imagining how virtual reality might change the world, Palmer had already set about inventing the technology to make it happen. So naturally, when *Ready Player One* was published, a lot of people began to recommend the book to Palmer because of its subject matter. When he finally read it, he told me that he found the novel's depiction of the potential of VR so inspiring that he immediately began to recommend it to everyone who came to work at the new company he was forming, Oculus VR. (He founded the company in July of 2012, less than a month after *Ready Player One* was released in paperback.)

One of the people who joined Palmer's new company was the legendary John Carmack, one of the cocreators of *Doom*—a game that served as another giant leap forward in the quest for VR. Carmack was also one of the inspirations behind the character of James Halliday. (I modeled Halliday's relationship with his partner Ogden Morrow after Carmack's partnership with John Romero, as well as the earlier tech titan team of Jobs and Wozniak.) It was incredibly surreal to learn that one of the game developers who had helped inspire my novel had now decided to work on virtual reality, and that he was citing that same novel as one of the inspirations behind his decision. I'd managed to inspire one of the game developers whose work had inspired me. It was the sort of thing every science fiction author dreams will happen.

Early in 2013, Palmer contacted me via email, to let me know what a huge source of inspiration my novel had become for everyone working on his team at Oculus. He told me that one of their conference rooms was named "The OASIS" after the virtual universe in my book. He also told me that they gave a paperback copy of *Ready Player One* to every new Oculus employee when they're hired. I was still trying to wrap my head around that when Palmer invited me to visit their offices for a book signing, and to experience what they were working on firsthand. I couldn't say yes quickly enough.

To hedge my bets, I set *Ready Player One* in the year 2045—over three decades into the future. But when Palmer and Brendan Iribe de-moed their prototype of the Oculus Rift for me for the first time in their tiny offices in Irvine, it immediately became clear to me that virtual reality was coming far sooner than I'd predicted. In fact, it was already here. I'd just seen my own science fiction become science fact—literally right before my eyes.

The future has arrived, I remember thinking as Palmer showed me his taped-together prototype of an Oculus Touch controller. *And we're already living in it.*

What follows is the fascinating story of how we reached this unique turning point in human history—the moment when we began using our technology to reshape the very nature of our existence, by fabricating an entirely new reality for our senses and ourselves.

This is the moment in our history when we began to make our dreams come true.

AUTHOR'S NOTE

IN FEBRUARY 2016, OCULUS/FACEBOOK GRANTED ME WHAT WAS ESSENTIALLY unlimited access to their employees for a book about the founding of Oculus, the magic of VR and the human drama that comes with trying to build the future. Over the course of the next two-plus years, I conducted hundreds of interviews and spoke every few days with key personnel at Oculus as they launched their first consumer product, attempted to lead a technological revolution and acclimated to life after a multibillion-dollar acquisition.

Shortly thereafter, as some of the events described later in this book transpired, my access abruptly came to an end. Fortunately, by this point, I had already obtained all the information that I would need to finish this book.

What follows is a story based on hundreds of exclusive interviews with Oculus/Facebook employees, many more with other pioneers in the VR industry, and a collection of more than 25,000 documents— coming from dozens of different sources—that provided me with an unvarnished, truly inside look at the history of a company, its people and a technological quest unlike any that's come before.

To capture the excitement, uncertainty and oddities of that quest, I have written this book in such a way that I hope readers will feel like they are in the room with these characters as the story unfolds and as their objectives, perspectives and relationships evolve.

Although, stylistically, this book is not an "oral history," my years of experience with that format (as well as with documentary filmmaking) helped guide a lot of the narrative decisions I made. Specifically, and as often as possible, I've tried to step out of the way and let the story be presented through the thoughts, feelings and words of those who actually lived it. As such, in an effort to authentically recreate those experiences on the page, this book contains a lot of dialogue. And since dialogue is rarely remembered verbatim, I wanted to briefly explain my process for crafting the dialogue in this book, which can be broken down to three categories:

1. **FIRSTHAND:** In the majority of cases, the dialogue that appears throughout the book is a representation of what firsthand sources recalled being said during a conversation to the best of their abilities.

2. **SECONDHAND:** In cases where those present could not remember exactly what was said in a given moment, or in situations where those present declined to speak with me (or declined to share certain details of a conversation), I occasionally—though rarely—relied on secondhand sources; but I did so only in situations where I had a good-faith reason to believe that the source would have specific knowledge of the conversation in question.

3. **TRANSPOSED:** In several cases—particularly when it came to long, or nuanced, conversations—firsthand sources could remember the substance of what was discussed, but not necessarily a word-for-word recollection. In these instances, I relied on feedback from those involved to capture the essence and meaning of those conversations, if not the exact words.

Where discrepancies in recollections occurred among direct participants, I have noted those divergences in the text.

PEGGY OLSON: You never say thank you.

DON DRAPER: That's what the money is for.

—MAD MEN (SEASON 4, EPISODE 7)

PROLOGUE

ON THE AFTERNOON OF MARCH 25, 2014, FACEBOOK CEO MARK ZUCKERBERG made a surprise visit to a SoCal-based start-up, where from the front of a small workplace kitchen, he proceeded to share some big news, discuss Facebook's grand vision, and geek out over a long-sought-after wearable technology that he believed would dominate the future.

"People describe it as, like, a religious experience," he said, addressing an audience of about fifty engineers, entrepreneurs, and dreamers. "They go into this world"—by putting on a pair of supercharged goggles—"and when they take it off they're, like, sad to be back in reality."

The tech that had Zuckerberg raving was a virtual reality headset called the Rift, and the engineers, entrepreneurs, and dreamers in that kitchen were the ones responsible for its creation. They were members of a young, cultishly popular start-up called Oculus, whose mission was to deliver the sci-fi-inspired dream of VR. "To *finally* make virtual reality happen!" Oculus's then-twenty-one-year-old founder Palmer Luckey liked to say. For gaming! For education and communication! For anyone who ever wanted to throw on a headset and step into a computer-generated world where anything was possible. And though this Holy Grail goal still remained a ways away, the Oculus team suddenly felt closer to achieving this than they ever had before—because as of about an hour ago, they had just learned that

Facebook would be buying their little company for three goddamn billion dollars.[1,2]

"It's unbelievable!" Zuckerberg continued.

It was, it really was. Just two years earlier, Palmer Luckey had been living alone in a camper trailer. And now—with Oculus becoming the fastest start-up in history to reach a multibillion-dollar exit—he'd be able to buy a house not only for himself, but one nearby for his mother and father.

What made the acquisition even more unbelievable was how out of the blue the whole thing seemed—especially to most of the team. Zuckerberg, for example, had only even visited Oculus's office once— two months earlier—when he came by to check out a demo called "The Room" (and, to the delight of many Oculus employees, he showed up with a to-go bag of McDonald's; Wow, they thought. Tech moguls: they're just like us!). At the time, nobody outside a small inner circle of executives even knew that an acquisition might be in play. And if they had, they might have tried to pump the brakes. After all, the folks at Oculus weren't a very Facebooky crowd. Their CTO, for example, didn't even have a Facebook account; their CEO had a piece of art in his office that was just the Facebook logo on a pack of cigarettes; and employees routinely mocked Facebook for being "lame," "poorly de-signed," "privacy averse," or "just plain parasitic."

Beyond the ideological differences, Facebook seemed an unlikely match for an even bigger reason: because for all intents and purposes, Oculus was a video game company—created to give players a way to "step into the game"—so selling to a console maker like Sony or Mi-crosoft made sense. Even selling to a Silicon Valley titan like Apple or Google made some sense (after all, Apple and Google had a history of building successful hardware products). But Facebook?

Zuckerberg had an answer to that very question when speaking to employees of his newly purchased company. "I think this has the potential to be—and you guys think about this too—to not just be the next gaming platform, but the next real computing platform."

He described how he believed that every ten to fifteen years a new computing platform takes hold, reaches a critical mass, and largely usurps its predecessors. Most recently, he cited, this happened with smartphones. "By the end of 2012," Zuckerberg explained, "there were a billion active people using smartphones. And I guess around 2012 or 2013 it started to overtake computers where everyone still had a computer, but smartphones started to be the primary way that people really used computing. I think that if we push this [virtual reality] really hard it has the potential to be this for this technology as well."

The more Zuckerberg spoke, the more Facebook's vision seemed to complement Oculus's. Any concerns that this might dramatically change postacquisition were assuaged when Zuckerberg said, "We have a good history, I think, at Facebook of buying companies and having them run independently; so that's what we're going to do here. We're not going to mess with the culture in any way . . . We're here to accelerate what you're doing. It's just awesome and you guys should all be so proud of what you've built so far."

By the time Zuckerberg opened up the floor to questions, he had seemingly managed to squelch any lingering concerns among the Oculus employees.

Well, except for one. An important issue—one that touched on an unnerving, underlying aspect of Facebook's existence that would not only wind up significantly impacting Oculus, but in the years ahead, would soon be wondered aloud by millions who were increasingly concerned about how Facebook operated, why it operated that way, and what this all meant for the future of privacy, social interactivity and even liberal democracy.

"Hey, Mark," began Chris Dycus, Oculus's very first employee. "Some people—not me, of course—but *some* people think Facebook is evil . . . so I'm wondering how that will affect the perception of Oculus."

Suddenly, the room fell quiet with anxiety. Save for a few choked-back chuckles of amusement. One of which came from founder Palmer Luckey, who was overcome by a singular thought: Chris Dycus has balls of steel! Seriously, that took major *cojones*—to ask Marky Z the question that literally everyone in this room is thinking. And though everyone in the room *was* thinking about the public reaction to this acquisition in some abstract way, Luckey was already dealing with tangible repercussions: from dozens of tweets along the lines of "FUCK YOU, you fucking SELL OUT" and "Thanks for killing the VR dream again" to proposed customer boycotts and even a few death threats.

Zuckerberg smiled, relieving the tension that had been building. And as he proceeded to laugh off Dycus's question—not in a dismissive way, but in an oh-well sort of way that seemed to ignore the severity of the already-apocalyptic online backlash—Luckey couldn't help but wonder just a bit: what did we get ourselves into here with Facebook?

But rather than plot out the possible permutations of what tomorrow might bring, Luckey decided that today—on this celebratory day; the day he sold his company to Facebook for more money than he could ever imagine—he would do something that he rarely ever did: he would stop looking forward and start thinking backward. Back to the beginning of how this whole crazy, extraordinary journey had started in the first place . . .

PART 1

THE
REVOLUTION
VIRTUAL

CHAPTER 1

THE BOY WHO LIVED TO MOD

April 10, 2012

UNLIKE SO MANY SILICON VALLEY SUCCESS STORIES, THE TALE OF OCULUS doesn't begin in a garage, a dorm room, or a small skunkworks lab. Instead, in a twist befitting the humble origins and pragmatic eccentricities of its founder, the tale of Oculus begins in a trailer.

More specifically: a beat-up, second-hand nineteen-foot camper trailer that, on the afternoon of April 10, was parked in the driveway of a modest, multifamily home in Long Beach, California. The bottom floor of this home belonged to the Luckeys—Donald, a car salesman, and Julie, a homemaker—and the trailer belonged to their nineteen-year-old son: Palmer. He had been living in there for nearly two years now, and based on how things in his life had been going as of late, Palmer Luckey seemed destined to keep living there for years to come.

From the outside, Palmer Luckey's trailer looked absolutely ordinary. Tinted windows, fiberglass shell, and corrugated once-white siding that had faded to beige. But inside, from end to end, it had been modified to fit his own desires.

The first thing to go was the bathroom. It just simply took up too much space, he reasoned. As convenient as having a bathroom would be, there was already a perfectly fine bathroom that he could still use in his parents' home. And if those facilities were occupied—as they often were by Luckey's three younger sisters—he could go try the public restroom that was located next to the Laundromat three streets away. So, the trailer's bathroom? That could go. So too could the trailer's kitchen. He didn't need a specific area devoted to preparing meals when his entire diet more or less consisted of frozen burritos, Mucho Mango AriZona tea and whatever he could afford at the Jack in the Box down the road. In fact, Luckey biked down to that Jack in the Box with such regularity that the manager gave him a special loyalty card entitling him to 15 percent off his meals (so special that, as Luckey would brag to friends, "it *can* be combined with other offers!"). Needless to say, Luckey wouldn't be hosting any dinner parties in the near future.

When neighbors passed Luckey's trailer, they would invariably feel a twinge of sadness—a guttural pinch of how-could-this-be. What had become of that kid with all the promise? That bright, bubbly home-schooled boy who had started taking college courses when he was only fifteen; who, last they had heard, was enrolled at Cal State, Long Beach, studying journalism, was it? But here he was, on this sunny Tuesday April afternoon—same as every afternoon, same as every morning and evening, too—holed away, doing God knows what in that monument of wasted potential.

Now as desolate as this situation might appear to be, one key detail cannot be ignored: Palmer Luckey *loved* living in that trailer. It felt, to him, like living in a spaceship; a feeling that probably had as much do with his love of science fiction books as it did with his Tonystarkian personality; that of someone who doesn't see the world as is, but rather who—defiantly, obsessively—sees things as they could be. Which is why Luckey took so much pride in how, after gutting it, he'd transformed that trailer into a makeshift laboratory.

The transformation began with the wide oblong cavity up front, which had been modified to fit Luckey's six-screen computer setup, and extended all the way to the back where (instead of a bathroom) a twin mattress was propped up by a series of component-filled boxes. In between these two ends was where Luckey conducted his hardware experiments; from control boards and soldering irons to lens equipment and power supplies, this space was overrun with an unlikely alliance of gear, gadgets, and tools. Scattered about—serving as both evidence and inspiration for the chaos that occurred in this trailer—were a handful of funky, helmet-shaped prototypes for a product that was unlike anything else being manufactured in the world. All in all, it resembled Walter White's mobile lab on *Breaking Bad*. But instead of being equipped to cook crystal meth, Palmer Luckey's trailer was optimized for building virtual reality headsets.

His obsession with virtual reality had begun three years earlier, when he was sixteen years old. At that time, the idea of an engineer building a virtual reality headset was not all that different from an archaeologist searching for the Holy Grail. To the current world at large, the quest for great VR was considered nothing more than a fool's errand. Unlike that mythological grail, however, virtual reality *did* exist, and had for quite some time.[1]

In 1955, cinematographer Morton Heilig wrote a paper titled "The Cinema of the Future," which described a theater experience that encompassed all the senses.[2] Seven years later, he built a prototype of what he had envisioned: an arcade-cabinet-like contraption that used a stereoscopic 3-D display, stereo speakers, smell generators, and a vibrating chair to provide a more immersive experience. Heilig named his invention the Sensorama and shot, produced, and edited five films that it could play.[3]

In 1965, Ivan Sutherland, an associate professor of electrical engineering at Harvard, wrote a paper titled "The Ultimate Display."[4] In it, Sutherland explicated the possibility of using computer hardware to create a virtual world—rendered in real time—in which users could interact with objects in a realistic way. "With appropriate programming," Sutherland explained, "such a display could literally be the Wonderland into which Alice walked."

Three years later, with assistance from a student named Bob Sproull, Sutherland built what is considered to be the first-ever virtual reality head-mounted display (HMD) system: the Sword of Damocles.[5] Unlike the Sensorama, the Sword of Damocles actually tracked the user's head movements (and, instead of film, placed users into a computer-generated world). But there was one major problem with Sutherland's breakthrough: his HMD weighed so much that it had to be hung from the ceiling (hence the name "Sword of Damocles").

Due to the exorbitant costs and perceived lack of consumer inter-est, this type of reality-defying research remained primarily in labs for the next twenty years. That changed in the late '80s with VR pioneer Jaron Lanier, whose VPL Research became the first company to go to market with virtual reality goggles. Although the company's flagship headset, the "EyePhone 1," was prohibitively expensive ($9,400), and required a multimillion-dollar workstation to power it, the sci-fi-like spectacle of VPL's products (plus the budding fame of Lanier, credited with coining the term "virtual reality") helped generate a cultural fascination for the technology.[6,7,8] Meanwhile, as consumer interest was piqued, the fervor among researchers also accelerated. This was aided in no small part by a company called Fake Space Labs, founded in 1991 to develop hardware and software for high-end scientific and government projects.

The hype for virtual reality seemed to multiply by the year. By the mid-'90s, VR was all the rage. Or, perhaps more accurately, there seemed to be unanimous consent that VR was *going to be* all the rage. But a decade full of flops (like Nintendo's Virtual Boy) and failures (like Virtuality, whose arcade initiative ended in bankruptcy) trans-formed the hype around VR into a cautionary tale.[9,10,11] When the '90s came to an end, virtual reality was no more than the butt of a what-ever-happened-to joke in the company of jetpacks and flying cars.

Given this fate, it may seem odd that a homeschooled, self-taught engineer like Palmer Luckey would have ever become interested in a subject matter as culturally radioactive as VR. Yet in an odd way, Luckey was uniquely qualified to try and resurrect this thought-to-be-dead technology due to his contrarian way of thinking, his love of retro-gaming (particularly '90s classics like *Chrono-Trigger*, *Pokémon Yellow* and *Super Smash Bros.*) and his life-defining passion for modify-ing hardware—or "modding" in the parlance of hackers, gamers and tech enthusiasts.

Specifically, Luckey was interested in a subset of modding called "portabilizing," which involved hacking of old game consoles into playable portable devices. Given that this was a highly technical, time-consuming hobby, there weren't all that many portabilizers out there. But there were enough that in June 2009, Luckey decided to cofound an online community called "ModRetro" that would cater to those who shared his niche interest.

With web forums, chat rooms and a motto of "Learn, Build, Mod," ModRetro sought to attract the world's best, brightest and most curious portabilizers. And for the most part, Luckey's community achieved that objective. But along the way, in the course of discussing projects and sharing work, something else happened: the purpose-driven relationships that ModRetro cultivated wound up becoming lifelong friendships.

"Man, ModRetro days were the best," Luckey would reflect years later. "We accidentally built a personal support community alongside the actual modding work—most of us were growing through similar periods in our life, and all of us were strange nerdy kids. For a lot of us, ModRetro was more important than any real-life group of friends."

Like many close friendships, those at ModRetro were imbued with a playful spirit of one-upsmanship. This type of banter didn't just keep the chat lively; it pushed everyone to try and be better; to build faster, smaller, and as cheaply as possible. To really up the ante, community members would endeavor to do all that with the most obscure piece of hardware they could find. Which, inevitably, is what led Luckey to eBay in search of some virtual reality headsets.

After making his first few purchases, tinkering around, and learning about the VR landscape, Luckey soon came to a critical realization:

As much as Luckey loved portabilizing consoles, he knew that even the coolest N64 portable would never change the way people live. But virtual reality, potentially, could. For example, he had read about something called "Bravemind," which was a virtual reality exposure therapy system that could help treat those suffering from posttraumatic stress disorder (PTSD).[12] Unlike previous treatments, where patients are forced to recall (and reimagine) traumatic scenarios, Bravemind enables therapists to virtually re-create those experiences—a city street in Afghanistan, a desert road in Iraq—and then walk patients through these re-creations under safe and controlled conditions.

Really, the only "limit" to the limitless possibilities of VR was computing power. The faster computers got, the better the graphics would be and the more real virtual worlds could feel. And if it felt real enough—if you could bestow users with a true sense of presence—then VR could achieve almost anything. It could reinvent how we communicate, educate . . . Luckey had to cut himself off. He wanted to be careful not to waste time getting lost in that stuff. But the bottom line was this: if technology existed that could allow *anyone* to be *anywhere* at *any time*, then not even the sky was the limit.

Luckey felt inspired like never before. As he tinkered away inside his trailer—pumping himself up with power metal music—he was reminded of something that visionary game programmer John Carmack had once said about virtual reality: "It's a moral imperative," Carmack had described, touting the ways in which VR could empower anyone—of any socioeconomic standing—to experience anything.

To try and unlock these uncanny possibilities, one of the first and most important things Luckey did was join an online community called Meant to Be Seen 3D (MTBS3D). With a focus on stereoscopic 3-D gaming, the forum wasn't exactly a perfect match for his VR interests. But back in 2009, when the number of VR enthusiasts worldwide numbered somewhere in the hundreds, this was the best option. MTBS3D didn't have the same chummy atmosphere as ModRetro, but that's not what Luckey was looking for. What he needed was a more cutting-edge, graphics-focused crowd, and that's exactly what MTBS3D proved to be.

Over the next three years—as Luckey put together more than fifty prototypes—the open dialogue and communal support on MTBS3D played a vital role in the evolution of his work. And along the way, in his quest to learn what had gone wrong before, he amassed the world's largest private collection of head-mounted displays. Since HMDs were so obscure and undesirable, these weren't the kind of things that tended to pop up on eBay or Craigslist. He'd check those sites compulsively just in case, but more often than not his geeky treasures would come from places like government surplus auctions or used medical equipment vendors. For example, one of his greatest hauls to date had come from a VA hospital in Kansas. The hospital was offloading old relics, and Luckey managed to snag an entire lot of Visionics headsets that had been designed for enhancing the vision of veterans.

Luckey's ability to hunt down obscure headsets was second to none.

But as far as hobbies went, this was not a cheap one. Rare HMDs could cost several hundred dollars, and the resources to repair them even more. Then add to that the cost of Luckey's own creations and experimentations, and suddenly this interest turned into a four-figure-a-month obsession.

To fund all this, Luckey often worked at the Long Beach Sailing Center, sweeping yards, scrubbing boats, fixing diesel engines. It only paid minimum wage, but the work there was something he could count on; his other go-to sources of revenue (i.e., walking dogs, or repairing busted phones) were a lot more sporadic. As a result, he was able to pull the trigger when items like the Fakespace Boom 3C popped up on eBay.

The Boom 3C was no ordinary HMD. Rather it was a heavy-duty, head-coupled display that weighed so much it had to be counterbalanced by a several-foot-long mechanical arm that moved with the user. All in all, it looked less like a piece of high-tech equipment than it did an unwieldly exercise machine. But setup aside, the Boom 3C had been capable of delivering some of the most immersive experiences in the '90s. Which is why, back then, it had cost upwards of $90,000. So when Luckey snagged one on eBay for less than a hundred dollars, he freaked out a little bit.[13]

Unsurprisingly, this steal of a deal needed major repair. Seeking advice, Luckey found an email address for the one person who knew more about Fakespace's Boom 3C than anyone in the world: Fakespace founder Mark Bolas, who was now teaching classes and running a research lab at nearby USC.

"Hi!" Luckey wrote, reaching out in June 2011. "I am a huge fan of your work, and as someone who loves the concepts behind VR and AR technologies, I greatly respect your contributions to the field. I have been wanting to get in touch with you for a while, there are a few things I wanted to talk to you about."

One of those things was the Boom 3C's color generator, of course, and another was a paper that Bolas had recently coauthored. That paper, "A Design for a Smartphone-Based Head Mounted Display," discussed how two iPhone 4 displays (one for each eye) could be used to drive an ultra-affordable HMD.[14] Luckey liked the solution that Bolas and his colleagues presented but thought he had come up with something even better, something that only required *one* phone and an external display. "I have several ideas which could be very easily and cheaply implemented," Luckey wrote, and then described a prototype he had already built that did exactly that.

It was a bit awkward, offering to try and help one of the very few VR experts in the world, but Luckey hoped that Bolas was the type of guy who would appreciate that kind of gumption. Especially because of the big ask that bookended his email: "I would love a low pay or unpaid internship at somewhere where I could get some experience," Luckey wrote. "If there is anything you can do to help me out, or point me in the right direction, I would greatly appreciate it."

Luckey knew that this was a long shot. But by the time he reached out to Bolas, he was beginning to suspect that those were the only kind of shots that he would get. Because the VR industry was tiny and, for what few jobs existed, Luckey was competing against much older applicants whose walls were adorned with prestigious college degrees. Meanwhile, here he was: a commuter student at Long Beach State whose only prior formal education consisted of some credits at the Huntington Beach community college. How could he compete? How could he even get his foot in the door? Luckey didn't know. But he truly believed that when it came to his singular obsession—building low-cost, high-immersion HMDs—there was no one in the world better at this than he. All he needed was a chance, someone willing to look past his age and résumé. Bolas proved to be that someone.

"I've been doing VR for 25 years," Bolas would later say to the *Orange County Register*. "He knew as much about the history of my products as I did."[15]

In July 2011, Bolas offered Luckey a lab technician position in his Mixed Reality Lab at USC's Institute of Creative Technologies (ICT). The twice-weekly job was everything Luckey hoped that it would be. Lots of busy work (cleaning, reporting, etc.), helping with student projects (short VR films), and organizing hardware that had been in the lab for years. The work wasn't glamorous, but Luckey was fine with that. All he cared about was being there and getting the chance to interact with people who shared the same niche interest as he. Of course, all the menial tasks of the job paid off when he could spend time in the warehouse-sized test chamber and experience what virtual reality could truly offer when money was no object.

On September 25, 2011, Luckey tried his best to describe what it felt like in an MTBS3D post titled "Truly Immersive (AKA 'Holy Crap This Is Real') VR Simulation."[16]

"WHOOSH!" Luckey wrote. "All of a sudden . . . your entire field of view is engulfed . . . [and] you seem to be standing on a post-apocalyptic bridge, what used to be a roadway that carries cars. Rust runs the lengths of the thick iron beams above you, and the road is littered with debris . . . You glance down, and quickly step back; your foot had been mere inches away from a sharp, rusty spike protruding from the ground, and your instincts want it as far away from your foot as possible . . . Up till now, you have been alone in the simulation. All of a sudden, though, you hear someone calling out from where you originally spawned, on the other end of the bridge. Now fully confident of the world you are in, you spring about 60 feet back to where you came from, meeting another avatar portraying a US Army soldier in full desert gear, carrying a large handgun. You salute, and reach out to shake hands . . . And then as quick as he came, the soldier thanks you for the help, and blips out of existence. You know in your mind that he was really just a software engineer controlling a virtual body, but your subconscious is having a pretty hard time believing that . . .: It sounds crazy, I know, but *The Matrix* is so much closer than we all think. . . . People need to experience this to believe it."

That last line—echoing Morpheus's famous words: "No one can tell you what the Matrix is, you have to experience it for yourself"—summed up one of the biggest hurdles Luckey faced in getting friends and family excited about virtual reality. This is why, when he wasn't at the lab or tinkering with his own prototypes, Luckey's favorite thing to do was share his work with others. And his favorite person to share it with was his long-distance girlfriend: Nicole Edelmann.

Luckey and Edelmann met at debate camp in 2009, with each on opposing sides of a heated policy debate. Unsurprisingly, their relationship did not get off to a great start. But later that day, Luckey noticed his opponent reading in a courtyard—her intense grayish eyes, that platinum bob of hair—paging through an issue of the Japanese manga *Lucky Star*. From there, things progressed in a more fortuitous direction. She, too, was being homeschooled. She also loved making things (albeit costumes, not electronics). And though she lived in Colorado, and he in California, they started dating about six months after that. A fact that Luckey liked to bring up on ModRetro. Sometimes just to brag, but usually in search of unusual romantic advice.

How Do I Ship Ice Cream on an Airplane

As some of you may know, I am no stranger to airplane shenanigans.

I am leaving on a trip this Thursday, and I need to somehow bring several boxes of ice cream, and keep them frozen... [because] I want to bring some to Nicole. The only place online sells a box for twice as much... WITH $35 SHIPPING! Overnight dry ice shipping is expensive.

I read that the TSA is trained to always leave professionally packed frozen seafood alone (Like, if you buy 2lbs of frozen crab from a fish place), and I was considering packing it up with some cold packs in one of those insulated lunch bag things, then wrapping it in brown paper, and printing out an official looking "Palmer's Wharf" label on a big adhesive sheet, to make it look like it was packed at a fishery.

Am I over-planning?

With the help of his friends (and a small helping of dry ice), Luckey was able to successfully transport his temperature-sensitive gift through the sky. But typically, to avoid the cost of a flight, he would visit Edelmann by car. So once a month, Luckey would drive out to Colorado in his red 2001 Honda Insight, watching the car's odometer pass 150,000 miles, then 175,000, and eventually over 200,000.

Occasionally, Edelmann would make the trip to see him in his trailer. Although she loved the boy who lived in it, she was not particularly fond of his "residence." In fact, the first time she visited Luckey's trailer she was so disgusted with the state it was in that she spent the whole trip cleaning it up (except for the dozens of empty Mucho Mango cans stacked up by the sink; he was proud of that collection and wouldn't let her toss them out). One day, Luckey believed, they would live together someplace nicer. But until then, the best thing he could do was transport her to incredible virtual spaces.

Well, okay, maybe he wasn't transporting her anywhere *that* incredible. His prototype headsets weren't capable of anything like the high-end, highfalutin stuff at USC. But Edelmann was still continually wowed by the experiences his inventions were able to give her. He was an unusual soul, this Palmer Luckey, and she liked this very much about him. "I'm the ground and he's the atmosphere," she would say, to explain their relationship. "And we need each other."

Edelmann didn't necessarily expect that Luckey's VR obsession would lead to anything, but she admired all the hard work he put in. Because one of the things she liked most about him was how—in an increasingly cynical, superficial, and shortcut-driven world—he seemed to be one of the few who still believed in the American dream. Central to that was his underlying ethos: don't complain about it, do something about it.

By 2012, this determination had led to the creation of a prototype that Luckey believed was almost good enough to share with the world: a headset way cheaper than anything else out there. His plan was to use Kickstarter, the popular crowdfunding website, to create a little campaign where he could sell an easy-to-assemble kit for VR enthusiasts like himself. It was a small population of people—like double-digits-worldwide tiny—but even so, Luckey was eager to produce a low-cost, high-performance option for his peers.

In his lab, he referred to his latest prototype—the sixth in a series of low-cost, high performance designs—as the "PR6." But thinking that wouldn't make a particularly attractive name, he decided to christen it the "Rift." Because, as he told his friends on MTBS3D, "the HMD creates a rift between the real world and the virtual world . . . though I have to admit that it is pretty silly."[17]

To make all this legit, Luckey knew he'd need to set up a company. And coming up with a name for that was a bit trickier because most of the names he initially considered were taken, and all the obvious ones—the ones with "virtual" or "VR" in the name—risked suffering the stigma associated with this technology. This was something he'd always known on some level, but it had become clear to him the previous Thanksgiving.

While in Kansas to celebrate the holiday with his grandfather, he managed to track down the company that owned most of the rights to that failed virtual arcade company from the '90s: Virtuality. At this point, Luckey had been toying with the idea of opening a modern-day arcade and was interested in maybe—if it was really, really cheap—buying the rights to the name "Virtuality." At an Olive Garden in Kansas, Luckey met with the rights holder to discuss this possibility. They wanted $150,000; not just for the name, but for a warehouse full of old Virtuality units, too.

"I just want the name," Luckey explained. "How much just for that?"

"Kid," he was told. "You don't want to use the name. Virtuality screwed over all of their suppliers before going out of business. If you go back into business under that name, you're probably immediately going to get sued."

In the following months, Luckey thought a lot about this advice. Not just the potential litigation associated with the name Virtuality, but how the toxicity of that name was indicative of the crippling baggage that virtual reality still carried. As such, Luckey realized it would be wise to try and distance his small company from the buzzwords of yesteryear. While running through possibilities for what to name this little entity, he recalled a conversation from a few months earlier, in which he was chatting with his colleagues at ICT. One of the researchers offhandedly mentioned that he had been "peering through the oculus." When asked what an oculus was, the researcher explained that it was simply a round window—any round window or opening. Luckey and his colleagues thought that was kind of funny. But even after they each put their fingers around their eyes—asking, jokingly, if that constituted an oculus—they all agreed that oculus was a supercool word and that someone should use it for something. Recalling those nods of agreement, Luckey was ready to put that word to use. He debated between using the word as is, or combining it with *lux*, the Latin word for "light," to dub the venture Oculux. But as luck would have it, Oculux was already taken by a medical lighting company, so Oculus it was.

With a company name and some semblance of a plan, Luckey had bought a domain (oculusvr.com) and planned to launch a website later that month. As he sat in his trailer, he typed up his vision for Oculus to put on the site:

> Unfortunately, virtual reality has risen and fallen many times, with a lot more emphasis on the latter portion. The tech has never gotten far enough to be truly convincing, and great VR hardware has been far out of reach for the average person . . . Until now.
>
> Oculus is my tilt at trying to change that. The tech has improved, and we can build hardware and software that is better, stronger, and faster than the old guard, companies that create niche, wildly expensive products. Don't get me wrong, these companies are important, and they have to solve some very tough engineering challenges to satisfy their customers. But the reality is that as gamers and dreamers, we have a different set of challenges to meet. Massive field of view to engulf your visual senses, low latency tracking to maximize presence, light weight and comfortable for long term use, and perhaps most importantly, prices measured in the hundreds of dollars, not tens of thousands. I have worked long and hard with a lot of brilliant people to try and meet those challenges, and now it is time to put it in your hands.[18]

Given the small size of the VR community, Luckey's reason for selling these Oculus Rift kits was hardly financial. There was just no way to strike it rich in such a niche market. But that was okay with him. And it was okay with Nicole. And perhaps, in a best-case scenario, if he could get any traction with this thing—if he were able to crack, maybe, one hundred sales—then perhaps having that under his belt might brighten his prospects for the future. Which, as he sat at his computer in that nineteen-foot trailer, was starting to become something of a concern. Because even though Luckey may have looked content in this moment—some eurobeat music now blasting, his soldering iron currently burning—he couldn't fully suppress a creeping sense of dread. Outside of this isolated, invincible space, a lot of things didn't seem to be going his way.

Take his job at the lab. He loved it. But because temp employees were supposed to be hired or let go after six months, and because ICT didn't have the budget to keep him on, Bolas had informed Luckey that he'd probably need to find work elsewhere sometime pretty soon. Except, again, that was going to be really hard due to his background. Although his time at ICT might open up some other possibilities, it didn't seem to be opening them up enough for the jobs that he was really interested in. Just recently, for example, he had rejected for a job at the VR-headset-manufacturer Sensics. And that was for a low-level lab technician position!

Maybe, Luckey reasoned, he'd be more hirable after obtaining a formal college degree. He was pursuing a journalism degree at California State University, Long Beach, and he had even been named the online editor in chief for the school's student newspaper. Luckey's "backup plan" (if you could even call it that anymore) was to become a tech journalist; if he couldn't build gadgets and gizmos for a living, then at least he'd get to write about other people who did. But after applying for a writing position, and then not even hearing back from the one site that seemed like a no-brainer—Hackaday, a blog that covered the type of modding he loved to do—Luckey had to face the facts: journalism might not end up being in the cards either.

Maybe, he thought, things would change *after* he graduated with a degree. But he wouldn't even get the chance to test that possibility: back in December, due to a glitch with the school's class-scheduling software, he'd missed his window to enroll for classes. Okay, he thought, I'll just take this semester off. But now apparently, going forward, a freeze had been placed on increases in enrollment. This shouldn't have really mattered anyway because, after that, he had applied as a transfer student to USC. Which not only was a better fit but, if he were a student there, then he'd be able to continue working at ICT. That wouldn't end up happening, either; because as Luckey had recently found out, he had been rejected from USC.

So, at that moment in 2012, Luckey had no idea what he was supposed to do next. Even though he was a through-and-through optimist, it was hard not to feel flustered by the tyranny of the status quo. But when the clock struck 2:28 p.m. on this fateful April afternoon, Luckey received a message from one of his childhood heroes that was about to change everything.

CHAPTER 2

CARMACK THE MAGNIFICENT

April 2012

IT WAS JUST SUPPOSED TO BE A MARKETING GIMMICK.

Six months earlier, in October 2011, id Software—the Dallas-based game developer best known for pioneering the first-person shooter genre—released a new game called *Rage*.

"A visual marvel," declared IGN, in their review of the game.[1] "A breakout achievement," hailed VentureBeat.[2] And, quite succinctly, "Carmack!" cheered the *New York Times*, referring to id's legendary technical director, John Carmack.[3]

Instead of taking a victory lap, Carmack—a thin, blond-haired forty-one-year-old coder whose work (and speech) often made him seem part machine—proceeded with his typical postlaunch ritual. "A little R&D period," he liked to call it. A little break from game development to think more broadly about the future.

Typically, these little R&D periods didn't lead to larger endeavors, but occasionally, there would be a venture to pursue. Such was the case in 2000, when after finishing *Quake III: Arena*, Carmack decided he wanted to learn about rocketry and soon after founded a company called Armadillo Aerospace (whose goal was to build a suborbital spacecraft capable of space tourism). Regardless of outcome, these were periods

Carmack greatly enjoyed and for this one, following the release of *Rage*, he decided to focus on virtual reality.

When asked why virtual reality, Carmack would say, "no particular reason." And while there may not have been a specific inciting incident, an almost equally accurate answer could have been "it was only a matter of time." Because in many ways, virtual reality was the unspoken end point of where his engineering efforts had always been heading.

AS A BOY—YEARS BEFORE HE'D PLAY HIS FIRST COMPUTER GAME—JOHN CAR- mack got his gaming fix with tabletop role-playing games like *Dungeons & Dragons*.[4] Though what he enjoyed even more than playing these games was overseeing them in the role of Dungeon Master. That enabled him—either from the rulebook or his imagination—to speak adventures into existence; and then when he grew bored with the loose restrictions of those rulebooks, Carmack moved to charting his own invented journeys on sheets of graph paper. Between that passion for world-building, and a penchant for fantasy or science fiction novels, it was clear from early on that Carmack preferred to spend his time inventing complex worlds or inhabiting those which had been invented by others. So naturally, he was drawn to the godly power and as-you-wish obedience of programming on computers.

Of all the things to program, Carmack's favorite soon became graphics. He loved how something as simple as binary code—just a mishmash of 1's and 0's—led to the creation of colors, images, and actions on a screen. But life behind the keyboard can be lonely. Like Palmer Luckey, Carmack found solace and purpose-driven friendship online, spending his teenage years hanging out on dial-up-accessible bulletin board systems—BBSs—where visitors could post notices, trade messages, and swap software. This exposed him to an incredible underworld of computer games; and eventually, while he was still in high school, Carmack set out to make a game of his own.

That game (*Shadowforge*) and his next (*Wraith: The Devil's Demise*) were both distributed by a small publisher, Nite Owl Productions. Neither game sold many copies, but just the fact that they sold any—that Carmack had created something good enough for others to spend their time playing—that was pretty damn awesome. And even more awesome: his work for Nite Owl got him hired to write a tennis game for a much larger, Louisiana-based publisher called Softdisk. There, Carmack met a handful of kindred spirits, three of whom he would end up starting his own game studio with in 1991: id Software.

At the time of id's founding, almost all the best-loved and best-selling games were being made exclusively for consoles. There was a good reason for this: with underpowered graphics, computer games just couldn't match the speed and splendor of those made for consoles. Take, for example, a side-scrolling console game like *Super Mario Bros.* When players decide to run Mario (or Luigi) across the screen, the "camera" is able to keep up, keeping our hero in the frame and doing so in a smooth and seamless manner. With computer games, however, this was not the case. If a character moved beyond the frame, this would lead him to an entirely new screen. That's just how it was, an understandable by-product of underpowered graphics, and this remained the norm until John Carmack came up with a technique called "adaptive tile refresh" that made it possible for personal computers—PCs—to perform smooth and seamless Mario-like scrolling. In fact, to prove *just* how Mario-like their games could be, the founders of id Software made a demo called *Dangerous Dave in Copyright Infringement*, which near perfectly re-created the first level of *Super Mario Bros. 3* (save for swapping out Mario with a spritely dude named Dave).

This breakthrough technique, adaptive tile refresh, became the centerpiece of id's first game: *Commander Keen.* Though that game sold pretty well, it still only sold pretty well "for a computer game." That was the qualifier that was always used back then; because compared to hit console games (like *Super Mario Bros*, which sold tens of millions of copies), or even just mildly successful console games (like *Hogan's Alley*, which sold over a million copies), the best-selling computer games (even those part of popular franchises like *Ultima* or *Zork*) rarely managed to crack a hundred thousand copies. So that qualifier existed—"for a computer game"—but it wouldn't last for much longer. Because after *Commander Keen*, id was able to change the perception of PC gaming with a decade full of megahits: like *Wolfenstein 3-D* (which sold over two hundred thousand copies), *Quake* (which sold over one million copies), and most notably *Doom* (which, at one point in the early 90s, Microsoft determined was installed on more PCs than their flagship Windows software[5]).

Games like *Wolfenstein 3-D*, *Quake*, and *Doom* made Carmack a rock star, and techniques like adaptive tile refresh, surface caching, and Carmack's reverse made him a legend within the gaming community. But there was also something else about him—something ideological in nature—that elevated Carmack from mere living legend to Gandalfian hero: a belief that openness, open sourcing, and technological transparency were critical to innovation. In a now-famous blog post entitled "Parasites," Carmack likened software patents to "mugging someone," explaining that "in the majority of cases in software, patents effect independent invention."[6]

These were not just empty words. Carmack lived by this credo. That's why he always publicly shared the source code for his games (after they had been released); that's why he regularly provided elaborate advice to hardware vendors (like Sony, Microsoft, and Nvidia); and that's why he readily divulged experimental findings and in-progress theories when delivering keynote speeches (particularly at QuakeCon, an annual celebration of id Software's games).

In fact, Carmack believed so greatly in the importance of this type of behavior—sharing, advising, divulging—that when he sold his company to ZeniMax Media in 2009 (and then signed a five-year-contract to work for his acquirer), he had special provisions written into that contract so that he could continue. That hadn't been easy to get in there. But the reason they acquiesced for the man fans called "Carmack the Magnificent" was because, over the course of twenty-one years, id Software had proven time and time again that Carmack's open and transparent approach worked. It had turned id's three iconic games— *Wolfenstein*, *Doom*, and *Quake*—into bona fide blockbuster franchises. And though—in October 2011—initial sales of *Rage* turned out to be less than ZeniMax had anticipated, there was still hope that it would lead to another world-class IP.

Right after the release, however, Carmack wasn't interested in thinking about sequels or spin-offs just yet. First, he wanted to enjoy the exploration of another one of his R&D periods. But what to explore? Well, around this time, many of the major electronics manufacturers were hyping 3-D TVs as the next frontier of television. Personally, Carmack was not too bullish on the odds of 3-D TVs going mainstream, but it was significant to him that both Sony and Microsoft had just released extensions that would allow their consoles to drive 3-D TVs. And somewhere along the way, while thinking more about stereoscopy—the process of presenting two images (one for the right eye, one for the left) in such a way to give the impression of three-dimensional depth—Carmack found himself wondering about virtual reality.

Carmack hadn't really thought much about virtual reality since the '90s, back when "VR" was one of the tech world's hottest buzzwords and a few VR companies licensed some of id's games. Nothing had come from that brief foray—either because the VR versions of those games sucked, or because the companies involved wound up going belly-up—but it had now been a couple decades since then, and Carmack was curious to see how much virtual reality had progressed. Surely, he thought, great strides had been made over the past twenty years! Computers, after all, were now hundreds of thousands times faster than they were back then. Not to mention all the progress that had been made with displays, sensors, and other relevant things. But after assessing the HMDs currently on the market, Carmack wasn't just disappointed by what was out there, he was flat out offended. In fact, with some of the headsets, the latency—meaning the lag between when someone *tries* to complete an action on-screen (like, say, firing a weapon) and when that action *actually* occurs—was actually a step *back* from the systems Ivan Sutherland built back in the '60s! Granted, those Sutherland systems were ultra-expensive with low-latency CRT displays . . . but still! How could this be? Had VR really flamed out so badly in the '90s that touching the technology was still perceived as toxic?

As of late 2011, the most popular HMD was probably eMagin's Z800 3-D visor, which cost about $900. Lightweight, and with a resolution of 800 x 600, the Z800 was certainly better than anything from the '90s. But the headset was flimsy, the tracking was lousy, and, worst of all, there was this:

> The field of view was terrible. With a FOV of only 40 degrees, it felt like looking at the world through a toilet paper tube.

How could a VR experience possibly feel immersive when you feel like you've got blinders just outside your eyes? The whole allure of VR was to actually feel *present* in a different place; and that just wasn't possible if you had the same FOV as a horse in the Kentucky Derby. This problem was not unique to eMagin's Z800 3-D visor; in fact that 40° FOV was just about the widest of what was out there.

Nevertheless, Carmack purchased several of these offensive headsets and spent some time trying to make them better. So Carmack used a "testbed" that he had already coded for unrelated graphics works—which, essentially, was just an playable scene from *Rage*—to experiment with the parameters of the VR headsets he had acquired. And eventually, he was able to make enough progress with his experiments that he felt what he would excitedly describe to colleagues as a "kernel of awesomeness." It was hard to put into words exactly what that kernel was, but it had something to do with finally, after all these years, feeling like he had stepped inside one of his creations. But again, with nothing more than a through-toilet-paper-tubes view, this sensation that he felt was barely a kernel.

So that was a big problem. And so, too, was this: Carmack had allocated only a few weeks toward researching VR and, as the calendar flipped to November, his time was almost up. As id's technical director, he needed to focus on tangible revenue streams; that is, to continue his current research, even in a minor capacity, he'd need to come up with a business reason to do so. This is how the "marketing gimmick" idea came about.

The following fall, id Software was planning to release a remastered version of *Doom 3*, which had originally come out in 2004. This remastered version—which would be called *Doom 3: BFG Edition* (BFG short for "Big Fucking Gun")—would, like most rereleases, struggle to earn buzz against a slate of shiny of new games. Unless . . . well, okay: What if, Carmack wondered, we paired *Doom 3: BFG Edition* with this weird novelty technology that people haven't really talked about in twenty years? What if, at the year's biggest trade show (E3), we demoed a virtual reality version of *Doom 3* for the press? It was, indeed, a gimmicky idea, but it seemed like an idea that might garner attention for an eight-year-old game.

This strategy gave Carmack a reason to keep VR on his radar. Now he just needed to find a headset that could deliver an optimal experience. Fortuitously, in November 2011, Sony released a "personal HD & 3-D viewer" that Carmack thought might fit the bill: the HMZ-T1. With cutting-edge OLED display panels, Sony billed the HMZ-T1 as a "wearable HDTV" that gave users "the equivalent of having your very own 150" movie screen just 12 feet away, in either 2D or 3D."[7] And while that may have been cool, it wasn't particularly useful for what Carmack had in mind. For one thing, the latency was unbearably high; and for another: it didn't come equipped with a tracking device— meaning a sensor capable of detecting the location of the user's head (so that—like in real-life—the screen you are watching does *not* move when you turn your head).

Nevertheless, a high-def, high-latency 3-D viewer still might have worked well enough for Carmack's pet project (it was a marketing gimmick, after all); but even for a relatively affordable HMD ($799), there was still a major, experience-killing problem:

> With a 45° FOV, Sony's HMZ-T1 was only slightly less terrible. The OLED screen made things more compelling, but it still seemed impossible to feel "present."

The FOV issue wasn't really something that Carmack could address. But he could do other things to his game itself—to the source code of *Doom 3 BFG*—to improve the VR experience of these HMDs. If this sounds counterintuitive at all—using software to improve a hardware experience—an easy way to think about this is with a television and a television program. Without the hardware (the television), you can't experience the software (the television program). Though much of the hardware experience is based on the quality of the hardware itself (i.e., the television's display, resolution, features, etc.), there *are* ways to film or edit the television program so that it optimizes certain aspects on-screen (i.e., clarity, contrast, consistency, etc.). And so with the exception of adding a tracking device to the few headsets that he thought might be decent enough to demo at E3 (this included the headsets from Sony and eMagin), Carmack found time over the next few months to try and make the software side of *Doom 3: BFG Edition* as "VR-ready" as could be.

His office, which now resembled a mad scientist's laboratory—partially dissected headsets everywhere—was proof of this ambition. As E3 approached, he continued to tinker toward something better. One possibility he found online involved using off-the-shelf optics—like a series of wide-angle lenses made by LEEP systems in the '90s—that, when coupled with an LCD screen, could potentially achieve a wider field of view. Exploring this possibility led Carmack to an enthusiast website called VR-tifacts, specifically an article entitled "LEEP on the Cheap."[8] In the comments section of that article, Carmack read about a hardware hacker who appeared to be doing some very interesting work in this apace. A hacker who went by the name "PalmerTech."

FROM: John Carmack
TO: PalmerTech

I would be interested in checking out, or buying outright, one of your high FOV prototypes if that is possible. I am going to be doing some private VR demos at E3, and it would be interesting to compare the relative merits of high FOV versus high resolutions versus high refresh rates

Reading Carmack's message from inside his messy trailer, Palmer Luckey could hardly believe his eyes. John Carmack, the man who had made several of his favorite games, wanted to borrow something that *he* had made? Despite a mild temptation to freak out, Luckey reminded himself that when it came to virtual reality headsets, he knew *way* more than probably anyone in the world—including even Carmack the Magnificent. So, trying not to sound like a drooling fanboy, Luckey replied, "I would be glad to lend or sell you one, whichever you prefer. I have been planning on selling it as a kit starting in June, but I can put one together for you before then . . . Are you interested in the 120-degree prototype or the 270-degree prototype?"

Now, as surreal as all this felt, it wasn't as if Luckey thought that this would lead to some sort of life-changing "big break." Sure, it felt incredible to be able to help out one of his heroes, but, really, what was the best-case scenario here? Probably that (if impressed) Carmack would post a nice review on the forum, which could potentially help Luckey sell a dozen more kits for his headset. And that was an outcome that would have thrilled him, a possibility that motivated him to finalize his plans for the Kickstarter and announce it on MTBS3D. But before doing so, he wanted to make sure that he could rely on the help of a few friends, most of all from Chris Dycus, a skinny, subdued-but-enthusiastic teenager who had been part of ModRetro since the early days and, by this point, was one of Luckey's best and most trusted friends.

"Dycus, I need your help," Luckey explained over the phone. "I want to make a Kickstarter for the VR headset thing. Would you still be down to help put some of the kits together?"

"Sure," Dycus replied. Partly because he wanted to be good friend, partly because he just loved building things and mostly because—quite frankly—he assumed Luckey wouldn't end up following through. In all the years they had been friends, Dycus could count on one hand the number of times Luckey *actually* finished one of his hardware projects. Typically what happened is that Luckey would get excited about a project; come up with a bunch of brilliant, creative solutions; and then before finishing up that project he'd get excited by something else.

This time, however, Luckey was determined to make it happen. This was virtual reality, after all—the most passionate of all his passion projects—and with his teenage years nearing their end, there was no more room for excuses. So on April 15, 2012, Luckey wrote up a lengthy post on MTBS3D:[9]

"Hey guys," he began. "I am making great progress on my HMD kit! All of the hardest stuff (Optics, display panels, and interface hardware) is done, right now I am working on how it actually fits together, and figuring out the best way to make a head mount . . . The goal is to start a Kickstarter project on June 1st that will end on July 1st, shipping afterwards as soon as possible. I won't make a penny of profit off this project, the goal is to pay for the costs of parts, manufacturing, shipping, and credit card/Kickstarter fees with about $10 left over for a celebratory pizza and beer. I need help, though . . ." After listing a few of the things he needed help with (a logo, ideas for the Kickstarter video, etc.), Luckey published the post on MTBS3D. He felt hopeful— hopeful that this might be "the kind of thing that jumpstarts a bigger VR community." But also: he suddenly felt very aware of the fact that he was only nineteen years old. No, that doesn't matter, Luckey reminded himself. Not online. On here—with an internet connection and keyboard at my fingertips—I'm it-doesn't-matter years old. All that really matters is that I'm going to do my part to try and resurrect virtual reality.

CHAPTER 3

A TALE OF TWO TRADE SHOWS

May/June 2012

IT TURNED OUT THAT CHRIS DYCUS WAS RIGHT. PALMER LUCKEY'S PLANS *DID* change. This time, though, it wasn't because he had gotten distracted with another project but rather because John Carmack's confidence in his vision had persuaded him to think bigger.

"This is a lot cooler than it looks," Carmack tweeted on May 17. along with a janky-looking prototype of the Rift. "Palmer Luckey's 90+ FOV HMD, soon to be available as a cheap kit." If that weren't life-changing enough, Carmack published a lengthy review later that day on MTBS3D—calling Luckey's prototype "the most immersive HMD of the five I have here" and noting that "If Palmer comes close to his price target, it will also be the cheapest."[1]

Luckey couldn't believe his eyes. A public vote of confidence from the Magnificent One himself. Privately, Carmack was even more ef-fusive, describing the Rift as "dramatically" better than every other HMD out there—"a completely different situation" that "blows every-thing else out of the water"—and going so far as to say that if Luckey let him bring the Rift to E3, those who received a demo would feel that they had "seen the future."

Weeks before the show, Carmack asked Luckey if he could demo the Rift prototype for "a few people at E3 in some private showings."

Luckey didn't even have to think twice. Of course it was okay! That would be a great way to start building buzz. Carmack agreed, though he expressed some concern that Luckey might get "short-changed" by any potential coverage. "I am going to do my best to not let anyone mistake the Rift as my work," Carmack wrote, "but I'm sure someone is going to get it confused next week. I am sensitive to the fact that the press has a tendency to over-attribute things to me because 'genius inventor' is such a convenient story hook."

Luckey understood and appreciated the heads-up. "I am not going to let it bother me if I can help it," he wrote. "I have seen plenty worse from lazy reporting. In particular, *Engadget* once wrote a story describing my project and using my pictures, but credited it to someone with a similar project who could be called a rival! That kind of bad tech reporting is what motivated me to get a journalism degree, but that is another story."

It would be a long while before Luckey ended up telling Carmack that story, but in the meantime he was just glad his childhood hero was going to try hard to make sure that credit went where it was due. And if that weren't already enough to get pumped about, Carmack also offered to give Luckey a copy of the testbed he had made. That, perhaps, might not sound like much—just a VR-ready sample level from *Rage*—but for Luckey, who at this point didn't have any software of his own (and openly admitted he was "not a software guy"), the testbed could serve as a go-to demo for his prototype. Which is explicitly why Carmack had offered it: to help Luckey generate "a ton of excitement, and go far beyond his original [goal of] 500 kits."

In addition to that, Carmack also put Luckey in touch with a pair of industry vets: Dan Newell, an engineer at Valve who wanted to preorder two prototypes of the Rift; and Mick Hocking, a senior director at Sony who wanted to preorder four prototypes and meet with Luckey after E3.

"I would love to help Sony bring VR to consumers in any way I can," Luckey told Hocking. And after some back-and-forth over email, the two agreed to try and meet after E3.

So, yeah, Chris Dycus ended up being right. But only because Luckey thought it would be wise to postpone his Kickstarter launch until after E3, just in case Carmack really did end up showing the Rift to a few people at E3 or just in case Newell or Hocking might be able to pledge some kind of support to Oculus.

Feeling inspired, Luckey began scripting an overview for his Kickstarter project: "My name is Palmer Luckey," he wrote. "I am a long-time stereoscopic 3D and virtual reality enthusiast, and currently work at the USC Institute for Creative Technologies. Over the years, I have developed a kind of fetish for head mounted displays (HMDs), collecting more than 40 different units from as early as the 80s."

After describing what inspired him to develop the Rift, and how rarely high-FOV headsets have gone beyond the research or military domain, Luckey got to the trickiest part of his pitch: appealing to enthusiasts without alienating everyday gamers. Ultimately, Luckey's goal was to get a future iteration of the Rift to millions of gamers, but now was not the time. The VR market was too fragile to disappoint that future audience, so he tried to be extremely clear.

"If you are just an average gamer, someone without significant skill in DIY hardware and software, do not get this product," Luckey wrote. "If you just want a high resolution TV that goes on your head, the SiliconMicroDisplay ST1080 and the Sony HMZ-T1 are both excellent products. If you have the DIY bug, though, and want something affordable that provides an image area more than 5 times as large as those products and actually gives you the sensation of being in a simulated environment . . . Then this is for you."

At $500 for the baseline model, this was a headset for hackers, for tinkerers, for fellow VR enthusiasts. And the momentum they built—enhanced by the open-source design—would enable this first wave of trailblazers to begin resurrecting VR for the gaming world at large.

ON JUNE 5, 2012, THE EIGHTEENTH ANNUAL ELECTRONIC ENTERTAINMENT EXPO kicked off at the Los Angeles Convention Center. Unlike other open-to-the-public trade shows, E3 is an industry-only event, which meant that with over forty-five thousand people in attendance, anyone who was anyone in the video game industry was there. So naturally, a nineteen-year-old nobody like Palmer Luckey was nowhere to be found.

In fact, Luckey was halfway across the country, in Boston, at a completely different trade show: SID Display Week. He had been sent there by ICT, who was looking for someone in the lab to check out the latest from places like Kopin, LG, and SiliconMicroDisplay, and then report back with a sort of state-of-the-industry report. As excited as Luckey was to walk the floor and see what those companies were working on, he had volunteered to be the one who went for another reason: not only would ICT pay for a hotel room—that's right, *h*otel with an *H*—but they were offering a per diem as well. Seventy dollars a day!

In an effort to save ICT money, Luckey chose a bed-and-breakfast that turned out to be a forty-five-minute, train-involved adventure away from the Boston Convention Center. For the first couple of days, the trek was a pain in the ass. But a few days into SID, Luckey stopped sweating the small stuff. Because on June 6, as he was walking the floor with his micrometer and camera—testing optical displays (which not every company was thrilled about)—his phone started buzzing with congratulatory texts and links to E3 articles.

Wait, Luckey thought between cellular vibrations. Wasn't Carmack only going to show the Rift to "a few people at E3 in some private showings?"

PART OF JOHN CARMACK'S GENIUS IS HIS ABILITY TO ADAPT ON THE FLY.

Which is why, from a sparse, white meeting room—little more than a desk, PC, three chairs, and a *Doom 3: BFG Edition* poster—Carmack decided to call an audible. Several, really. He had started the day with three different headsets, each one meant to show off a unique strength in VR. There was Sony's HMZ-T1 (which had the best resolution), eMagin's Z800 (low-latency) and the Oculus Rift (wide field of view). But before giving his first demo, Carmack realized that, ultimately, all he'd want to do is hustle visitors through the other headsets so that they'd get to the Rift. Because despite its flaws—which had been minimized through Carmack's modifications, as well as the Hillcrest-made head-tracking device he added—that's the one that would cause people to "get it." They got it so much that he kept inviting more members of the press to come see what he started describing as "probably the best VR demo the world has ever seen."

Midway through that first day, there was so much buzz for "the Rift" that Carmack, who had only been scheduled to be at E3 for one day, decided to extend his trip. And it wasn't just the press who was impressed. So, too, was Sony—who Carmack suggested should hire Luckey—and even Robert Altman, the chairman and CEO of Zeni-Max Media (which had acquired id Software in 2009).

"What would it take for us to make these ourselves?" Altman asked. "Just go to China and make a whole bunch of these. Sell them with our games."

ZeniMax had never manufactured hardware—nor had id—but Carmack was becoming so smitten with VR that he was intrigued by this possible trajectory. "I would be happy to lead a charge on that," he replied. "But it's going to be more work than you might think. It would be a full-time project. We would have to bring on more people. But say the word, and I'm game."

All in all, things couldn't have gone better. Except for one minor hitch: many of the gushy articles coming out of E3 had headlines like this one from *PC Games*: "Why John Carmack's Rocket-Powered Goggles Won E3."[2]

THE MISATTRIBUTIONS BOTHERED LUCKEY. SO MUCH SO THAT HE CONTACTED the author of that *PC Games* article to say, "My name is Palmer Luckey, and as the designer and builder of the HMD that Carmack was using, I would really appreciate it if you would give me some credit." Regardless, the more coverage he read about "John Carmack's" Rift, the more eager he was to return home and finally launch his Kickstarter. Before doing that, however, he still had a few more days in Boston. This gave Luckey time to meet with Laurent Scallie, a stylish Frenchman he knew from MTBS3D who was hoping they could meet up while both were in town.

Scallie had been fascinated by VR ever since 1992, when he first saw *The Lawnmower Man*. He then moved to the US so as not to miss out on what he—and millions of others—believed was a soon-to-boom industry. This, of course, did not turn out to be the case; but despite the market crash, Scallie remained one of the few true believers who stuck with VR. And he did so by pivoting from the out-of-home entertainment market and into military simulation. This proved a difficult transition, until demoing a shoot-aliens-with-lasers game that he sold for a retired four-star general.

After this demo, the general opened up about his experiences in Vietnam, about how he had led young soldiers into battle and watched them fall victim to the horrors of war. He talked about how he had tried so hard to train those soldiers for what they were about to face, but things were just different when you were actually there, when you were on the ground in enemy terrain. And sadly, there was just no way to prepare for that . . . until Scallie opened his eyes to the potential of VR simulations. In fact, the general came to believe in VR so much that he soon became the lead investor in Scallie's new venture: Virtual Edge, whose mission was to provide "gaming-based simulation solutions for the military and government users."

Over the next ten years, Scallie's simulations helped train thousands of soldiers. What began with crude graphics and isolated combat experiences evolved into a networked virtual reality environment where thirty-two soldiers could train together. From head to toe, each soldier was fully tracked so that additional elements of realism could be added to the simulation. Like gloves (so that soldiers could gesture with hand signals and aim virtual weapons accurately), steering wheels (so that soldiers could drive VR Humvees), and force-feedback vests (so that, when hit, soldiers could feel the impact and pain).

With Scallie's focus on military simulations, it had been several years since he thought seriously about VR going mainstream. Not until he met Palmer Luckey on the MTBS3D forums.

"I think what you're doing is amazing," Scallie explained, as he and Luckey met for lunch in the food court at the Boston Convention Center. "And the timing is just perfect."

"Yep," Luckey replied. "Consumer VR is pretty close to viable. That's a little heady, isn't it?"

Scallie agreed that these were indeed interesting times, but he wasn't quite as bullish about how quickly VR could reach consumer quality and pricing. In the meantime, he believed that the market was ripe for "more of a professional deal" than the build-it-yourself Kickstarter Luckey had planned. More of a prosumer-level HMD, Scallie explained. And he suggested that he and Luckey work on this venture together, along with a wealthy friend "who just sold his company" and might be the "perfect guy" to provide funding. Especially if Luckey could get Carmack involved in such a venture.

"How does that sound?" Scallie asked.

"THAT SOUNDS AWESOME!" LUCKEY SAID A FEW DAYS LATER, THOUGH NOT IN response to Scallie. Instead, over lunch at the Huntley Hotel, Luckey was replying to Mick Hocking and a couple other Sony execs who had just made what seemed like a dream job offer:

- A full-time position with Sony Computer Entertainment Europe
- The role would entail leading an R&D lab at Sony's studio in Santa Monica
- Tentative start date: August 1, 2012
- Starting Salary: $70k

In a sense, Luckey would be doing the same stuff he'd been doing in his trailer these past few years; except now he'd have the resources of a consumer electronics giant behind him and he'd actually be getting paid to do it. Just imagine what Nicole would say! Or even his parents! Yet as amazing as all this sounded, one little thing prevented Luckey from saying yes on the spot: the Rift.

By accepting Sony's offer, he'd essentially be giving up control of his baby. Which at the moment—since he was basically being hired to work on his baby for Sony—didn't really seem like that big a deal. But what if Sony later pivoted away from VR? Or what if they relocated him to another team? The answer, of course, was that that's what the money was for; and, for the most part, Luckey was okay with that deal. But still a small part of him wavered—a small part that spent the rest of that lunch thinking about advice that his grandpa Larry had given him as a boy.

"Money is freedom, and freedom is happiness," Grandpa Larry had said then and repeated many times since. The advice resonated with Luckey, who was forever seeking work—be it scrubbing ships or repairing phones—so that he could afford the freedom to work on projects that he truly loved.

Working for Sony would enable Luckey to do work that he loved, not to mention the fact that running a lab would be a million times better than the typical gigs he did. Nevertheless, he still wanted to weigh what he might be giving up. So Luckey asked Sony for a few days to think about their offer.

That was fine. Sony wasn't expecting an answer on the spot. Given that they were offering the moon to a teenager with limited educational and professional experience, they had to be confident that he would accept. And they likely would have been correct in that assumption, if not for one thing: a guy named Brendan Iribe, who was about to enter the picture.

CHAPTER 4

THE SCALEFORM MAFIA

June 15, 2012

TO THE FRONT OF BRENDAN IRIBE, THIRTY-THREE, WAS AN IMMACULATELY BLUE swimming pool.

To his back: a panoramic view of West Hollywood.

And to his left and right—tanning beside him on the rooftop of the Mondrian Hotel—were four fellow serial entrepreneurs: Sven Dixon, Nate Mitchell, James Bower, and Greg Castle.

"What about Mike?" Dixon asked. "He gonna join us?"

Mike was Mike Antonov, a brilliant, Russian-born programmer whose code was so good that Iribe called it "voodoo magic." And for the past thirteen years, Iribe had been finding ways to productize that magic, ever since he and Antonov founded their first start-up . . .

START-UP #1: SONIC FUSION

Brendan Iribe and Mike Antonov first met in the fall of 1998, as freshmen at the University of Maryland. Housed in same dorm—their rooms across the hall—and both computer-obsessed coders, it might have seemed like they were destined for fast friendship. At the time, though, Iribe was already moving too fast for something as trivial as friendship. Because in addition to carrying a full load of classes, he was also juggling a full-time job.

"At Quatrefoil," he explained to Antonov, during one of their earliest interactions. "I'm the lead programmer and product designer on this project to build this tech museum in San Jose."

"You're a programmer?" Michael Antonov asked, revealing a thick-but-cheerful Russian accent. "Me too! So cool."

Iribe looked up, getting a better look at the gawky, bespectacled neighbor who would occasionally stroll into and out of his room. "Nice!" he replied, and then disappeared Antonov from his mind and returned to his work.

Instead of taking the hint, Antonov continued to linger around the room. Although he had a gentle presence and frequently flashed a bubbly Muppet-like smile, there was something unnervingly tentative about the way Antonov moved: he lurched instead of walked and his arms swung with the rigidness of a robot. These mannerisms were likely a by-product of Antonov moving to the US when he was fourteen years old, coming to a country he didn't know and struggling to befriend kids whose social circles had already been sewn years earlier. Instead of trifling with that, Antonov found solace in programming.

"I started getting into computers a tiny, tiny bit in Russia," Antonov said. "Because a friend of the family brought this . . . it's called 'Spectrum.'"

"I know the Spectrum!" Iribe said, nodding along without looking up. "Good computer."

"Yes! But you had to boot up with a cassette player. Do you remember?"

Iribe chuckled in agreement.

"So I played some games and did a little bit of basic programming. But that was taken away because it wasn't ours. So it was really in high school here where I learned for real."

First, on a TI-85 calculator. Where, amazingly, Antonov built for himself a falling puzzle game like *Tetris*. And then later honing those skills on a computer of his own—a Packard Bell—that Antonov had been able to get a great deal on at Montgomery Ward (discounted so deeply "just because it had been the store's floor model!").

Although Iribe didn't seem to share Antonov's enthusiasm, Antonov felt that perhaps a small bond had been forged. They both knew Spectrum for Chrissake! So before returning to his room, Antonov boldly exclaimed, "Hey! I have an idea: let's make a game together!"

"Look," Iribe replied. "You seem like an awesome, supersmart guy. But I gotta get this thing done for Quatrefoil and I'm superstressed. We need to ship it soon and—look, I just can't make a game right now. Why don't I come by in a couple weeks?"

But by the time he did, the gawky Russian kid no longer lived across the hall.

On a campus of thirty-five thousand students, it seemed unlikely they'd ever cross paths again. But a few months later, while driving down Route 1, Iribe thought he spotted Antonov walking along the side of the road. So he pulled over and rolled down the window to see if it was indeed his long-lost neighbor. "Mike?" Iribe asked. "Is that you, Mike?"

Antonov nodded, recognizing Iribe immediately. "Hello, Brendan!"

"What happened to you?" Iribe asked, excitedly waving him over.

"I was in a double," Antonov explained, "but there were three people in it. So they moved me to different dorm. I wanted to stay, but they made me get moved."

Iribe nodded. "Where are you headed?"

"I'm going up to the bank. The SECU bank."

"That's pretty far. Why don't you get in and let's talk? I'll give you a ride."

As the two caught up, and Antonov talked about a job he'd recently taken—which involved UI (user interface) design and computer vision for handwriting recognition—Iribe realized that Antonov must actually be pretty great with computers.

"I'm still working for that company Quatrefoil," Iribe explained. "It's a tech museum project and I could *really* use some help."

Antonov remained relatively neutral on the idea until Iribe started talking about graphics. Mike really liked graphics. But between school and his current job, he wasn't necessarily looking for anything else to do. "I'll think about it."

Whenever Iribe needed to wait for someone to "think about it," he always tried to be there for as much of the thinking process as he could. This is why Iribe ended up continuing the conversation back at Antonov's dorm room.

"Tell me more about you," Iribe asked. "What have you programmed before?"

"I've made some games," Antonov replied and then walked his new friend over to the computer to show him what he had made.

It was a shoot-'em-up spaceship game. The graphics were crude and the game itself wasn't too original (it was basically a clone of *Galaga*), but after Antonov showed off a few more games, Iribe realized that was the point. Antonov wasn't trying to make something to commercialize; he was simply making his own versions of classic games for himself to play.

"You made all these?" Iribe asked, particularly impressed by Antonov's *Tetris*.

"There are more," Antonov replied. "I can show you."

"Wait. Hold up. Who else worked on these with you? Who did the art?"

"Oh, I did everything *by myself*," Antonov replied with a mixture of pride and defiance. "I built the tools to do the art *by myself*. I laid down each pixel *by myself*."

As Antonov showed off some of the tools he had built, Iribe became incredulous. Antonov hadn't even used Photoshop, or DirectX, or any of the Microsoft libraries that would have made the process so much easier. Instead, Antonov had literally built every tool from scratch.

"Mike," Iribe said. "This is *crazy*. How'd you learn to do all this?"

"I taught myself. I have these books." Antonov picked up a black-and-white composition notebook from the floor and handed it to Iribe.

Expecting a computer book, Iribe was confused. Even after opening the notebook—filled with row after row of really small, fine print—it took him a few seconds to figure it out: Antonov had transcribed an entire computer book. "Why didn't you just get the actual book?"

"It's expensive!" Antonov explained. "But at the library, I can copy assembly instruction set details line-for-line for free. Plus anything else I need."

Antonov then pointed across the room to an entire stack of notebooks and, right away, Iribe had three thoughts:

- This guy's a genius!
- I've hit a gold mine here.
- Maybe we should do something *much* bigger than the tech museum.

"Mike," Iribe eventually said. "I think we should start a company together."

Iribe didn't launch into a full-fledged pitch right then and there, but over the next few months it became clear that not only was Iribe serious about starting a company with Antonov, but he also already knew what their start-up should do.

"Do you know the company Maxis?" Iribe asked.

Of course Antonov knew Maxis. They made *SimCity* and a bunch of other cool *Sim*-related games. And they had just been acquired by Electronic Arts for $125 million. They were a *big* deal! Did Iribe think they should do a company that made simulation games? Did he think that one day *their* company might actually make them rich? Rich enough to one day have a million dollars . . . each?

Iribe laughed. It was funny and endearing, the way that Antonov said "million dollars." His tone hushed and his face scrunched as if just uttering the possibility would jinx any chance they had at riches. This wasn't totally shocking since Antonov had grown up poor. But then again it wasn't like Iribe had been raised with a silver spoon. He was an only child, the son of a single mother, and though they always managed to get by, it was never without an uphill battle. And so, befitting his world-class ability to weave optimism into confidence (and noticeably in direct contradiction to Antonov's sensibilities), Iribe could talk about millions—hundreds of millions, even billions—with a straight face. Because why the hell not? That money was out there, just waiting for someone to take. Why not him? Why not us? Why not Brendan Iribe and Michael Antonov?

Iribe's idea for a business was not quite what Antonov expected. It wasn't a game studio like Maxis, but rather an idea that had come to Iribe while listening to a Maxis engineer talk at the 1998 Computer Game Developers Conference.[1] That engineer was talking about an in-house tool he had created at Maxis—a graphics-based windowing system that was used to build games like *SimCity*—and as Iribe listened to this engineer talk, he started thinking about how great it would be if such a tool existed for all game developers.

"That's what we should build," Iribe explained to Antonov. "A windowing system to compete against Windows and Mac. Initially, we'll make it for developers, but eventually it'll be for everything and everyone."

When asked the obvious question—how can a pair of college freshmen compete against the likes of Apple and Microsoft?—Iribe would say, "Microsoft was started by two guys; Apple was started by two guys; and here's the two of us. If they can do it, we can do it. And the reason ours is going to be better is that it'll offer more functionality and cross-platform play."

In retrospect, Iribe would concede that his vision was "way too ambitious." But even so, he never regretted the size of his reach. And at the time, Iribe genuinely believed that he and Antonov—together coding a high-end, vector graphics engine they dubbed "GFC"—would be able to outduel any and all Goliaths that stood in their way. A start-up, Sonic Fusion, was born.

Yet even with things coming together, there was still too much work for Iribe and Antonov to handle by themselves. As it turned out, Antonov's roommate, Andrew Reisse, poked his head into the conversation. "You know . . ." Reisse said in a drawl. "I could . . . help."

Given that Reisse's coding skills rivaled those of Antonov, the founders of Sonic Fusion were thrilled to bring him on. And they were also thrilled to recruit a raw-but-talented pixel artist named Sven Dixon, whom Iribe had worked with during a teenage internship at Alien Software.

After receiving a small seed of financing from Iribe's mother and uncle, the quartet found a small basement office in Laurel, Maryland. To anyone who stepped into that office—past the building's trash compactor and into the dank light—the place was atrocious. But to the guys, it was something else. It was *their* office, for *their* own business; and to have all that at such a young age was just kind of magical.

To fully embrace the magic, Iribe decided to drop out of school, and then tried to persuade Antonov to do the same.

"My parents would never let me!" Antonov protested. "My family is all scientists. They all have PhDs. To them, education is everything."

"You don't need their permission," Iribe replied.

Maybe that was true, technically speaking. But Antonov felt that he owed his parents a piece of the decision. Especially his mother. His mother the biologist. Ever since he was a boy, she had put him above herself and always prioritized his education.

"She would take me to the lab whenever they would let her," he recalled. "And she's the one who taught me English. Not to say she knew it very well herself, but she got me studying English very early and then made all this effort to have me go into a special school in which they taught English. She always gave everything to me."

Iribe understood. His mom, who at times had raised him on her own, had also tried to give him everything. But grateful as he was, that didn't entitle her to his decisions. And soon enough, he was able to convince Antonov to see things his way (though, as a compromise to his parents, Antonov agreed to enroll in some evening classes).

To keep costs down, Iribe and Antonov moved into Iribe's childhood home and lived with Iribe's mother as they tried to get Sonic Fusion off the ground. Needless to say, it wasn't an easy time. Nothing embodied this struggle—this quest that always felt *just* out of reach, just *one* break away—more than the pile of uncashed employee paychecks that mounted ever upward on Iribe's desk.

Each check—made out to either Iribe, Antonov, or Dixon—represented a week that one of three had not taken a salary because there just wasn't enough money in the bank.

What made the struggle even more difficult for Iribe and Antonov was that, throughout this entire time, they were receiving almost unanimous praise for what they were doing. "It's really a case of having the right idea at the right time," explained tech analyst Laura DiDio to the *Baltimore Sun* in a 2003 article about Sonic Fusion. And yet, still they struggled.

Until 2005, when Iribe struck a deal with Firaxis Games to license GFC for the latest iteration of their flagship series *Civilization*. This was it! That big break had come! So Antonov made some tweaks to the core system and Iribe was "lent" to Firaxis "to integrate GFC and program the UI on Civ4." That fall, *Civilization 4* shipped on time to rave reviews; but instead of celebrating this milestone, Iribe pulled Antonov aside and said the hardest thing he'd ever had to say: "We need to kill our baby and start over."

"We just can't continue in this direction," Iribe explained, almost in disbelief at his own words. He loved GFC and thought it was a beautiful piece of code. "I'm extremely proud of it. But it's just too big. It requires way too much support. There's just no way we can scale it."

"But . . ."

"I know. Believe me, I know. But I'm out there, trying to sell and support this thing and all the developers think it's too complicated. They all just wish they could use Flash."

By Flash, Iribe meant Macromedia Flash, a graphics-based animation platform that became very popular around this time because of how smoothly and speedily it worked on the internet. And with internet usage drastically rising during this period, betting on Flash seemed like a pretty good gamble. That's why, as much as Antonov wanted to fight Iribe on this, he knew that his partner was right. And so, after six years and over three hundred thousand lines of code, they decided it was time to restart their start-up.

START-UP #1.5: SCALEFORM

Back to the drawing board, Iribe found an open-source Flash project online. Over the next six months, Antonov and Reisse used that as a starting point to build what Iribe had envisioned; a high-performance middleware package that could play Flash in 3-D video games. They named this new product "GFx" and, ready to scale their business, they renamed their company too: Scaleform.

The risk soon paid off, beginning with a call from Cevat Yerli, the CEO of Crytek. "We've been playing with your product," Yerli told Iribe, "and we're finally at a point where we're ready to use it."

Not only was this incredible news—Scaleform's first customer!—but, oh, what a customer it was. Crytek, a German-based studio, had only published one game to date; but that game, *Far Cry*, was one of the best games of 2004. And they were already at work on a second called *Crysis*.

"Wonderful," Iribe replied with an even voice. "This would be for *Crysis*?"

"Not just *Crysis*," Yerli replied. "We'll use it for everything going forward. For ten years."

Iribe could hardly believe his ears. Ten years? That was unprecedented for a middleware licensing agreement.

"For ten years," Yerli continued. "But we want it for free."

Iribe quickly gamed this out in his head—no revenue versus the value of partnering with a hot studio—and after a long pause, he agreed to the proposition. "But *only* if you promise not to tell anybody."

"Okay," Yerli replied.

"And you'll need to pay us for annual support."

"Sure."

Yerli also agreed to issue a press release and talk to the press about this ten-year licensing agreement. "Scaleform shares Crytek's commitment to delivering cutting edge experiences to gamers," said in an interview with Gamasutra. "Using our new state of the art Cry-ENGINE 2 and Scaleform GFx, Crysis will deliver an unrivaled cinematic experience."[2]

Within days of the PR blitz, Iribe received a call from Ray Muzyka and Greg Zeschuk at BioWare. Not only did they want to use GFx for their big-budget sci-fi shooter *Mass Effect*; but they also wanted to lock down a license for the next nine games after that.

After Crytek and BioWare, other developers followed. Within a year, Scaleform had about twenty licensees. Within two years, they were over fifty. And by 2008, they'd hurdled past a hundred. As a result, Scaleform became a top-tier middleware company and Iribe was now, in June 2012, able to enjoy some of the finer things in life. Like $150,000 cars, $50,000 watches, and weekends like this one—"West Hollywood Weekends" Greg Castle called them—where Iribe and his old Scaleform friends would check into a pricey hotel (usually the Mondrian) and for either twenty-four or forty-eight hours, they'd enjoy the spoils of what they'd accomplished together.

"What about Mike?" Sven Dixon asked. "Is he gonna join us?"

"Nah," Iribe replied. This wasn't really Antonov's scene.

"But he'll probably be joining us for dinner," Nate Mitchell added. Then just to Iribe: "Assuming you're able to confirm with the kid."

Nate Mitchell had joined Scaleform in the summer of 2008. Before that, he and Brendan Iribe had been hearing about the other for years from Paul Iribe—Brendan's cousin and Nate's childhood friend—who would constantly point out how similar the two of them were and suggest that they work on something together. But both being busy and not really having anything in common except computers, they didn't meet. Not until Mitchell's junior year at Dickinson College when, with summer approaching, he was searching for an internship in the video game industry. After striking out with all the big companies, Paul Iribe once again suggested, "Dude, my cousin!"

Immediately, Iribe and Mitchell hit it off. As with all great relationships, it's difficult to pinpoint the how and why. But certainly, a large part of it stemmed from the thing that had pushed Paul Iribe to introduce them in the first place: their similarities. Both were good (but not great) programmers; both were boyishly handsome, geeky but extroverted and suave enough to find their bearings in any conversations; and both were utterly, utterly relentless. No task was too big and no schedule was too tight; no matter the circumstance, Iribe and Mitchell would find a way to get things done.

Iribe—being nine years older and the CEO of a now-thriving middleware company—had way more experience than Mitchell, but he saw in Scaleform's intern a younger version of himself. Someone he could count on, someone he could mold. So much so that by the end of the summer, Iribe pleaded with young intern to skip his final year of college to come join Scaleform full-time.

"I can't!" Mitchell said. "I have to go back. My parents, you know?"

"All right. I guess. But please come right back after you graduate, okay?"

As would soon often be the case, Mitchell wound up exceeding Iribe's expectations. Not only did he return right after graduation, but Mitchell also spent his winter break working there. Scaleform just felt like home. He had a great mentor in Iribe and great engineers in Antonov and Reisse, and he was constantly surrounded by a bright, young roster of friends who felt like family. It was everything Mitchell could have hoped for . . . well, except for one thing: all his life, Mitchell had wanted to start his own business. With Scaleform, of course, that would never be the case. He hadn't been there from the start. But whenever the guys—his brothers, really, that's what it felt like—whenever they would talk about the "good old days," he'd always feel a pang of yearning.

Even though Mitchell hadn't been there to go from nothing to something, he was there as Scaleform grew from good to great. So great, in fact, that in March 2011, Iribe and Antonov sold the company to Autodesk for $36 million. Then, in a move that surprised his friends, Iribe decided to resign two months after the acquisition.

START-UP #2: GAIKAI

With Autodesk's resources and Scaleform's stable of assets, Iribe felt like the company he had founded was now fixed for success, and the time had come for him to *build* again. So Iribe said good-bye to his home state of Maryland and moved out west to become the chief product officer for a promising start-up called Gaikai.

Gaikai—based in Orange County, California—was a cloud-based gaming service devoted to delivering console-quality streaming experiences onto tablets, mobile devices, and Smart TVs. "It's like Netflix, but for video games," Iribe would explain to his friends. "On-demand gaming, right to your TV. No console needed, just a gamepad." And as for what Iribe would be doing there, that was best summed up by a recent 2012 article in *The Verge*: "He's here to help build an SDK [software development kit] that not only lets TV, tablet and set-top box manufacturers integrate with Gaikai's network, but game developers too."[3]

Several things attracted Iribe to Gaikai—innovative technology, a disruptive strategy, and the backing of top-tier VCs like NEA and Benchmark Capital—but what most sold him was the man whose vision it had originally been: David Perry, the prodigious game developer best known for 16-bit games like *Cool Spot*, *Earthworm Jim*, and *Disney's Aladdin*. Growing up, Iribe had worshipped Perry; and now heading into this new phase of his career, inspiration from one of his childhood heroes was exactly what Iribe needed.

"We're going to change the world!" Perry told Iribe.

Changing the world is an ensemble effort, which is why, within months of joining Gaikai, Iribe set out to bring over "his guys." Specifically, he wanted Antonov, Reisse, and Mitchell.

Of the three, Antonov was the easiest to recruit. Because like Iribe, ten years in the same business had made him ripe for a new start. Reisse was a little tougher woo, but Iribe knew the key to his friend's heart was freedom. Personal freedom was part of it—the ability for Reisse to continue being his quirky self and to work at his own quirky pace (i.e., taking off a day each week to rest his aching fingers)—but it was a threat to Reisse's professional freedom that had him concerned.

"So . . ." Reisse told Iribe, "Autodesk wants me to sign . . . that silly agreement."

All recently acquired Scaleform employees needed to sign it. The agreement itself was pretty standard, but Reisse—who cared deeply about the freedom and rights of programmers in the workplace—objected to things like noncompete provisions and intellectual property transfers, both of which were covered in the agreement he was being asked to sign.

"Well," Iribe said, his eyes lighting up, "if you come out here, you won't have to sign any of that. I'll make whatever you want work. You know me."

Reisse was in, Antonov too; all that remained was Mitchell. Which from the get-go, Iribe knew would be a challenge. Because unlike Antonov, it actually made sense for Mitchell to expand his role at Autodesk and begin filling some of Iribe's shoes. But that initial challenge became a whole lot tougher when in the midst of Iribe's effort to put the band back together, Mitchell had been offered a job at Riot Games doing UI development for a soon-to-be-superpopular new online multiplayer game called *League of Legends*.

"It's basically my dream job," Mitchell explained to Iribe over the phone, a long-distance call from France—from Paris—where Mitchell had some meetings scheduled for the following week. But he had flown in before the weekend so that he could spend a few days savoring the historic city; except that was not to be because Iribe then made it his mission in life to recruit Mitchell away from this so-called dream job.

Financial incentives. Career development. Being a part of something from the beginning. For the next three hours, Iribe put on the full court press, and then he continued for another three hours the next day. As Iribe laid out visions of Mitchell's future and waxed nostalgic about their past, it was impossible for Mitchell to focus on the present. The Louvre? The Eiffel Tower? The Arc de Triomphe? One by one, these historic landmarks faded into mere set dressing as Mitchell ambled through Paris, wrestling with existential questions.

All this made it that much more difficult for him to decide when a third challenger entered the ring to recruit his talents: Greg Castle. This is the same Greg Castle who—just moments earlier, during their June 15, 2012, West Hollywood weekend vacay—snatched some poolside fries out of Mitchell's now-tan hands. Be careful, Castle joked, because he might be back for more!

Of all the guys in this Scaleform crew, Castle was the one Iribe considered his "best and oldest friend." Which was sort of ironic, because Castle hadn't joined Scaleform until 2010. Even so, Castle *had* been there since the beginning, since the University of Maryland days. In fact, it was during Antonov's "hiatus"—that time when he went from Iribe's across-the-hall-neighbor to rooming with Andrew Reisse—that Castle and Iribe really hit it off.

Unlike Iribe, Antonov, and Reisse, Castle didn't know the first thing about coding. But he loved technology, entrepreneurship, and sales; this, in turn, made him the perfect pal for the nonengineering side of Iribe's personality, the part that loved dealmaking, risk-taking, and marketing the future. They shared these passions in such a genuine way that at the end of freshmen year, when Iribe told his friends that he'd be dropping out, Castle wasn't just one of the few to actually be supportive; he was also pretty damn impressed. So much so that while he did not have any money to invest himself, he introduced Iribe to some of his more well-heeled friends who invested early in Sonic Fusion. That was a gesture that Iribe would never forget—not when Castle left Maryland for a finance job in New York, nor when he left the US to start a restaurant in London—and it was a gesture he was able to pay back, years later, when he formally welcomed his best and oldest friend into the Scaleform family.

Castle had been brought on to run marketing, advertising, and PR for Scaleform, and postacquisition became a marketing manager for Autodesk. When Mitchell was in Paris, deciding between two great opportunities, Castle had gotten in the mix to try and convince him not to choose either.

"Stay at Autodesk!" Castle told Mitchell.

"Come to Gaikai!" Iribe would counter later. Oddly (but beautifully) in the midst of this back-and-forth, Iribe and Castle still hung out that weekend, respectful not to mention their respective efforts to lure Mitchell—other than to chivalrously say: "Best of luck to you, sir!"

The following night, Mitchell made up his mind and he called Iribe to share his decisions: "Sorry, I'm gonna go with Riot Games."

"I think you're making a mistake," Iribe told him.

"I know you think I'm making a mistake!" Mitchell shot back.

"Look," Iribe said, all the persistence in his voice now gone. "I understand. But I want you to remember one thing. You can always, always, always hop down if it's not good at Riot. Blast out and come down here, come work with me. You know, I'm only forty-five minutes away."

By this point, it was past midnight in Paris, so Mitchell thanked Iribe—seriously, man—hung up and went to bed. Finally, a decision had been made. Finally, a good night's sleep would be had. Or so he thought. "Nate," Iribe said a few hours later, after waking up Mitchell. "I thought about it and, just, no. You have to come to Gaikai. I'm sorry, you just have to."

"I already made up my mind," Mitchell replied. "Don't do this. Please."

"Nope, sorry. You're coming to Gaikai. Do that for me. Just come to Gaikai for six months. And if you don't like it, Riot will still be there. Trust me."

"You don't know that."

"I do. I really do. You think that in six months they're not going to remember how valuable you can be? Not gonna happen. And if it does, then I'll call them, and I'll keep calling them, until they remember. So just give me six months. Okay?"

"Okay," Mitchell replied. As soon as the word came out of his mouth, he felt a pinch of frustration. Not because he regretted the decision or because he felt bullied into making it, but because it felt so right that he didn't know what had taken him so long.

After successfully recruiting Mitchell—or "closing him" to use one of Iribe's favorite phrases—Iribe set his sights on closing an LA-based start-up called Wikipad.

START-UP #3: WIKIPAD

Like many gadget aficionados, Iribe first heard about Wikipad at the 2012 Consumer Electronics Show when the founders (James Bower and Matthew Joynes) demoed an education-oriented, glasses-free-3-D prototype tablet.[4] But unlike most of his fellow aficionados, it wasn't the 3-D aspect that piqued Iribe's interest; instead what really caught his eye was the tablet's detachable, snap-right-back-on game controllers.

This, Iribe thought, would be *perfect* for Gaikai. Because the problem with traditional tablets (like Apple's iPad or Google's Nexus) was that if you wanted to use them to play games through Gaikai's streaming service, then you had to use the tablet's default touchscreen controls, or you had to go out and buy a compatible game controller. But the beauty of Wikipad, as Iribe saw it, was that you immediately had this ready-to-rock gaming tablet. "The ultimate video game tablet," Iribe called it.

Initially, Iribe had envisioned some sort of partnership with Wikipad. But as he spent more with Bower, and as he further fell in love with the idea of what this tablet could be, Iribe ended up making a sizable personal investment and becoming a cofounder of Wikipad. And then in May 2012—the same month that Antonov, Reisse, and Mitchell moved from Maryland to California and all officially started at Gaikai—Gaikai announced a partnership with Wikipad to create the "World's First Gaming Tablet."

Now, one month later, from the rooftop of the Mondrian, Iribe found himself thinking about an entirely different hardware start-up—

one that he had heard about over an unexpected phone call he had received just one week earlier.

START-UP #4: OCULUS?

Like in *Back to the Future*, when Marvin Berry calls up his cousin Chuck to tell him about that new sound he might be looking for, Laurent Scallie had called up Iribe to tell him about a business opportunity that might just be the one. The big one. "You gotta meet this kid Palmer Luckey!"

"Who?" Iribe asked, exiting his silver Audi R8 in the Gaikai parking lot.

"Palmer Luckey," Scallie replied. "He's doing this VR thing and it's really working. You gotta meet him, Brendan. VR's gonna really work this time."

Iribe loved Scallie's enthusiasm, he always had. But unfortunately—as a short montage of Scallie making similar VR proclamations played in Iribe's mind—it was hard to take seriously. VR, really? It's been thirty years and it hasn't worked. But then Scallie said something that gave Iribe pause: John Carmack was involved.

"I don't know the nature of his involvement," Scallie explained. "But he was demoing Palmer Luckey's Oculus headset at E3."

If Carmack's excited about this, Iribe thought, then I should at least take a look. So as soon as he got into his office, Iribe started reading stories about this kid named Palmer Luckey. And about how John Carmack had demoed his VR set at E3 to reviews like these:

- "It's beyond thrilling . . ." (Eurogamer)[5]
- "Truly astounding" (CNN)[6]
- "The level of immersion was unlike any other gaming experience I've ever had, and that bodes well for the future if Carmack or someone else can take the tech to the next level." (PC Gamer)[7]

After reading through these stories, Iribe was overwhelmed by a thought: I don't want to be the guy who misses out on the thing that changes the world.

"And if this is real," Iribe said, grabbing Antonov and Mitchell, and pulling them into his office to show them what he'd just read "then we should try to get involved."

Antonov and Mitchell were still skeptical—echoing exactly what Iribe had thought just fifteen minutes earlier: VR, really?—but, like him, they saw a possible opportunity and wanted to explore this further. So Iribe asked Scallie for an introduction to Palmer Luckey; and then he tried to arrange a dinner with the young inventor. Which is why, beside that immaculately blue rooftop pool, Iribe kept checking his phone. Hoping to hear back from Luckey, hoping to lock down plans for the evening. Luckey replied that, yes, he could meet Iribe and his friends at 9:30 at STK Steakhouse.

"So tonight," Iribe explained to Greg Castle after hearing about the dinner, "I just made plans to meet with this guy named Palmer Luckey. He's this supersmart kid who dropped out of college to build virtual reality headsets. This is something that Nate, Mike, and I might want to get involved with. You down to join us?"

"*Virtual* reality?" Castle asked—playfully, skeptically, lovingly.

"You gotta see this," Iribe said, pulling out his phone and playing a video of John Carmack demoing a prototype of the Oculus Rift at E3.

Castle was impressed. It was cool, very cool. But what wowed him most was that in the fourteen years they had known each other, he had never seen Iribe so excited about a business opportunity. Truth be told, Castle wasn't totally sure why his best and oldest friend was *this* excited.

"This thing," Iribe started, his eyes brightening with visions of five, ten, twenty years down the road, "it's going to be the future."

CHAPTER 5
STK

June 15, 2012

UNLIKE THE SMOKY WOOD AMBIENCE AND DARK LEATHER DECOR TYPICAL OF most upscale steakhouses, the STK on La Cienega—with its dramatic gold lighting, faux crocodile tiles, and towering wall of bleached steer horns—took LA swank to a level that Palmer Luckey had never seen before. Which is why, when Luckey stepped inside—wearing flip-flops, cargo shorts, and a gray Atari T-shirt—he received stares from just about everybody nearby.

"Palmer!" Iribe boomed, coming over to collect his guest. "You made it."

"Brendan? Oh hey!"

Sensing that Luckey was out of his element, Iribe smiled and gave Luckey a friendly it's-not-as-bad-as-it-looks pat on the back. "Follow me," Iribe said and then led him across the crocodile tile floor and over to a circular booth in the corner, where Luckey was a bit surprised to see five other guys who—at first glance—all looked to him like they could have passed for clones of Iribe. Then one by one, they introduced themselves:

MIKE ANTONOV: Software architect at Gaikai. Cofounded Scaleform with Iribe.
NATE MITCHELL: Senior product developer at Gaikai. Before that a Scaleform guy.

SVEN DIXON: Designer at Scaleform/Autodesk. Known Iribe since forever.

GREG CASTLE: Former finance guy. Currently marketing manager at Autodesk.

JAMES BOWER: CEO of Wikipad. Cofounded with Iribe.

And then, of course, Brendan Iribe. "We just started Wikipad," Iribe explained, making it sound like starting up companies was as easy as vacationing in nearby cities. That was kind of the whole point of this evening: to convince Luckey that starting a company wasn't a big deal.

Given this objective, Iribe couldn't have asked for a better ensemble. Four guys whose faith in him had paid off handsomely when Scaleform was sold. Plus another, Bower, whose start-up Iribe had financed. Forces combined, these guys were a walking, talking advertisement for Iribe's expertise and character. All of which was set against the backdrop of shiny, beautiful people. Even if that wasn't exactly Luckey's scene, Iribe's crew couldn't help but think the whole situation felt like a scene out of *The Social Network*, the one when Sean Parker dazzles Mark Zuckerberg and Dustin Moskovitz with a glimpse of what their lives could be. Except in this situation—as Castle would later put it—it was "six sharks and a deer caught in the headlights."

Although what Castle next noticed, and soon the others did as well, was that this kid was no deer. Because the minute Luckey opened his mouth and started talking—building momentum with each sentence; segueing, tangenting, and bantering with great aplomb—it was clear to everyone around the table that there was something magical about Palmer Luckey.

"Is it true you sent Carmack your one and only prototype?" Iribe asked. "No paperwork? No money? No nothing."

"Yep!" Luckey replied. "I mean . . . when Jesus asks to borrow your clothes, your boots and your motorcycle, you say yes."

It took the guys a moment to get Luckey's *Terminator* reference, but even during those seconds the verdict was already in: they loved this kid. He was confident without being arrogant. He was intelligent but still intelligible. And through the night—whether talking, listening, or just picking at Shrimp Rice Krispies—there was this exhilarating and invigorating is-he-really-only-nineteen? air of genius to him.

"And you dropped out of college?" Castle asked. "Is that right?"

"Hey!" Iribe chimed in. "Nothing wrong with that."

Luckey chuckled. "Well, I *don't* think there's anything wrong with that. But my situation was a little . . . complicated. It's a long story, and unfortunately not a very interesting one. What it boils down to is that Cal State Long Beach screwed me and I can't really blame them. Basically, there's this thing called 'The Long Beach Promise,' which is a system set up between the California State universities and the local community colleges. It was designed to keep the universities from being too overcrowded by encouraging students to use the community college more. And then if you had a certain grade point average, they 100 percent guaranteed that you'd be able to transfer into a Cal State. So that was a really cool program. Up until the Long Beach Promise was suspended, right around when I needed it. So what a promise it was!"

"And where were you before that?" Castle asked.

"I was homeschooled for most of my life."

"That's awesome," Mitchell said. "Was it by yourself? Or with siblings?"

"Let's see," Luckey said, running through the chronology in his mind. "It was my sisters for a lot of it. But they ended up going to private high school. A lot of it wasn't my mom teaching me. There's a lot of, basically, self-driven learning courses. I was mostly doing it myself. But it only works if you're a certain type of person. Like my sisters were not into that. They were not self-motivated or self-driven. So they couldn't keep up with the work and with the testing. So they had to go to a school system and I was lucky enough to escape that."

"Well done," Iribe said, laughing. "Did you ever feel like you missed out on stuff? By, you know, not going to high school?"

"Yeah, it's a good question," Luckey replied. "But not really, no. Because, one, I had a lot of friends outside of school. Two, I was involved in a lot of extracurricular stuff. And, three, I had the internet and I had online communities. Plus, you know, I always had my own projects to work on. But having nothing to compare it to, I don't know, I guess it's possible I missed out on some valuable life experiences."

"Nah," one of the guys said. "The only thing you missed out on was stupid high school drama."

"Yeah, maybe," Luckey said. "But internet drama is better anyway!"

After Luckey recounted some of his favorite online sagas, the conversation shifted to his father. Was he also an inventor? Is that how you got started on all this?

"No," Luckey explained. He was a car salesman at an Acura dealership in Cerritos. "And my dad's probably one of the best car salesmen in California. Like he actually has won awards."

This information was a big hit with the guys around the table. Where does your dad work? What are his favorite cars to sell? What is it about him that makes him such a good salesman?

"He knows how to sell people," Luckey said with a big smile on his face. "He knows how to convince them that *you need to leave in a car today*! I used to go to work with him and learn all the tricks. You know, like 'Hold on, I'm gonna go talk to my manager and see what I can do.' And then I'd go into the manager's office with him and we'd all just joke about stuff that was totally unrelated. Then, after a few minutes, my dad would come out and say, 'Good news, I was able to get you below invoice!' So, yeah, it was really fun. And my dad is what they call a 'closer.' Like if one of the other employees is struggling to close a deal, they bring in my dad and he gets it done. His job is to finalize deals. For a small cut, of course!"

"That's awesome," Iribe said, seeing a connection to himself that instantly made him feel closer to Luckey; and, he hoped, vice versa. "I like that: the closer."

"Yeah," Luckey explained, "but being a closer also meant that you're closing up the dealership every night because you have to be there until every last customer leaves. I remember, when I was a kid, I was, like, 'Dad, why do you got to work every single day?' And he told me [in an older man's voice] 'That's what you gotta do, because you're a man, and a man needs to provide for his family.' So I asked if he liked doing his job. 'Most of the time,' he said. 'But there are lots of times when I hate it—when I hate it a lot. But you still gotta do it whether it's a good day or a bad day.'"

Eventually, the conversation shifted to what had brought them there that evening: virtual reality. And Mitchell asked the question they'd all found themselves wondering at some point during dinner: *If virtual reality really is as awesome as you say it is, and as awesome as John Carmack says it is, then—with all due respect!—how the hell did a teenager end up being the one to invent this awesomeness?*

Luckey blushed. What Mitchell was essentially asking was: Why are you the Chosen One? What makes *you* so special? The truth was that Luckey didn't have a good answer.

"My obsession started when I was sixteen years old," Luckey said. "But I had been curious about VR way before that. Except it was in the same way that millions of people were curious about VR; like it was just this crazy, cool sci-fi thing that surely would never happen in my lifetime! I thought of it the same way I thought about time travel."

His admission seemed bizarre, until everyone at the table realized that they were all guilty of thinking the same thing prior to learning about Palmer Luckey. Deep down, they'd all known that VR didn't really occupy that same sphere of apparent impossibility as time travel. After all, these guys had lived through the '90s and all tried out headsets before; shitty ones, sure, but headsets nonetheless. "I'll own that," Iribe admitted through laughter. "But what I want to know is if *anybody else* was thinking seriously about VR. Like outside of the hobbyists you met on that forum, are there any, you know, companies or whatever that might be serious about VR?"

Luckey explained that there *were* a handful of companies that made VR-related hardware, but that none seemed to have any interest in selling something affordable and consumer oriented. "Definitely none with a wide field of view," Luckey said. "And definitely none with a focus on gaming. The closest thing, though not even really at all, would probably be Valve."

"Wait," Iribe said. "Valve? As in *Valve* Valve?"

Technically speaking, Valve was a game company that was best known for developing a handful of instant classics (such as *Half Life*, *Counter-Strike*, and *Dota 2*) and operating the world's biggest game distribution platform (Steam). But to call Valve "a game company"—even to do so with superlatives—still missed the point. Unlike Nintendo or Blizzard or any of the other truly beloved game makers, there was a mythology to Valve that made it seem less like a company and more like a religion. For these reasons and many more, Valve was uniquely able to attract many of the most talented designers, developers, and engineers—including, very recently, a prominent programmer named Michael Abrash.

Within the gaming industry, Abrash was a legend. Not amongst gamers, per se—not like John Carmack—but amongst those who worked behind the scenes in the industry. Some knew of him from his books on coding—books like *Zen of Assembly Language* and *Michael Abrash's Graphics Programming Black Book*—while others knew of him from his work at id (where he co-created *Quake* with Carmack), at Microsoft (where he was the graphics lead on Windows NT; and then later returned to help launch the Xbox) or at Rad Game Tools (where he co-wrote the Pixomatic software renderer).

"When he was at Rad," Iribe said, "I'd go up to his booth and talk to him every couple years at GDC—just being, like, in awe of what he had pulled off with that Pixomatic thing. And he did it all by himself, basically."

"How well do you know him?" Luckey asked.

"Not that well," Iribe said. "Just from talking to him at his booth—I would just walk up and be like a young nerd in awe of super godfather nerd."

The reason Luckey asked was because of what Abrash was now working on at Valve: wearable computing. And specifically—as was now publicly known via his recent posts on a blog called *Ramblings in Valve Time*—Michael Abrash was researching augmented reality.[1]

"Wait," Castle said. "What's that? Augmented reality?"

"The easiest way to explain it," Luckey said, "is basically: Virtual reality makes you feel like you're some place totally different and the entire environment is synthetic. Augmented reality places some digital elements into your view of the real world. So that's really cool, but it's like ten years away from being viable in any legitimate way."

"Do you anticipate that Valve will produce a headset of their own?" Antonov asked. "For VR or for AR?

"Actually," Luckey explained with a beat of pride, "the guy I've been in touch with over there"—meaning Dan Newell, via John Carmack—

"wants me to make them an AR-version of the Rift. But the short answer to what I think you're really asking—like: is Valve going to be a "competitor"—is I seriously doubt it."

Luckey told the guys that—as far as he was aware—Valve saw VR as a "stepping stone" to AR and that, really; that their goal was to bring games to AR and AR headsets (not to make the headsets themselves); and that, believe it or not, Michael Abrash—this man they all revered!—was "particularly susceptible" to simulator sickness.

"Also, interestingly enough," Luckey said, "I actually applied for a job there a couple months back. In April, I think. But I got rejected. They said Abrash said they were looking for people with 'a proven track record of self-directed work.' Oh well!"

By the time their food arrived, Luckey still wasn't sure he could tell these guys apart in a Generic Business Dude lineup, but he had to admit they weren't what he expected. These guys were thoughtful, considerate, and, most surprisingly, genuinely passionate. By the time they started talking about Luckey's plans for virtual reality, he felt as though he could at least trust them enough to have an honest conversation.

"So my plan," Luckey explained, "well, prior to the past couple weeks, was to do a Kickstarter for prebuilt kits. I mean, I basically already know everyone with any interest in VR. So it's very doable to . . . basically, I was gonna bring some people together, have a pizza party, and build these things. Probably like a hundred or so."

Iribe tried his best not to look at Luckey like the kid was insane. But that was sort of impossible because if that's all he was planning to do with it, then he was insane. Of course he was leaning toward Sony if that was the alternative. "Dude," Iribe finally said, with a gentle shake of the head. "No. Just no. You can't do that."

"Well," Luckey said, flashing a smile, "*obviously* my plan has evolved."

"Are you speaking to anyone else besides Sony?" Nate Mitchell asked.

"John said that id—or I guess, really, their parent company ZeniMax—might be interested in giving me some funding. But it didn't sound like a very good deal."

"Are you guys still on good terms?" Iribe asked.

"With John?" Luckey clarified. "Yes, very good terms. With the businesspeople? I'm not really sure. I'm honestly not even sure what they want, other than to hold off on the Kickstarter. But the guy there—Todd Hollenshead—he's been really nice, at least. And they haven't uninvited me from QuakeCon, so I guess that means we're still good!"

The guys laughed.

"So there's that," Luckey continued, "and then there's one other possibility. My boss at ICT—well, technically, I don't know if he is my actual boss—anyway, Mark Bolas, who you might have heard of, he's very well connected with the Disney Imagineering Lab in Burbank. And he told me that, if I wanted, he could probably get me on the team over there. Working on VR."

"No, just no," Iribe repeated with a smirk on his face. "As cool as that sounds, you'd be better off doing the pizza party option. At least that way it's yours."

"It is hard to say this in the exact way," began Michael Antonov, "but this way—that Brendan is describing—it is the best feeling in the world."

Antonov's comment earned nods from around the table.

"You need to be able to control your own destiny," Iribe elaborated. "Otherwise, what's the point? I know it's cliché to call it 'your baby,' but that's really what it is. You can't let someone else take your baby. How can they possibly raise it better than you? All they see is a chance to make some money; they'll never be able to see what it is you see."

Luckey weighed these words. "I agree with you," he said, "but then let me ask you, and I don't mean this in an offensive way, isn't that what you want me to do? To work with you?"

"*With*," Iribe said, without hesitation. "I want to work *with* you to help you grow your vision and ensure that you have the resources to make it happen."

"Sure," Luckey replied. "Except—"

"—Except for a million things," Iribe said, cutting him off. "And maybe one time we'll talk them all out. But—and I mean this so sincerely; just trust me here, please—that's not even what this dinner is about. Obviously, I think it would be awesome if you let us help you make this thing *huge*, but what really, really matters—so much more than working with us or not working with us—is that you work on this yourself."

"What about Laurent?" Luckey asked. He didn't have any particularly strong desire to work with Laurent Scallie, the man who had introduced him to Iribe; but Luckey figured that if he and Iribe ended up moving forward, Laurent would probably want to be involved.

"Don't worry about Laurent," Iribe said. "This dinner isn't about Laurent, it's about *you*. Making sure that you understand your options and you do what's best for you."

"I want to do what's best for VR," Luckey replied.

"Of course!" Iribe agreed. "That's what we want too! All I'm trying to say is that opportunities like this are rare. Trust me. You only get a shot like this once in your life."

By midway through the meal, Luckey was convinced. This really was, most likely, his one and only shot. So of course he couldn't allow himself to be a spineless sellout! He needed to see this thing through to the end. But there was just one problem with that: he was practically broke. How do you turn down a $70,000 salary when you have less than $700 in your bank account?

"Here's the thing," Luckey said, with a hint of vulnerability. "Of course I want to run my own company. That's always kind of been my dream. But can I really afford to pay myself nothing? With an offer for so much money on the table?"

"Palmer," Iribe said, with a restaurant-sized grin. "You realize that if you start the company, you can pay yourself whatever you want."

Luckey's lit up in a way that revealed his age. "Oh my God. You can really do that?"

For the next hour or so, the conversation shifted to all the wonderful things that you can do when you run your own business. All the fun you can have. All the money you can make. And, in fairness, all the hours of stress that your mind and body can only barely handle. But that didn't deter Luckey. He was already conditioned for that kind of life anyway. And yet, as much as can-do spirit dominated the conversation, the evening's most important moment came when Iribe told Luckey the one thing he couldn't do.

"You can't sell a kit," Iribe emphatically stated.

"What do you mean?" Luckey asked. "You think I should wait and then sell, like a finished product or something?"

"Oh no, sorry," Iribe said, swiftly shaking his head. "But it's good you brought that up because it gets exactly to my point. Let's say you *did* suddenly have a totally polished, cool-as-all-hell HMD that anyone in the world could buy and experience VR at an affordable price. Then what?"

Luckey didn't quite know where Iribe was going with this, and by the looks of it, neither did his Chorus of Entrepreneurs. "Okay," Luckey said. "Then what?"

"Then *nothing*!" Iribe exclaimed. "Why would anyone want to buy this thing? Sure, the hardware is cool, but you've got no software. Or maybe, somehow, you've magically managed to convince a publisher, maybe even a *huge* publisher, to make a game for your hardware: well, that's still only one game. That's not a compelling reason to spend a few hundred dollars."

"I agree," Luckey said. "There needs to be a whole ecosystem of software."

"Exactly," Iribe said. "And what's the best way to accomplish that? An SDK."

By "SDK," Iribe meant "software development kit," which was an all-in-one suite of tools that gave users the ability to build comprehensive applications for a specific platform. This was actually part of what Iribe was pitching for Gaikai—an SDK that made it easy for developers (devs) to create content for Gaikai's cloud-based platform. And so Iribe was suggesting that Luckey do a similar thing: create an SDK that would empower devs to create VR content and bundle that with his Oculus headset.

"Can you imagine how much content would be created if you packaged your hardware with an SDK?" Iribe asked.

He was so fired up about this idea, that Luckey almost didn't have the heart to say what he said next: "Most of the VR headsets out there, they already come with their own SDKs. Like I know eMagin has one . . . Vuzix has one . . ."

Iribe appeared completely undeterred. So much so that Luckey was impressed. "I don't know too much about those, but what I'm talking about is a *super* easy-to-use, well-documented SDK that'll integrate with *all* the big game engines."

"Like Unity and Unreal?"

"Yep."

"Oh shit . . ." Luckey replied. Now he was the one who felt fired up.

"And not to sound arrogant," Iribe said, "but Nate, Mike, and I are probably the best people in the game industry to create such a thing. I mean, we've got *incredible* long-term relationships with Unity and Epic [the company that made Unreal]."

By the time Luckey got back to his trailer, he was already brainstorming ways to turn down Sony. Iribe and his crew had convinced him that he needed to see things through with Oculus—though he wasn't yet sure how much he wanted them to be a part of that. They seemed like smart, passionate guys, and they clearly had experience turning ideas into companies; but it felt weird starting a VR company with guys whose passion for VR was less than one week old. Then there was the fact that every argument they had made against working with Sony could be applied to working with them. As if Iribe had known what he was thinking, Luckey received an email from his new potential partner:

FROM: Brendan Iribe
TO: Palmer Luckey
SUBJECT: Oculus

Hey Palmer,

Just wanted to send a quick note and say how much everyone enjoyed meeting you and hearing the story.

We're fully on board to help you build out the "vision." Pun intended;-)

We're also excited to help make it even bigger than you imagined . . . And in the end, it's all about you, your vision, your company, your product.

You'll hear me say many times, it's going to be a really fun ride!

Let me know what time you're available tomorrow. Girlfriend first of course.

Talk soon,
—BRENDAN

Shortly thereafter, Luckey replied:

FROM: Palmer Luckey
TO: Brendan Iribe
SUBJECT: Oculus

I really enjoyed the meeting as well, thanks to you and all your friends for convincing me out of my rationalizations! Can't wait to get into this, very enthused.

I will let you know, thanks for being understanding about the time thing.

Best,
PALMER

CHAPTER 6

PIVOTS, PROTOTYPES, AND PARTNERSHIPS

June 2012

"WHO IS OUR AUDIENCE?" MITCHELL RHETORICALLY ASKED IRIBE THE FOLLOW-ing morning, as the two brunched at Habana, a Cuban joint in Costa Mesa. They were still starry-eyed from their meeting with Palmer Luckey the night before, so much so that even before knowing whether Luckey was going to turn down Sony's offer and move forward with Oculus, they decided they were ready to get to work. "Who is our audience?" Mitchell repeated. "And what are the proof points that will get our audience to care?

The answer to that first question, they determined, was threefold: "hard-core" gamers looking for a new way to experience their favorite games; game developers interested in experimenting with VR and/or adding VR support to their old/new games; and average consumers interested in trying the latest in virtual reality at an affordable price.

To get those audiences to care, Iribe and Mitchell began compiling a list of things they could do prior to build value for Luckey's project:

- Create integrations with popular game engines (i.e., Unity, Unreal, CryEngine)
- Provide visual explanations of the hardware's technical specifications and advantages

- Highlight Palmer Luckey's unique backstory
- Get endorsements from trusted industry experts like John Carmack

"Who else do you think we can get?" Mitchell asked, referring to endorsements by heralded game developers like Carmack.

Iribe had close relationships with a lot of developers. Getting endorsements would likely just be a numbers game; meeting with a lot of devs, as they called them, in a short period of time and seeing who might be willing to lend some of their credibility to this project.

For hours, Iribe and Mitchell brainstormed like this. Then after brunch, hours turned to days as they continued plotting ways to amplify Luckey's vision for Oculus, from a Kickstarter campaign to all the beautiful uncertainties that lay ahead.

"Nate," Iribe said, calling Mitchell from the road the following week. "I figured it out: this has to be marketed as a devkit."

By "devkit," Iribe meant "developer kit." And what he really meant was that instead of selling DIY headsets (as per Luckey's original plan), or even preassembled units (for those who didn't want to build headsets by themselves), Oculus would sell preassembled units to only those who built games: the developers.

For console makers like Sony, Microsoft, and Nintendo, getting devkits to developers in advance of a new product release was par for the course. But it was very much a behind-closed-doors process in which devkits were only offered to "select partners." And even if you were lucky enough to make that list, the devkits themselves were usually incredibly expensive[1]—for example, early devkits for Sony's PlayStation 3 sold for nearly $20,000[2].

So what Iribe was proposing was uncommon in the gaming space. But it actually made a whole lot of sense, as both a way to tap into the booming indie development space, and also a way to try and spare Oculus from the hype cycles that have plagued VR headsets of the past. Because, at least in rhetoric, what they'd be selling *wasn't* a headset. It was a devkit—just a devkit. That would keep user expectations low and also buy Oculus time to ramp up to a consumer product.

"I love it," Mitchell told Iribe. "We should totally do devkits."

Conversations like these continued throughout the week after meeting with Luckey. The more time Iribe and Mitchell spent architecting what Oculus could become, the more excited they each got. A new company! A new journey! There was just one problem: they still didn't have an answer from Luckey. And worse, Luckey's phone appeared to be off. Everything went to voice mail.

As Iribe would soon learn, this was not unusual for Luckey. The kid just didn't like carrying a cell phone; partly because he was a bit of a conspiracy enthusiast (and didn't want to be tracked), but mostly because he just didn't want to be available to everybody who wanted to talk to him whenever they wanted. Besides, he was almost always available online.

Which, of course, meant that Luckey could have replied to Iribe over email. He should have, he knew that, and he felt guilty for not doing so sooner. But the truth was that Luckey still wasn't sure if he wanted to move forward on his own, or with Iribe. Sensing this might be the case, Iribe kept reaching out and eventually persuaded Luckey to come meet him in person at a Starbucks in Long Beach on June 22.

"You made it!" Iribe said with a bright smile, gesturing for Luckey to sit down.

"Actually," Luckey said, "I'm gonna get something to drink."

Iribe reached for his wallet, which made Luckey uncomfortable. He hated taking from people without giving in return. He kindly waved off Iribe, walked over to the counter, and ordered a drink. While waiting for it to be made, Luckey glanced at the ATM receipt in his pocket: $306.68. All the money he had to his name. He wondered what Iribe would have to say about such a meager balance. Would he recoil in disgust? Would he remember what it's like to have a three-figure net worth? Luckey stuffed the receipt in his pocket and returned to the table.

"What did you get?" Iribe asked, bright as can be.

"Just hot chocolate."

"Not a big coffee guy?"

"I don't do caffeine."

"Really. So what do you do when you need a jolt?"

"Snap out of it!" Luckey playfully replied, earning a laugh from

Iribe. "Sometimes sugary stuff. I keep caffeine in my glovebox. Just in case. Emergency road trip fuel."

"You're an interesting guy, Palmer Luckey."

Luckey smiled and took a swig of his chocolate.

"All right, Palmer," Iribe began, trying not to begin too firmly. "It's been a week . . .

Luckey apologized for the delay and for occasionally going "off the grid."

"That's okay," Iribe explained. "I get it. But you officially turned down Sony, right?"

This was correct, technically speaking. On June 20, Luckey had emailed Sony to say, "Sorry to bow out at this stage, but I was made an offer I could not refuse by people who can help me develop my current HMD on my own." That was that, Luckey thought. But then shortly after that, Sony emailed Luckey to offer him twice the money: $140,000.

"Look," Iribe said. "This is a big decision you are making. So, actually, I *respect* that you're taking your time. But I really think we've gotta do this together. It's going to be awesome."

Luckey didn't reply.

"I promise you, it's going to be great. Nate's already been coming up with all sorts of ideas for the Kickstarter and Mike is psyched to get started on the SDK. Those guys want this. They believe in this, they believe in you. But they're starting to get a little frustrated that we're still sort of in this limbo."

Luckey nodded. He really was sorry about that. There were few things he hated more than wasting people's time and, as he now realized, it was time to stop wasting theirs. "You know, Brendan, I've been thinking about it a lot and I've decided I'm gonna do it on my own."

"Mmmmkay," Iribe said, digesting the news. "Why?"

"I really like everything you're saying, and I appreciate all the advice you've given me. But you're talking about it like it's this big thing. But I know the community and it's just not like that. I think I can get my friends together on this Kickstarter, and it's only going to sell probably a hundred or two hundred of them, and I'll get my friends together and I'll buy them all pizza and they'll help me build this thing."

Iribe pursed his lips and considered Luckey's response. After several beats, he shook his head and replied, "No."

Luckey's head flinched back in confusion. No? No to what?

"That's not acceptable, Palmer. That's not a great idea, I'm telling you. You can do it with us and it's going to be so much bigger. Trust me. You need to reconsider."

Luckey laughed, which was maybe an odd response.

"I think I have a good idea," Iribe said, nodding with confidence. "How about we table this for a minute and go get something to eat?"

"I like that idea."

So Iribe and Luckey got up, left Starbucks, and walked across the street to low-key spot called BJ's Pizza Grill. They grabbed a booth by the window, ordered food, and then picked up the bumpy conversation. In a new venue, it no longer felt as uncomfortable. "I know sometimes I can come on strong," Iribe said. "I'm sorry if that's the case. I don't want to pressure you into doing something you're not 100 percent psyched about; and even though I know in my heart that if you work with us you'll get everything you want and more, I'll stop pushing on that. All I really want to know right now is when can I see the prototype?"

"Soon," Luckey confirmed.

"Soon?" Iribe echoed. "I mean, I'm ready to start a company with you—to take this enormous leap—and I still haven't seen the thing. That's a little crazy, no?"

Luckey couldn't disagree. Wanting to put thousands of dollars into something you've never seen or tried yourself; that did indeed seem crazy.

"What's really going on here?"

"There's nothing going on!" Luckey replied. "I'll get you a demo soon. I just need to buy a few new parts, but there's a problem with my credit card."

"Got it."

"It's not a payment issue or anything like that."

"Look," Iribe said, now digging something out of his pocket. A checkbook. "I'm gonna cut you a check right now. How much would it cost for you to build me a headset? Actually, it'd be great to also get one for Nate, Mike, and Andrew. So how much do you need to make four?"

"I don't know exactly," Luckey said. "I can check the prices when I get home. But I recently did four for Sony and it came out to $3,700. That included shipping though."

"Great," Iribe replied, starting to fill out a check. "I'm going to give you $3,700. No strings attached. If you disappear after this and I never hear from you again, I get it. But if that's not what this is about, then please go buy parts. We want to at least see a freakin' prototype!" Iribe signed his name and handed the check to Luckey. "Okay?"

Luckey stared at the check, entranced. This wasn't about the money exactly, but the generosity and curiosity of the guy who had given it to him. "Maybe," Luckey began, sounding clearer than he had all day, "we should do this together."

"Yes!" Iribe replied. "Yes, Palmer, let's do this together."

Luckey and Iribe smiled and shook hands. Then, as the burst of excitement subsided and they felt more like just two guys eating pizza—permutations of partnership running through each of their heads—they both couldn't help but wonder: So how, exactly, is this relationship going to work?

CHAPTER 7

FREEDOM IS HAPPINESS

June/July 2012

"I NEED TO UPDATE YOU ON SOME STUFF," LUCKEY TYPED TO HIS TRUE-BLUE buddy Chris Dycus over Skype the day after his meeting with Iribe.

"What stuff?" Dycus asked, assuming it would probably be something related to their beloved online forum, ModRetro.

"Sony offered to hire me after E3," Luckey said. "$140k salary."

"Holy crap," Dycus replied. Here he was, just a senior in high school—graduating in a few weeks and then taking community college classes in the fall—and his best friend—who was only one year older—was being offered the job of a lifetime.

"I turned them down," Luckey said.

"What?! Why?!?!"

"I was going to go with them, but then I met these venture capitalists with crazy amounts of money. And they said that they would make sure I have money to start my own company . . ."

"Wow," Dycus replied. "That is awesome."

"Yeah. So skip college and come work with me."

"You being serious?"

"Yes.

"I'm gonna need a few more details before I can decide something like that," Dycus said. "Like: what would be my daily duties?"

"You would be helping me to fabricate prototype units, doing some slight design revision."

"Will you be on the same level as me, for the most part? Doing similar things? I just don't wanna feel alone, though I'm sure I'll be fine no matter what."

"I will be focusing more on the hardware design side," Luckey replied. "Since I know VR like no other. But we will be working closely together. Even if you cannot stick around after the summer, I at least want to get you around for a couple of weeks."

A couple of weeks? Sure, Dycus could commit to that. Building prototypes sounded like fun (and as an added bonus, it'd get his mom to stop bugging him about finding a summer job). But there was just one problem. "I still have no car," Dycus reminded Luckey. "And my sexy, sexy legs aren't quite toned enough to bike there."

"After I get my first paycheck," Luckey said, "I'm gonna use that money to move out. So you can stay with me for sure. At least until September. Sometime around early September, I am proposing to Nicole."

"Oooh. Nice!"

"Anyway, let me know when you finish high school. The way things are going, it will probably ramp up in Mid-July. And if you like things, maybe you can ditch college. Unless your parents would kill you . . ."

"If I'm already enrolled and ditch, yeah, I'd be dead."

Luckey understood. A few weeks earlier, he had told his parents that he was going to take some time off of school to start Oculus. They did not like this idea, and had even started threatening to sell his trailer. "But I am going to take a year off for this," Luckey told Dycus. "If it succeeds, then I put it off more. If not, oh well, I will have had a lot of fun and made some money . . ."

TO CELEBRATE THE FOURTH OF JULY, LUCKEY WAS PLANNING TO TAKE EDEL-mann someplace special for fireworks. But first he had business to attend to: demoing the Rift for Iribe, Mitchell, and Antonov at the Hilton Hotel in Long Beach.

By this point—as agreed to at BJ's Pizza Grill—Iribe was already in. To what extent was still a question, but the size of his investment and how much business counsel he'd be able to provide would come down to a combination of how he felt about today's demo and how many of his after-hours at Gaikai he wanted to devote to this Oculus thing. Mitchell, not having profited nearly as much as Iribe (or Antonov) from the Scaleform acquisition, wasn't in a position to make a significant financial investment. But as had been the case throughout his life, Mitchell made up for whatever resources he lacked with an insatiable degree of effort and enthusiasm.

That's why, even before any papers were signed, Mitchell had already devoted all his free time to figuring out how they could launch a Kickstarter campaign that would surpass any that had come before. In particular, he'd been racking his brain to try and figure out a way to destigmatize virtual reality for all the skeptics out there. And a *lot* of big-time skeptics were out there. In here, too, as Antonov didn't yet know what to make of this VR start-up that Iribe was already using "we" to talk about.

That seemed impulsive to Antonov. But that had always been one of the big differences between him and Iribe. Whereas Antonov liked to learn everything he could first and then take a few weeks to vet his thoughts, Iribe had a particular way of deciding things quickly. He liked taking risks and had no problem jumping into things. So it didn't really surprise Antonov that Iribe had already written Luckey that $3,700 check and agreed to pay for the production of a high-quality Kickstarter video. But what did surprise him was that Iribe had already started to plant the seed that they should leave Gaikai and go run Oculus. Come on, Antonov had replied each time Iribe brought it up. Let's at least see the device first!

"There he is!" Iribe exclaimed, as Luckey ambled into the lobby, carrying a white bucket filled to the brim with electronics.

"Okay," Antonov said, grinning at the sight of Luckey and his bucket. "This is certainly a start-up!"

"Shall we get started?" Iribe asked, slinging his arm over Luckey's shoulder.

"Yes," Luckey replied. He booted up his laptop, fiddled with a bunch of wires, and then connected them to his headset, which, without a head strap, was more like a lightweight, gold-bar-shaped, foam-core viewfinder. When Luckey had the setup ready to go, he asked Iribe to turn off the lights.

"What are you gonna show us?" Mitchell asked.

"It's a testbed that Carmack made. But I won't spoil it," Luckey explained. "Who wants to go first?"

"Mike," Iribe suggested. "Mike, Mike, Mike. You first."

Antonov pressed the prototype up to his eyes and quietly experienced his first fifteen seconds in virtual reality. "Uh, I kind of get it," Antonov said, removing the device.

Iribe and Mitchell quickly glanced at each other with pale looks on their faces.

"It's pretty neat," Antonov continued. "But it's a little bit blurry."

"Oh!" Iribe exclaimed. "I think you need your glasses, Mike."

Lying there, beside the bucket, were Antonov's glasses. He forgot he had taken them off. So Mitchell picked them up, passed them over, and Antonov tried again. And this time, it didn't take him fifteen seconds to react. "Ohhhhh wow!" Antonov shouted. He was looking at a colorful room with pipes. There was a remarkable sense of depth. It really, really looked 3-D! And as he turned left to right, up and down, and then back and forth again, everything was in the right spot. "This is *very* compelling!"

Mitchell tried it next. And, oh boy, compelling was an understatement. This was . . . magical, mesmerizing, momentous. It was like he was there—actually there!—like he had been zapped inside of a video game. Then, as if that weren't already enough, Mitchell saw something else that got his heart racing: opportunity. For years, Mitchell had wanted to start a business of his own. Scaleform, with its entrepreneurial spirit, had scratched that itch a little. Gaikai, too. But in both cases, Mitchell felt like he was late to the party. For once, he wanted to be there from the beginning, to start a start-up. With Luckey's invention, he saw a chance to make that happen.

Finally, Iribe received the prototype and held it up to his eyes. From the very first moment he peered into the device, the hair on the back of his neck stood up. A hundred years from now, Iribe thought, we'll all have virtual reality goggles like these. "Wow," he said, in disbelief. "So let me ask you: How much of this is yours—like, *your* invention—versus stuff that Carmack did?"

"Oh," Luckey said proudly. "It's all mine. I mean, the testbed you just tried was made by John. I can't take any credit for that. But the actual HMD? That's my prototype."

Iribe, Mitchell, and Antonov smiled at Luckey's ingenuity.

"The only hardware that John 'added' was an IMU," Luckey said, referring to an "inertial measurement unit," which is small, sensor-filled chip used to track the orientation and rotation of an object in space. "But I say it like that—like 'added'—because the only reason I didn't send him an IMU was because he didn't need one. He was already using a Hillcrest FSRK-USB-2 for his experiments with the HMZ, the Z800, and all those other HMDs."[1]

"Can you say again the name of this IMU?" Antonov asked.

"FSRK-USB-2," Luckey answered. "Actually, it's a slightly modified version of the FSRK-USB-2 that Carmack was able to get because, well, he's Carmack. So this one, it's a 125 Hz tracker, but he got Hillcrest—the company that makes it—to make custom firmware so it runs at 250 Hz. I don't know how much they'd charge us, but the trackers cost about a hundred bucks and do pretty solid 3DOF tracking."

With three blank stares now looking back at him, Luckey realized that this required some additional explanation. "Sorry!" he apologized. "So DOF is short for degrees of freedom. Which refers to the types of movements you can do in your virtual environment. Probably the simplest way to put it, at least for now, is that there there's 3DOF and 6DOF. And the more DOF the better, okay? With 3DOF, you can track head movement; so, you know, you can turn your head left or right and look around in the virtual world. But that's all you can do—meaning you *can't* move around. To be able to move around (or even just lean forward) you need 6DOF. That'll give you "positional tracking" so that you can actually walk around the virtual world. Basically," Luckey ended by saying "we *want* 6DOF, but we *need* 3DOF. Because without head tracking, your HMD is pretty much just a giant TV that's really, really close to your face!"

The guys laughed.

"Not *totally*," Luckey added, "because you could still display stereoscopic images that are the same size and scale of the real world. But you get the idea . . ."

"Totally," Iribe said, still laughing. "I think it's fair to say we need head tracking! And, wait, how much did you say that Hillcrest tracker cost? A hundred bucks?"

"Yeah," Luckey replied. "That's the retail cost, at least. But I have no idea how much we'd be able to get it for if we buy them in bulk."

"And if we wanted to go with a different option, are there other tracking devices that are good? Other ones that you've worked with?"

"Definitely. I've worked with a lot. I've worked with Polhemus magnetic trackers and Virtual I-O inertial trackers from the old days. And I've worked with more recent hardware like the Spacepoint Fusion, PhaseSpace, and Razer Hydra."

This gave Iribe a thought. "Could we make our own?"

"Sure," Luckey nodded. "It would take some time, obviously. But it's not unreasonable. In fact, a friend of mine on MTBS3D is working on an open-source tracker. He sent me one of his prototypes and I was even thinking about including that with my Kickstarter kits. Well, before I met you guys and the scope changed a bit."

"Go big or go home!" Iribe said.

"So," Luckey said, pointing to his prototype. "Does it live up to the hype?"

"Guys . . ." Iribe said, but he didn't finish the sentence right away. He couldn't. His mind was bursting with grand and jagged thoughts. He bought himself a few seconds by walking across the room to turn the lights back on. "Guys," he said again, "do you have any idea how huge this is going to be?"

"I see many challenges," Antonov said. "*But* it is very compelling. And if it is compelling, then it is worth doing."

"Oh man," Mitchell said, still in a sort of what-just-happened daze. "I cannot even begin to express how thrilled I am to take this thing to the next level."

"Agreed," Iribe said. "Let's go build a company that changes the world."

"I'M SO HAPPY FOR YOU!" LUCKEY'S GIRLFRIEND, NICOLE EDELMANN, SAID later that night, as the two of them gazed upward from the beach—watching fireworks set off from aboard the *Queen Mary. Pop—pop-pop-pop*—and bursts of red, white, and blue electrified the sky. "So what happens next?"

"So much," Luckey replied, his eyes glowing with patriotism. "But the number one thing for right now is to ramp up for the Kickstarter campaign. We're going to launch it on August 1."

"That's so soon!"

"I know. But Brendan's not worried. He thinks we can pull it off."

"You trust him," Edelmann said. The way she said it sounded almost like a question, but before Luckey replied he realized that it wasn't at all. Instead, it was his girlfriend—the love of his life, the one who knows him best of all—noticing that the way he talked about Brendan had changed. Any anxiety he'd had about working with him was gone.

"Yeah," Luckey replied. "I do trust him. I mean, he's still a suit. But other than that, he's a good guy. And I think he'd make a good CEO."

"You don't want to be CEO?" Edelmann asked.

Although it was a little hard to admit aloud, Luckey knew that Oculus's best chance to succeed didn't involve him running the company. If push came to shove, he felt he could handle it. Largely because that same grandpa who had taught him that "Money is freedom and freedom is happiness" had also instilled the importance of responsibility and financial accountability. But managing one's own finances was very different from managing an entire company. And while he was tempted to find out *how* different—by following in the footsteps of famous hackers-turned-tech-moguls like Steve Jobs, Larry Page, and Mark Zuckerberg—he didn't want to let ego guide his decision making. "I realize that the thing to do is to try and be the God, emperor, and CEO of everything," Luckey told Edelmann. "But running a company is not my aspiration. There's a lot of operational aspects to running a company that I don't want to be in charge of—like I do not want to be in charge of making sure everyone gets paid on time. And that the bills are all paid. So if Brendan (or somebody else) could do that—make sure shit gets done on time, the bills are all paid—then that means I get to spend time working on what I *actually* want to do: work on VR."

From what little time Luckey had spent with Iribe, Luckey could already tell that he was well-organized, good at managing relationships and had a talent for generating enthusiasm. And though Luckey would often give him a hard time for being a "Suit," he knew that Iribe wasn't a typical Suit. For one thing: Iribe didn't *actually* wear a suit. For another thing: Iribe *actually* seemed like an honest and caring individual.

That said, Luckey was smart enough to acknowledge the possibility this was all an act. That, at his core, maybe Iribe wasn't actually any of those things. But, even if that were the case, Luckey reasoned that if Iribe was a good enough actor to trick him, then he's probably good enough to trick others.

"I think he'd be a great CEO for Oculus," Luckey explained. "But he just got to Gaikai. And he has his friends there. Although I kind of get the sense that they're not happy. But who knows?"

As the fireworks grew even bigger and louder, Luckey put aside his thoughts about Oculus. No point in thinking about virtual reality when the one you've got in front of you is so damn good. And it was. She was. All of it was.

God bless America, Luckey thought. God bless this wonderful country of ours.

CHAPTER 8

THAT FATEFUL PROMISE

July 2012

"WHAT ARE YOU IN THE MOOD FOR?" NATE MITCHELL ASKED, PEEKING OVER HIS menu to see if Palmer Luckey was ready to order. This was the first of several meetings that Mitchell and Luckey would have at Pei Wei Asian Diner locations in the area. Yet despite the many rendezvous, it would be the last time that Mitchell asked him because it was clear, from the recited precision with which Luckey ordered, that the kid's order would always be the same: pad thai, regular-size protein combo with chicken and shrimp, double chicken, double shrimp.

The repetition of Luckey's order not only conveyed to Mitchell that Luckey was a creature of habit, but—as evidenced by his still reviewing the menu each time—it revealed that Luckey at least made an effort to avoid being blinded by his own familiarity. But perhaps most important, Mitchell learned to never expect a short answer from Palmer Luckey. No matter the scope of the question, the kid would almost always respond with some sort of eloquent elaboration. The inquiry could be as technical as asking about the algorithm for distortion correction or as casual as wondering what the "protein combo" entailed.

"The protein combo?" Luckey asked. "That's where they use half your meat allotment for one meat and half your allotment for another. Because usually you only get one. Chicken, steak, shrimp, tofu, you know? There's no inherent value to the combo—it's really just a matter of taste; what are you hungry for?—but if you get the pad thai (or any noodle dish, really), I highly recommend the double-meat option."

"Oh yeah?" Mitchell asked with a soft smile. He loved listening to the kid rant—how he could just go on and on about almost anything with such an absurd level of specificity.

"Yeah," Luckey replied, nodding. "Because if you buy that . . . the thing with the pad thai is that there's too many noodles. Compared to the protein, that is. But for only two bucks you can double the meat and you end up with a pretty good mix of protein to carbohydrates."

"So let's talk Kickstarter . . ." Mitchell eventually said.

Although Mitchell had never run a Kickstarter campaign before, it was a role that, in a way, he'd been groomed to play; originating from the influence of his father, David Mitchell, a partner at Greer Margolis Mitchell Burns & Associates (GMMB), a DC-based communications and advertising firm. Since its founding in 1983 GMMB has been hugely influential in the political sphere, from helping to secure victories for presidential candidates (like Bill Clinton and Barack Obama) to producing iconic awareness campaigns (like Click It or Ticket and Save Darfur).[1] Given this backdrop, it should be no surprise that, as a boy, Nate Mitchell had a reverence for campaign messaging that other kids his age reserved for cartoons and action figures. As such, a career in political advertising had always been on Mitchell's horizon.

Within days of receiving that first demo at the Hilton, Nate Mitchell and Brendan Iribe determined that, for Oculus to succeed, they had to do more than just sell a product. This is true of selling any disruptive technology to consumers (and, of course, is true of all advertising to some degree as well). But what made this proposition even more difficult for Mitchell and Iribe was that unlike most disruptive innovations—be it a car, television, or smartphone—the Oculus Rift wasn't an immediate upgrade over something similar that had come before. There was no horse and carriage to point to and say: Wouldn't you prefer to ride in a car? Nothing like that could be said because nobody had ever tried to sell a mass-market virtual reality headset before, at least not in the past twenty years. Furthermore, comparing the Rift to those failed headsets from the '90s would only serve as a reminder of false hope.

The physics and psychology that went into product messaging were things that Luckey had rarely thought much about before. Mitchell anticipated this—the kid made things; he wasn't concerned with the narrative around their making—but, to his increasing delight, Luckey took to these discussions with poise and enthusiasm. Given how the first of their Pei Wei rendezvous had started, with Luckey assuring Mitchell that he'd already shot a Kickstarter video they could use, Mitchell was impressed.

"Tell me about what you filmed."

"Sure," Luckey said. "So I set up this old JVC camcorder in my trailer and . . . actually: have you ever seen those old Ross Perot infomercials from the 90s? The ones where he would sit down beside a flip chat and just go through the points?"

Mitchell looked horrified.

"No, no, it's good!" Luckey plead.

Mitchell laughed. "Well, what exactly did you talk about? What was the messaging?"

Immediately, Luckey launched into a passionate speech he had memorized. "If you don't know what this is, you're not the target audience. This is really intended for people who are on the MTBS forums and, if you're not, if you're following or if you've been following Oculus.com, or if you've been following the prototype I gave to John Carmack, or perhaps the work I've been doing for the Institute of Creative Technologies, then you are probably some type of game developer, software developer, or hardware enthusiast who is interested in virtual reality and believes it has a future. And here is the kit that you can assemble. I've pulled together . . . oh, I should probably mention that off to the side, just off camera, I had a bunch of components and stuff so that I could reach for them, hold them up, and explain. You know, like 'These are some of the laser-cut plastic components that all slot together using normal machine screws and bolts' or 'These are some of the optical elements and mounting rings; they're held in with a retaining O-ring.' You kind of get the image I'm painting?"

"Yeah," Mitchell replied, trying to remain upbeat and friendly. "I have a pretty good idea of the picture you're painting. But, look, I'm just gonna be honest: that's not gonna work."

"Are you sure? Do you want to at least see it?"

"No, it's not about that. Honestly, I'm sure what you shot is pretty good. But if we're going to make this happen in a real way, we need to knock it out of the park. We need to make a video that's worthy of your creation. That's really the best way to put it. Something to get people as fired up and inspired about the Rift as the three of us were at the Hilton. I hope you don't think I'm being offensive."

Of all the things Luckey had said during the meeting thus far, it was what he was going to say next that interested Mitchell most. In a way, this was the first make-or-break moment for Oculus, as it would provide insight into how much of a collaboration Luckey really wanted this to be.

"You know what?" Luckey replied. "I think you're right. I'd started doing this on my own. I'd kind of decided I was going to do this as a Kickstarter project and had this version of it in my head. But you and Brendan have convinced me that there is an audience larger than the audience I had previously anticipated. It would make sense to cater to that audience."

Mitchell was stunned by Luckey's response. He actually considered getting up and giving the kid a hug. But it was too soon for that kind of camaraderie. Rather, he pushed the conversation forward with a compliment. "Although I don't think what you said in your video is the right message, I'm more confident than ever that you are the right messenger. When you're talking about something you care about, you have this magnetic presence. I assume that people have told you that before?"

"My girlfriend says I am larger than life. But, you know, she's my girlfriend."

Mitchell laughed. "I'm serious, though! There's a theatricality to your cadence."

"Oh, well, actually—did I mention this before?—I used to act?"

"Really? Tell me. I need to know this stuff!"

Luckey told Mitchell about his prior life as an actor, about how, in his early teens, he'd been in some local plays and a few short student films. How, at one point, he even had an agent. "Technically," Luckey continued, "she's probably still my agent. I don't know. It was never very lucrative. But I'll tell you what *was* lucrative: modeling."

"Wait," Mitchell said. "You were a model, too?"

"I was a fit model. Do you know what that is? Nobody knows what it is. That's probably how I got the job. But basically a fit model is someone that a clothing designer, or a retailer, uses to check the fit of a particular size. Essentially, your job is to be a human mannequin."

Before moving the conversation to virtual reality and Luckey's vision for Oculus, Mitchell remembered something that he'd been meaning to ask for a little while now. "So . . . what's the deal with the trailer?"

"What do you mean?" Luckey replied.

"Well, I don't know many nineteen-year-olds who live in trailers—let alone gutted-out trailers parked in the driveway of their childhood homes . . ."

Luckey laughed and then gave Mitchell the whole sordid history. "So my trailer wasn't really *my* trailer; my parents bought it. They got it about two years ago on Craigslist, or Autotrader—it was one of the two, I think—for somewhere in the neighborhood of fifteen-hundred-dollars."

"Ah, okay. So it was, like, a gift?"

"Um, not exactly. It was more like: 'Palmer, you're seventeen and we don't want you to live at home and be dependent on us forever. So we're forcing you to move out!' But at the same time, they wanted me to go to college, and they reasoned that there was very *little* chance of me going to college if they just threw me out entirely onto the streets. Let me put it another way: it was not so much about kicking me out as it was about sending a message."

"And were you . . . I guess . . . cool with that?"

Luckey didn't answer right away. "I was initially not super psyched about it. But then I realized that I had way more freedom to do whatever I wanted living in a trailer."

"But you're still living only like ten feet away from your parents!"

"It makes a huge difference!" Luckey replied. "You can set up shit however you want, nobody bothers you. Or you can just be like 'I'm hungry. I'm going to Jack in the Box.' And you don't have to go downstairs and have Mom say: 'Where are you going? No, don't go there—don't waste your money on that crap!'"

"That makes sense," Mitchell said. "Sounds cool. Maybe I can swing by some time?"

"Well," Luckey said, "that's gonna be a problem. Because they weren't all that thrilled when I told them I was dropping out of college to start Oculus. And so that finally came to a head a couple weeks ago, and they wound up selling the trailer."

"Wait. So where are you living now?"

"Technically, I'm kicked out. But they were like: if you're willing to sleep in the garage—on top of these boxes, and underneath the cabinets—then that's okay . . . for now."

"Seriously?"

"I know: it's ridiculous!" Luckey said, now laughing as he spoke. "But don't worry! I mean, sure, it's musky and damp; and, sure, I'd be toast without that dehumidifier at the foot of my twin mattress. To be fair: from their perspective, I'm their loser-internet-son, dating his loser-long-distance-girlfriend, who just dropped out of college to start a company that will sell a product that—in the entire history of the world—has only ever crashed and burned."

Mitchell started cracking up so hard he nearly fell face-first into his food.

"So yeah," Luckey said, now cracking up too. "That's the deal with the trailer!"

"Dude," Mitchell said, "we gotta get you out of there! I mean, I don't have much space, but you're welcome to come stay with me."

Luckey appreciated the offer, but said he was planning to get a place of his own towards the end of the month: a studio apartment in Seal Beach. "It's pretty expensive—$1,049 a month, if I recall correctly—but it comes fully furnished and has a ton of benefits: Free gym, free donuts, a hot tub, tennis courts . . ."

"Wait," Mitchell said, eventually cutting Luckey off. "So if you weren't moving to that sweet setup with the free donuts, and things hadn't really taken off after you connected with Carmack, what would you have done? Like: where would you be living?"

"Well," Luckey said, "I would have taken the job at Sony."

"Ah, that makes sense . . ."

"But I did have a backup plan. My backup plan was that I would sell some of my Bitcoins, purchase my *own* trailer, and move to the Inland Empire and live in a trailer park there."

"And keep making virtual reality headsets?"

"Something like that . . ."

On that note—and over the course of several Pei Wei meetups throughout July—the guys got to talking about VR. "So I checked out your Oculus website, and I loved what you wrote about Oculus being 'your tilt' to bring VR to the 'average person.' And that stuff about gamers and dreamers? That was gold! But I want to dig even deeper."

"Oh!" Luckey said, thinking of something else. "So there's this comic that someone on 4chan made and it totally nails the game industry. Have you ever seen it? It's that one that starts with a game dev saying: U CAN DO ANYTHING!!"

Mitchell had not, so Luckey explained it to him. The comic was only six panels long and it begins, as Luckey had mentioned, with a game developer announcing that his upcoming game will be so amazing—so filled with incredible features—that players will be able to do . . . anything! In response to this wonderful news, fans freak out about the possibilities. Excitement builds—Anything? Anything!—until one of the megafans finally gets his hands on the game, installs it, and is greeted by this message: WELCOME TO THE TUTORIAL . . . YOU CAN'T DO ANYTHING.

Mitchell laughed. "Yeah, that pretty much sums up the hype cycle for every game."

"But we can't do that," Luckey said. "I don't want to do that. Not just because it's a shitty thing to do, but because that would be bad for VR."

This was not the first time Mitchell had heard Luckey make a comment like this—about something being "good" or "bad" for virtual reality—and he found it utterly charming. Who talked like that? Who *thought* like that? Mitchell, in his capacity of campaign manager, couldn't have asked for better putty with which to craft a candidate.

"After the devkit," Mitchell began, "when we launch the consumer version of this thing, what do you envision the controller should look like?"

For the past few months, Luckey had been giving demos with an Xbox 360 gamepad. This made sense, since the Xbox 360 gamepad was considered one of the best console controllers ever made (and with a current install base of over seventy-five million people, it was incredibly familiar). "But that won't be good for VR," Luckey said. "A gamepad is fine for now, but when we launch for real, it's critical that we have some kind of wands."

By "wands" Luckey meant trackable hand controllers. In its simplest form, he explained, this would "essentially be like an Xbox 360 controller split in two. One for each hand. But with awesome tracking so you can move your hands freely in VR."

"So kind of like the Nintendo Wii-mote?"

"Exactly. Because think about it: What's the first thing people do when they get into VR? They look down. They want to see if their hands are there. So that hand presence is critical if you want to trick the brain." Wands, Luckey explained, would only be the start. After that, they'd need to create an input device capable of simulating what it feels like to open and close your hand—for grabbing, gripping, pinching, and so on—and then eventually they'd need to come up with something that could track individual finger movements as well. To Mitchell, this all sounded awesome. But was it realistic? Absolutely. As proof, Luckey cited an input device that Jeremy Bailenson, the director of Stanford University's Virtual Human Interaction Lab, currently had at his facility. "It tracks your position *two hundred* times per second," he said. "And it's accurate down to one-quarter of a millimeter."

This got Mitchell's mind flowing with possibilities: grabbing the side of a mountain (for a climbing game), gripping the seams of a baseball (for a sports game), ripping open beautifully wrapped gifts (for a virtual Christmas morning). Mitchell's excitement jumped even higher when Luckey started talking about other inputs they might want to consider down the line: recoil simulation guns, omnidirectional treadmills, and galvanic vestibular stimulation devices.

"What's that one?" Mitchell asked.

"Galvanic vestibular stimulation? It's basically a technology where you use electrical impulses to stimulate the inner ear, to stimulate orientation or acceleration. I've put together some GVS systems myself, actually. It's, uh, not particularly safe . . ."

The guys spent a little longer talking about futuristic technologies. After that, the conversation shifted to what the two of them, along with Iribe—sometimes in person, sometimes weighing in from the road—would primarily discuss during these Pei Wei meetups: messaging.

Early on they talked a lot about the vernacular. How to distance themselves from the negative stigma associated with "virtual reality." At one point, they even considered staying away from the term completely in favor of "immersive gaming." But that felt too vague. So instead, for a while, Iribe and Mitchell liked the idea of heavy messaging around the word *3-D* (they even bought the domain Oculus3D.com), the goal being to capitalize on the buzz around 3-D movies and 3-D TVs. But persistent resistance from Luckey—"Guys, I'm telling you, the whole 3-D hype bubble is crashing"—brought them back to where they started: virtual reality. At which point they realized there was no use running from it. They were better off steering into the skid and, instead, coming up with a way to redefine how the term was perceived.

When thinking abstractly about "virtual reality," Luckey, Iribe, and Mitchell soon realized they *all* had one positive perception of VR in common: the promise they had made to themselves.

Ever since they were kids, all three guys had yearned to set foot inside of the video games they played. But year after year—no matter how good the graphics got, or how immersive the gameplay became—that thin glass screen always separated them from the video game world. All their lives, like most gamers—especially those who also loved sci-fi—they felt an unspoken promise that they'd one day be able to step into the game.

"Step into the game?" Luckey asked, digging into his double-chicken, double-shrimp, protein combo pad thai.

"Yup," Mitchell explained. "That's the tagline Brendan and I came up with. And we're feeling real good about it." Not only were they feeling good about that—Luckey, too, nodding with his mouth full—

but the tagline (and the promise that had inspired it) led Iribe and Mitchell to the conclusion that Oculus was only going to succeed one way: by starting a revolution to galvanize people around an incredible mission—an unlikely, fifty-years-in-the-making mission—of finally delivering that fateful promise of virtual reality.

This, they realized, was Oculus's mission.

This, they realized, was Oculus's responsibility.

CHAPTER 9

MOUNT UP

July 2012

THE NEXT STEP? ASSEMBLE A TEAM. SO LIKE DANNY OCEAN PUTTING TOGETHER the perfect, ragtag crew of thieves, Brendan Iribe spent July flying around the world in search of key hires and strategic partners for Oculus's VR revolution, starting on July 8 with a visit to the man who some referred to as "the Guitar Hero."[1]

JACK MCCAULEY

"Anyone can make a prototype."

This was something that Jack McCauley was fond of saying. He didn't mean it disparagingly—after all, he'd spent a third of his adult life (and about half his childhood) building things from scratch—but he said it as often as he did to contrast the oft-forgotten difference between a prototype and a product. The difference between making something for an audience of one, and making something that could be replicated, manufactured and shipped to strangers around the globe. The ability to do the latter required a different sort of talent. An endangered talent, really, in an increasingly software-driven world.

"Brendan?" McCauley asked, standing in the doorway. Tall, with a square jaw and lightly salted black hair, McCauley gave off a stoic vibe. Standoffish, almost. But there was something about him—maybe his relaxed shoulders, maybe his bright blue eyes or maybe the fact the looked much younger than his actual age (fifty-two)—that seemed to indicate he was more affable than he immediately appeared. "Brendan Iribe?"

"Yeah!" Iribe enthusiastically replied. "It's so great to meet you."

"It's great to meet you too," McCauley said, waving Iribe inside.

If Iribe had any doubts that this was the guy Oculus needed, they were gone before McCauley even walked him upstairs. The first floor of McCauley's studio looked like a lost wing from the Rock & Roll Hall of Fame with at least a dozen, spiffy-looking guitars mounted on the wall, a couple of guitar kits lingering beside them, and myriad news clippings, colorful posters, and framed platinum discs. These items were not relics from some former career as a rock star, but rather trophies of McCauley's talent as a hardware manufacturer. From afar, these guitars looked like gorgeous handcrafted instruments; a closer inspection would reveal that they were actually plastic video game peripherals. Plastic video game peripherals—designed and manufactured by McCauley—that were critical to the success of what was perhaps the most unlikely multibillion-dollar game franchise of all-time: Guitar Hero.

"There was this arcade game in Japan," McCauley explained with a soft smirk on his face. "I reverse-engineered it and made a demo. We were hoping to sell, maybe, a hundred thousand units. We ended up doing twelve million. Then we sold the company to Activision so that we could raise the capital to produce more. Ended up selling fifty million of those things. It worked out well for everybody. Incredible for Activision."

It was through one of McCauley's former Activision colleagues— Greg Deutsch—that this meeting with Iribe had been arranged. Deutsch, a lawyer whom Iribe knew (and trusted) from his Scaleform days, had agreed to help out a bit with the legal formation of Oculus. He'd also suggested that McCauley would likely be a great fit to handle Oculus's hardware manufacturing.

"I swore that I would never do games again," McCauley told Iribe, which momentarily sunk his heart. "And then you called me and the first thing I thought was: Virtual reality? Oh no, here we go again. But you seem like a nice guy and Greg said some nice things about you."

"I appreciate that," Iribe said, speaking now with a magnetic flow of energy and enthusiasm. "And look: whatever you're thinking—

whatever you're feeling—I went through exactly the same thing. Trust me. I mean: virtual reality? If you weren't skeptical about this, then I'd probably be skeptical of you!"

McCauley loosened up. He liked that Iribe was somehow able to make him feel calm, especially since he often felt withdrawn. "Because I grew up in Holland," McCauley explained, "but my parents moved back here when I was ten and I was shocked by the culture. So I just worked on things, built things, on my own."

"What kind of things?"

"I got a toy fire engine from my dad, when I was like six or seven. I took it apart and hacked into it and made it do something different. And I started doing that with cars and motorcycles and all kinds of stuff. Making it do the things I wanted them to do. I also started experimenting with chemicals. I read somewhere that you could make synthetic heroin and I thought: I could make a lot of money doing that."

"Really?" Iribe asked with a pleasant disbelief.

"Oh yeah," McCauley said. "So I bought all the chemicals, everything to do it, but I never did it because I realized: this could harm people. So I threw away all the chemicals. But I made explosives. I built bombs and cannons and all kinds of things as a teenager. I didn't really use them, but I was just interested in the process of making them. Then I started building cars and bikes." McCauley waved his hand toward a car engine docked at a table in his lab. "So I've worked on a lot of crazy things, Brendan. A lot. But I've never worked on virtual reality."

"So few people have," Iribe replied. "That's the amazing thing. Nobody's working on this stuff, except for this kid Palmer Luckey—he's, like, this crazy-smart VR enthusiast; I can't wait for you to meet him, you're gonna love him—he's just been building all these headsets, doing this stuff for years, and he's come up with this thing that actually works. That's how he got hooked up with Carmack and, oh man, I'm just stoked for you to . . ." Iribe trailed off as he fiddled with his MacBook laptop to make sure everything was properly connected to Luckey's prototype.

"Brendan, you're a great marketeer," McCauley said. "When you walked through the door, it was like: I don't know this guy, he seems like a scammer. But you're not. You're a very, very good pitchman. I can tell by your enthusiasm that you think you're onto something."

"I don't want you to take my word for it," Iribe said, handing him Luckey's headset. "Here, try it for yourself."

McCauley felt a visceral thrill. Years ago, he reasoned, it would cost hundreds of thousands to manufacture a headset like this. But now, because of technological advances in several areas—from computing and displays to gyroscopes and accelerometers—this thing in his hand could be made for only a few hundred bucks.

"So?" Iribe asked after McCauley removed the headset. "What do you think?"

"This is really, really something," McCauley said. "This is really going to change things. Provided, of course, that you guys don't fuck it up."

DILLON SEO

Growing up, Dillon Seo wanted to be a superhero. Such an aspiration was not unusual for a five-year-old boy, but, unlike his childhood friends, Seo had the semblance of a plan. Realizing that he didn't inherently possess any powers (like Superman) or have the wealth to build something comparable (such as Batman or Iron Man), he aspired to be a biologist so that he could create a medicine that would turn him into a superhero. But, like most childhood dreams, this one got lost somewhere along the way to adulthood. Until July 11, 2012, when Brendan Iribe happened to be passing through Korea.

Iribe was in Seoul to disappoint LG and Samsung. He had to inform them that Gaikai would be selling to Sony and that, by closing this deal with one of their competitors, Gaikai would therefore not be able to license out their cloud-based solution. It was a trip that Iribe knew he had to make in person, and one he hoped might be brightened by a last-minute meeting with Dillon Seo.

Iribe and Seo had known each other since October 2008, when Scaleform, looking to expand globally, hired Seo to be their business director for Korea. Seo quickly proved to be a key part of that expansion. By the time Scaleform was acquired by Autodesk in 2011, Seo's region was responsible for about 25 percent of the company's global revenue. That staggering percentage was not only a tribute to Seo's salesmanship but also to the size and import of PC gaming in South Korea—a county where, of its fifty million residents, half are estimated to play online games regularly.

In 1997, on the heels of the Asian financial crisis, the South Korean government made a concerted effort to improve the country's internet infrastructure. Specifically, this involved heavy investment into ADSL (asymmetric digital subscriber line), which, unlike conventional voice modems, used copper telephone lines to enable faster data transmission. With this superspeedy service in place, internet usage proliferated faster in Korea than anywhere else in the world.[2,3] By 2004, for example, 70 percent of Korean households had broadband internet access (compared to only 25 percent in the US that same year). This pervasive connectivity also gave rise to "PC bangs" in South Korea.[4] These are LAN-based, pay-by-the-hour gaming centers where patrons could easily, and socially, play the latest MMO (massively multiplayer online) games; from Western favorites like *EverQuest* and *StarCraft* to Korean-made hits like *Lineage* and *MapleStory*.[5]

Like most Koreans his age, Dillon Seo (thirty-five on his way to Iribe's hotel) took pride in his home country's technological foresight, and its role in birthing an unparalleled PC gaming industry. But what he admired even more—and participated in firsthand—was Korea's continual commitment to build that industry.

In April 2007 Seo had become the global business manager for the Korea Game Industry Agency. As a government employee, affiliated with the Ministry of Culture, Sports and Tourism, Seo's job primarily consisted of building relationships, conducting market research, and helping to organize Korea's annual trade show (G*Star). By virtue of his days working for the government, his years at Scaleform, and his present role as a product specialist and sales executive for Autodesk, Seo knew the Korean game market as well as anyone in the world. Which is why Brendan Iribe was so optimistically delighted when he heard a knock on the door of his hotel room.

"Thank you for coming!" Iribe said, ushering Seo inside.

"Of course," Seo replied, eager to see whatever it was Iribe planned to show him. "How did your meetings go?"

Iribe rolled his eyes and shrugged off whatever residual *ugh* he was still feeling. "This," Iribe said, pulling out one of Luckey's prototypes, "is the future of gaming."

That's a pretty strong claim, Seo thought. Especially for something that looks like a high school project. "What the heck is this?"

"Close the curtains," Iribe said. Seo followed these instructions as Iribe booted up his laptop. Then, after connecting the device to his laptop's HDMI and Micro USB, Iribe handed it over to Seo. "Try this out."

Seconds later, Seo found himself inside what appeared to be a circus tent. It was hard to tell exactly what it was because he was moving all around. No, not just moving; he was flying!—soaring up and down through the scene in Carmack's testbed demo. The graphics were low-resolution (with a relatively strong ghosting effect), but Seo felt magnificent. Like a superhero. Until he looked down and his knees buckled with vertigo.

Seo pulled off the headset and found Iribe silently laughing.

"I *felt* the height," Seo said. He then looked down at his hands and saw that they were sweating a bit. "What the hell kind of experience is this?"

Iribe reconvened with the laughter, slightly audible this time.

"This is a very, very interesting sensation," Seo explained, still sounding more like he was talking to himself than Iribe. "I started to forget about where I was and started to feel like I was inside of that tent. And that was a very, very strange feeling."

"Well, Dillon, I'm going to start a company. Oculus. We're going to deliver that feeling you just experienced to the world. We're going to deliver virtual reality. And we're going to launch with a Kickstarter campaign in August."

"I think you will find a lot of success with this."

"I think so too," Iribe replied. "But we need the right team to make this happen. And for virtual reality to be a thing, we *really* need to make this happen." Iribe paused to make sure he caught Seo's eyes. "Korea's got such a strong PC gaming community. It's critical we reach that audience. So why don't you come and join us?"

Now it was Seo's turn to laugh. He had a wife, a young son, and a steady paycheck at Autodesk. There was even talk of a promotion.

Autodesk had been suggesting that Seo could move to Singapore and become director of the whole Asian market. "Brendan," Seo replied, "right now, the company pays me handsomely. And, you know, I am outperforming my expectation."

"Just think about it," Iribe said. "Okay?"

While Seo deliberated about joining Oculus, Iribe got some good news from his first recruit: Jack McCauley. On July 14, Iribe received an excited email from McCauley that began: "I did not sleep well last night as I was excited about the Oculus project."

McCauley was in, back to the industry he swore he'd left behind. And though his enthusiasm was through the roof, one thing vaguely concerned him: Iribe's plans for Oculus. McCauley had been part of numerous start-ups and they all ended one of two ways: disaster or acquisition. For once, he wanted to work with a company that was in it for the long haul. And Iribe was happy to confirm that this was their goal. "To be clear," Iribe wrote on July 14, "we're not going to flip Oculus quickly, we're building a large independent company that's going to revolutionize the game industry and pioneer a new world of VR."

Adding experts like McCauley (and possibly Seo) was key. But that key wouldn't unlock anything important without a game engine to help developers create content for the Rift—which is why Iribe next headed to Cary, North Carolina.

EPIC GAMES

In April 2010, Tim Sweeney, the CEO of Epic Games—the makers of hit game franchises like *Gears of War* and *Infinity Blade*—publicly remarked that Scaleform GFX (the flagship product of Iribe's former company) was "the clear leader in user interface design for casual and triple-A games."

Now, two years later, Iribe was back at Epic; hoping once again to get another glowing endorsement from Tim Sweeney. But this time, for Palmer Luckey's VR headset.

"Tim!" Iribe said, greeting Sweeney in Epic VP Mark's Rein's office (where Iribe would be giving demos to a handful of employees).

In addition to the endorsement from Sweeney, Iribe had arrived in North Carolina with one other key objective: get Epic excited enough about Oculus that they would be willing to work closely with the VR start-up and help integrate their Unreal Engine into Oculus' SDK.

The Unreal Engine was a game engine first developed by Epic Games in 1998 to power their first-person shooter *Unreal*. The game itself did very well (spawning a bevy of sequels over the year), but its engine—the Unreal Engine, notable for its easy-to-mod architecture and being the first to integrate AI, collision detection, networking, and rendering into a single engine—that was the real game changer. With the Unreal Engine and its ensuing iterations released in 2002 (Unreal Engine 2), 2005 (Unreal Engine 3), and 2012 (Unreal Engine 4), smaller studios and amateur developers could simply license the engine for a small fee (or, in some cases, for free) instead of investing all the time and money it would take to build their own game engine from scratch.

In addition to powering some of the most beloved games since 1998 (like *BioShock*, *Borderlands*, and *Mass Effect*), the Unreal Engine had become so popular and easy to use that even nontraditional developers (like the military and the FBI) were licensing it to create training simulators. By the time of Iribe's meeting with Bleszinski in 2012, Epic's Unreal Engine was one of only two game engines that anyone in the industry even seriously considered anymore; the other was made by Unity, a Copenhagen-based game engine company, which tended to cater to smaller developers.

As such, Iribe knew what a coup it would be if he could persuade Epic or Unity to support Oculus. If just one of them believed in the Rift enough to equip their engines for virtual reality, then almost overnight thousands of developers could start building games in VR. Hopefully, Iribe thought, CliffyB really meant what he had tweeted.

"I saw you tweeted about Ouya," Iribe said. "The reaction to their Kickstarter . . . I mean, it's just . . . *crazy*."

Bleszinski couldn't help but nod in agreement. There was just no other valid reaction. It *was* crazy.[6,7] Ouya's Kickstarter campaign went live on the morning of July 10. In just eight hours, on the heels of over eight thousand backers, the campaign surpassed its fund-raising goal of $950,000.[8] Even crazier: Ouya would end up raising $8,596,474 (from 63,416 backers) by the end of its thirty-day run.[9]

Iribe knew it was unrealistic to envision Oculus's Kickstarter campaign achieving that same kind of runaway success. Not only was

Ouya's $99 product significantly cheaper, but it was also something that gamers had no trouble believing; as they always did, the challenges only served to inspire Iribe.

"So," Iribe said, handing the prototype to CliffyB, "you ready for your first taste of virtual reality?"

Bleszinski nodded, pushed the prototype to his face, and, within seconds, delivered exactly what Iribe had hoped for: a full-fledged CliffyB freak-out. Holy Shit. *Ho*llllly Shit! No way, no way, no way! Bleszinski was in such disbelief that he couldn't even finish the demo.

Not only did Bleszinski feel compelled to publicly share his excitement—tweeting "Today I got to play with a prototype Oculus VR headset, exciting! VR is coming back! They're Kickstartering soon . . ."—but he arranged for everyone in Epic's core crew to try out Luckey's prototype. From founder Tim Sweeney and VP of engineering Daniel Vogel to devs like Mark Rein and Steve Polge, CliffyB wanted to make sure that as many of his colleagues as possible bear witness to the holyshithollllllyshit he'd just experienced. And Iribe was able to set up demos for dozens of them, which he excitedly relayed to the Oculus team on his way to the airport.

FROM: Brendan Iribe

DATE: July 16, 2012

Good news—I just came from visiting Epic where just about their entire dev team (several dozen of their best, including CliffyB, Sweeney, Vogel, Rein, Polge, etc.) all tried and LOVED the prototype. We'll launch with Epic support now. They're going to give us access to UE3 and UE4 right away, at which point, it's up to Mike, Andrew and crew to integrate.

Palmer's working on id/Carmack/Bethesda, Valve, and Mahjong [*sic*]? As well. I'm going to reach out to Crytek, Unity, Hawken and a few others now. Very important to take the time to nail the video . . .

THE LMU CREW

"Sound?"

"Speeding."

"Camera?"

"Speeding."

Satisfied with the answers, Win Bates called, "Action!" And with his word, on July 18, filming of the Oculus's Kickstarter video officially commenced. Opening, of course, with none other than Palmer Luckey.

"My name is Palmer Luckey, and I'm a virtual reality enthusiast and the designer of the Rift," Luckey said, delivering the lines Mitchell had written to an eyeline just off camera.

Unlike most engineers, who preferred a role behind the scenes, Luckey liked being on camera. It wasn't so much that he liked the limelight as that he liked how the limelight forced him to raise his game; when the camera was rolling, he knew he needed to be "on." And being on was a good thing.

McCauley had graciously offered to let the team film at his studio in Livermore. And although Luckey knew, for sure, that this was leaps and bounds better than his trailer, what was gained in production value was lost in comfort. A man's lab was his sanctuary and this, well, this was another man's sanctuary.

Bates, the director, detected the mild sense of discomfort. He hadn't yet directed many films himself—he was in his final year at LMU School of Film and Television—but had seen enough to know this was almost always the case with docu-style projects. So he and his crew (of classmates from LMU) did their best put Luckey at ease, praising the kid's natural presence and reminding him not to worry because, as Bates would say, "so much of it will be pieced together in post anyway."

Further encouragement came from Nicole Edelmann, whose cheery face invigorated Luckey. As did the nods of encouragement from Iribe, Mitchell, and Antonov, who were already coming to view Luckey as something like a younger brother. Feeling more comfortable by the minute, Luckey started to hit his stride.

"Games are something I'm really passionate about. And even more than playing games I'm passionate about bringing games to the next level," he explained on camera. "I was interested in stereoscopic displays, I was interested in head mounts, and the problem was that there was nothing that gave me the experience that I wanted: the Matrix. Where I could plug in and actually be in the game. And I was sure that somewhere out there, there was something that I could buy. And the reality is there's nothing. I set out to change that with the Oculus Rift."

There was no word-for-word script that Luckey was given to follow, but Mitchell and Bates had plotted out a list of beats they wanted to hit. The trickiest points to hit—the ones that would require the most finessing in post—were those about the tech itself. Fortunately, this was where Luckey felt most comfortable. An ever-present optimism in his eyes and voice enabled him to retain the attention of his audience when he said things like, "The magic that sets it apart is immersive stereoscopic 3-D rendering, a massive field of view and ultra-low-latency head tracking."

"Have you *seen* this place?" Edelmann asked Luckey during one of the filming breaks, referring to McCauley's studio. "I don't think I've ever been so close to someone so rich before!"

Luckey chuckled. "Yeah, Jack's done pretty well."

"Pretty well? I just went into the garage," Edelmann said, "and there was some sort of a Ferrari kit kind of thing in there!"

Just a few months ago, Edelmann was working at an old-timey photo booth in Colorado and now, on this whirlwind journey, her boyfriend was being filmed for a video about his new company. Not just filmed, but *starring*! And it was all happening at this place—this fancy, fancy place. Was this the life that she and Palmer were headed toward? The whole thing was so surreal, and somehow her boyfriend was taking it all in stride. Luckey wanted to give her this sort of life— something that'd make her forget all about that trailer—and now, for the first time, such a possibility felt . . . not impossible.

Given how far he'd come in such a short amount of time, Luckey thought nothing could knock him down. But then something Iribe said to McCauley sent him crashing back to Earth: "If Wikipad misses Black Friday, we're pretty much fucked."

There was nothing cruel or pernicious about this comment. Luckey had known from the start that Iribe was involved with other start-ups (like Wikipad) and that Oculus, really, was just one of the companies in Iribe's portfolio. If the Kickstarter went well, then it might become a big part of that portfolio—perhaps even big enough to lure Iribe away from Gaikai—but whether or not that happened, this was a sobering reminder that Iribe had other options. Whereas for Luckey, it was Oculus or bust.

"We're getting involved with Oculus now," Iribe said a little later, speaking to the camera, "because we see an incredible opportunity here for game developers to experience something new."

"A great SDK is simply about a great developer experience," Antonov added, "where they get a piece of software that is really efficient and skilled for all their needs."

As Iribe and Antonov bantered, Luckey rejoiced. Each vote of confidence (even from members of his own team) played a critical role at this early stage, mitigating murmurs of an impostor syndrome he would feel from time to time. One thing that helped Luckey instantly feel more like himself was his bright red "Ryu headband," which he named after the beloved, hadouken-shouting *Street Fighter* martial artist. Luckey had gotten it at a *Street Fighter* launch party a few years back, but its origin wasn't what made it special. Its significance actually came from a photo—of him head-banded and smiling like an idiot with TV personality Adam Sessler—which his ModRetro friends took endless joy in modding. One friend cropped out Luckey's head, tinted it blue, and pasted it onto the body of Captain Planet; another added a red headband to the screaming face in Edvard Munch's famous painting; and yet another blew up the image, cropped out the eyes, and indicated where to cut so that ModRetro members could create (and then wear) their very own "Palmer Tech Mask."

Needless to say, the headwear held a special place in Luckey's heart. And even though everyone on the set looked a bit squeamish when Luckey put it on, he felt the video wouldn't be complete unless the Ryu headband made an appearance.

Finding a spot to fit *Street Fighter* Luckey into the final cut was the least of Win Bates's concerns. More pressing was editing the short film together in only about ten days, especially considering that every few days Iribe would inform the director and his editor (a classmate, Luc Delamare) that they would need to weave in another new endorsement. Including, possibly, Luckey's former boss at USC, Mark Bolas.

MARK BOLAS

Although Luckey had only worked at USC's ICT lab for less than a year, Bolas was someone whom he had come to admire and trust. Bringing on someone who Luckey felt comfortable with was also important to Iribe, especially as the company continued to add more and more people Luckey had never even met before. So, Iribe believed, hiring someone Luckey had a history with would give him an anchor for the choppy waters ahead. And even more important, Mark Bolas was one of the foremost virtual reality experts in the world.

Unfortunately, with that expertise came the baggage of his own experiences trying to commercialize VR. He had, in his own words, "been on that roller coaster before." And with a daughter soon headed to college, he wasn't particularly interested in abandoning academia—and the possibility of tenure—to go on another wild goose chase.

As such, Iribe and Mitchell made every effort possible to convince Bolas that Oculus was not another goose chase. At USC, which could afford a six-figure setup like the one Luckey wrote about on MTBS3D, Bolas was focused on pushing the limits of possibility; whereas, at Oculus, the focus would be on the possibilities of creating a consumer product. Or, to steal an analogy that Steve Jobs must have used at the dawn of Apple, the goal here wasn't to build the most powerful mainframe possible; it was about empowering the masses with an affordable personal computer.

"I genuinely believe that we can accomplish that," Mitchell explained to Bolas and his Fakespace cofounder, Ian McDowall, during a last-minute, marathon dinner the three had in mid-July 2012. Mitchell, who was up in Berkeley to visit his girlfriend, had received a call from Iribe earlier that day asking if he might be able to ditch his lady friend and head over to SFO airport.

"Where do you need me to fly?" Mitchell had asked.

"Nowhere," Iribe had told him. But Bolas and his partner were at the airport with some time to kill. "They'd like to meet you."

Prior to this airport meeting, Bolas frankly hadn't known what to expect. Bolas was a seasoned veteran in an industry that was barely an industry and certainly had very few veterans. As such, Mitchell wouldn't have been shocked if Bolas's mentality was along the lines of "You mangy kids don't know anything about VR and you seriously want to start a company?!" So, with that comment, Mitchell felt a sense of mutual respect. The dinner ended on an optimistic note, with plans to keep the dialogue going.

By this point, now that Bolas had a pretty good idea of what Iribe and Mitchell brought to the table, much of that conversation continued with Luckey. And given the bond he felt with Bolas, Luckey frequently made it clear how much he wanted him to come join Oculus. But, as the days counted down to August, Bolas remained unsure.

"He keeps wishy-washing back and forth," Luckey told Iribe and Mitchell. "But if you asked me to bet, I would bet on him joining Oculus."

Luckey's hunch seemed to be on point when Bolas agreed to re-cord an on-camera endorsement for the Kickstarter video. "The Rift is taking years of virtual reality research and putting it into a package everyone can use," Bolas stated from a recording booth the guys rented at LMU.

While on campus, the guys checked in with Win Bates to see how the video was coming together. Excellent, Bates raved, and then showed off some of the cut, which included an endorsement from CliffyB that Luckey had not yet seen. "So I recently had a chance," CliffyB began, "to check out the Oculus headset. And needless to say: I'm a believer." CliffyB had become such a believer that he wanted to invest $100,000 in Oculus.

The guys had also recently heard back from Dillon Seo, Iribe's contact in Korea. Seo had decided that he no longer wanted to be Mr. Nobody, that he didn't want his young son to look at him and see just another salaryman. So instead of being just another face in the crowd, Seo wanted to try and become that which he had once upon a time dreamed to be: a superhero. Or, at the very least, someone who was willing to take a chance.

In addition to gaining attention in Korea, Oculus was also gain-ing traction with the press. This was largely due to the help of a public relations duo, Jim Redner and Eric Schumacher, who had done some work for Gaikai and were willing to help Oculus get out of the gate. As they tried to convince the media that this no-name start-up was for real, Iribe and Mitchell tried to convince Jon Malkemus, Gaikai's art director, to make some motion graphics for Oculus's Kickstarter video. He was slammed with work from his "actual job . . . [cough, cough, Brendan?]" but willing to fit in some animation during his off-hours. And just like that, some semblance of a team was forming.

"What about Laurent?" Luckey asked Iribe, inquiring about the guy who had first put them in touch. "I got the sense that he wanted to be involved."

"He did," Iribe replied. "He does, I guess. But it's not a great situation."

Iribe explained how he loved Laurent, considered him a true friend, but just didn't think there was a place for Laurent at Oculus. Again, he loved the guy, but why did they need Laurent Scallie?

Excluding Scallie seemed a bit cold to Luckey, but who was he to argue? Iribe had been able to reel in CliffyB, Jack McCauley, and Dillon Seo. They seemed pretty close to recruiting Mark Bolas, too. But as Luckey and Iribe were trying to sign Bolas—to bring him on board as Oculus's chief scientist—Mark Bolas was busy trying to get Luckey to sign an agreement of his own.

UNITY

"Your timing is just perfect," David Helgason said, welcoming Iribe to Unity's Headquarters in Copenhagen. "As I mentioned over email, all three founders are here today."

In addition to Helgason—Unity's then CEO, outfitted that day in a silver dress shirt whose collars popped so high they bumped his ears as he walked—the company's other founders were Joachim Ante (Unity's CTO) and Nicholas Francis (CCO).

"Fantastic," Iribe said, as Helgason brought him to the conference room, where he found Ante and Francis waiting there with friendly faces. Yet behind the good cheer, all three founders harbored serious doubts about virtual reality. The only reason they had even agreed to this meeting was out of respect for Iribe. Ever since they first crossed paths, back in 2009, Helgason had been a big fan of Iribe. "Brendan's relentless," Helgason often remarked. "He knows how to bend the world to his will in a very positive way."

"I can only imagine what you must think of virtual reality," Iribe said, "but I promise you that I'm not here to waste your time. I genuinely believe that VR is going to change the face of gaming; and, at Oculus, we've made it our mission to do exactly that. To deliver the promise of true immersion."

To a different audience, those words may have sounded overly ambitious. But to the three men in that room, it was familiar, transporting each of them back in time. Back to a dark, cluttered ground-floor apartment that they all remembered so well (albeit not always so fondly), where ten years earlier, younger versions of themselves—a trio of Danish Steve Jobs Fanatics—had just started a game company, Over the Edge Entertainment, that most agreed would fail. On top of all the risks typically associated with such an endeavor (cost, development cycle, etc.), these guys had made the bold decision to focus exclusively on Mac-based games, a segment that, at the time, made up less than 1 percent of the video game market.[10] Or to put it even more starkly: of the one hundred best-selling games in 2002, none were made for the Mac.

Initially, what they had in mind was a spy game set against the Israeli–Palestinian conflict. It wasn't quite meant to be educational, but the goal was to utilize spycraft—which would require players to learn about, and empathize with, the opposing side—as a means to explore that conflict in a deep and refined manner.

That game never materialized, and the one that did (*GooBall*) would inevitably struggle to find an audience. During *Gooball*'s development, however, the guys found themselves less interested in the game that they were building, and more interested in the engine they were building to run it. They didn't necessarily think their engine was "better" than the engines already out there (like Epic's ultrapopular and just-released Unreal Engine 2), but they did feel its focus on usability and workflow made things a lot easier on developers. Especially the developers with less experience and limited resources: the have-nots of the developer world. Those who could never afford something like Unreal Engine 2 (which cost about $400,000); those who couldn't even afford to *know* if they wanted to be a developer. These were the people they wanted to reach. These were the ones they wanted to empower. And so, out of failure, Unity was born; as well as new mission: to democratize game development.

One year later, in 2005, Unity 1.0 was unveiled at Apple's World-wide Developers Conference (WWDC).[11] Although Unity 1.0 was no runaway success, the engine—which cost $1,499 for professionals and $249 for indie developers—was exactly what the three founders envisioned it to be: proof-of-concept and a promising first step toward a game development revolution.

The next few years were touch and go, but the trio managed to cobble together something close to a break-even business. A few employees were hired, features were added to the engine, and salaries were paid on time (more or less).

"We were working insanely hard," Helgason reminisced to Iribe, as he began prepping the conference room for his demo. "We were consistently behind on everything. And the hole kept getting deeper and deeper. Luckily the business was growing, but it was slow. And then . . . the iPhone."

The iPhone was first released in June 2007. At the time, the Unity guys didn't see this as the watershed event it turned out to be, but they were intrigued enough to develop a version of Unity that would enable publishing applications for the new device. In December 2008, mere months after the launch of Apple's App Store, Unity iPhone was released. And suddenly, following six years of struggle, Unity's trio of democratizers found themselves on the kind side of a tipping point.

In 2009, the number of developers using Unity hit 10,000. In 2010, that number hit 170,000. And by Iribe's July 2012 visit to Copenhagen, the number of developers using Unity had just surpassed one million. But the power of Unity went beyond sheer numbers; it was as much about the end-to-end ecosystem that Unity had created.

What had started as a Mac-based toolset that supported Mac-based platforms soon grew to support more platforms than any other engine. This was good, the trio thought—now developers didn't need to put their eggs in a single basket—but it still wasn't good enough. Because even though Unity empowered any developer to become his or her own game studio, bigger studios still wielded many non-engine-related advantages. Specifically when it came to the details.

For example, imagine a virtual forest: something 3-D, populated by squirrels, with trees as far as the eye can see. Well, every tree (every squirrel, every asset) in that virtual forest had to be created by someone. This meant that the bigger studios had more resources and could build better assets. They also had better in-house tools (i.e., scripts, shaders, special effects, etc.) to maximize how those assets could be implemented. So, in October 2010, Unity released a solution to this inherent inequality: the Asset Store.

Unity's Asset Store was a marketplace where developers could buy and sell just about any game-making ingredient that might be needed. Trees and squirrels, shaders and special effects; developers could now buy what they wanted or sell what they made—and by the thousands they did just that.

In March 2011, Brady Wright's EZ-GUI and Sprite Manager netted him over $15,000.[12] In March 2012, Michael Lyashenko's NGUIs made more than $25,000 (thanks, in part, to the Asset Store's "Madness Sale").[13] These are cherry-picked examples, of course, but the takeaway was clear: Unity was relentlessly committed to bringing quality development tools to the masses in whatever way they could. This was a sentiment that Iribe hoped would work to Oculus's advantage as he handed the prototype to David Helgason, launched the testbed demo, and eagerly awaited his reaction.

"This is really shitty!" Helgason announced, in a cheery tone that didn't match his words.

"Wait, what?" Iribe asked, even more puzzled as Helgason began to chuckle.

"The hardware keeps drifting," Helgason explained. "It's really shitty. *But* this is so compelling to me. I want to stay in here. This is a great place to be!"

Helgason removed the headset with a sheepish grin and then passed it along to Ante and Francis, who reached a similar conclusion: there are challenges that must be solved, but man alive this thing is extraordinary. Iribe told them how he needed Unity on his side so that independent developers didn't get left behind in the coming VR wave.

Iribe also kindly asked if he could film Helgason talking about his experience so that it could be added to the Kickstarter video. "CliffyB will be in the video too," Iribe added, noting Helgason's uncertain look. "And you wouldn't necessarily have to say that Unity would be supporting the Rift. I mean, that'd be great. But this is mostly just about you describing your experience."

Helgason remained uncertain—he was the CEO, after all, and couldn't just toss out endorsements willy-nilly—until, one by one, the CEO watched his employees walk into the conference room, try the Rift, and walk out utterly elated. To the extent that the reality of virtual reality seemed clear: Brendan was going to bend the will of the world once again, and Oculus was going to resurrect VR.

"Okay, Brendan," Helgason said, "I believe I'm ready to record that video testimonial."

"Awesome!" Iribe replied, "Are you sure?"

"I am not going to regret this, am I?"

"No, no, this is going to be great."

"Well, I've gotta say," Helgason soon began, "I just tried the Oculus prototype and it was such an immersive, amazing experience that we pretty quickly—like within an hour—decided to get behind this project."

With only five days to go until the Kickstarter launch, Iribe couldn't help but reflect on how far Oculus had come in such a short amount of time. They looked almost like a . . . real company. Wow. Not bad, considering that it had been only three weeks since he and his crew tried Luckey's invention for the first time.

Even so, Iribe knew that one thing would put them over the top: Valve. If Valve vouched for Oculus, then Oculus would be seen in an even brighter light to the 100+ million people who used Valve's platform.

The good news was that—through a recommendation from Carmack—Luckey was able to get a meeting with Valve. The bad news—at least to Iribe in Copenhagen—was that he wouldn't be able to attend. Nor would Mitchell, who had just left for a vacation with his stepmom and dad to China. This meant that Oculus's most important meeting would come down to the company's least polished founders: Michael Antonov and Palmer Luckey.

CHAPTER 10

VALVE!

July 26–27, 2012

ON THURSDAYS, VALVE ORDERED IN LUNCH FOR THEIR EMPLOYEES. ON THURS-
day, July 26—the day that Luckey and Antonov were supposed to meet
with Valve—they would be ordering in hot dogs from a place called
Dante's Inferno. And though Luckey did not need another reason to be
psyched about visiting Valve, he *was* looking forward to trying one of
Dante's gourmet dogs. But he'd never get to try them due to an issue
at the airport.

"Excuse me, sir," a TSA agent said to Luckey, after flagging something in his carry-on. "Can you please explain what this object is?"

"It's a virtual reality headset," Luckey replied.

The agent gave Luckey a dubious look and then shuffled him off to
the side where somebody would come by to see him shortly.

Although Luckey knew that there was never a good time to be
stopped by airport security, right now—mere days before the Kickstarter, at a time when every minute felt precious—this just felt like
a kick to the groin. Especially as his flight was called to board, and
especially as Antonov, carrying his own headset, slid through security
with no problem.

Luckey's VR headset *did* look suspicious. Any VR headset probably would have caught their attention (well, except Antonov's, apparently), but this just-built prototype—with exposed PCB boards and

a spaghettic mess of wires—looked less like a consumer device than a high school science project gone awry.

"Part of the reason it looks so weird," Luckey explained to a new TSA agent, "is because I used the shell of an old VFX1 headset. They were popular in the '90s. Not *that* popular, obviously. But—"

The agent signaled for Luckey to stop. "Why are all the wires just out like that?"

"Because it's not totally finished. And it's easier to work on that way."

"So it'd be possible to modify the function of your device while up in the air?"

"Well . . . yeah. Yeah, I guess it would."

As soon as the words flew out of his mouth, Luckey knew he had screwed up. Who tells the TSA they're carrying an electronic device that can be modified in midair? Idiot! Now he was probably going to miss his flight and ruin the chances of working with Valve and—worst of all—let down his new partners; who, by this point, had figured out that he had, shall we say, a laissez-faire attitude with time. He had a habit of showing up for meetings a little late, extending end-of-day deadlines into middle-of-the-night hours, and just generally moving at his own easy-breezy Palmerian pace. But, of course, there was a huge difference between jumping on a call five minutes late (or happily working five more hours into the night) and missing a key, company-critical meeting, especially when that meeting was with Valve.

To a large percentage of gamers—particularly those who, like Luckey, primarily gamed on PCs—being invited to Valve was like being invited inside of Willy Wonka's Chocolate Factory. Not only because they made great games (or because 100+ million people used Valve's PC-based Steam platform to buy and play those games) but because the eccentric and mysterious ways in which they did things was unlike anybody else in the industry.

To start with, Valve had a flat organizational structure. Nobody had any bosses! For people like Luckey and Carmack—ambitious self-learners who often felt stymied by structure—a flat organization sounded too good to be true. And for years, many suspected that was probably the case. That Valve was progressive was an understatement. A recently leaked copy of Valve's handbook for new employees, which was seventy-four pages long and subtitled "A fearless adventure in knowing what to do when no one's there telling you what to do," provided outsiders with an illustration-filled inside look at what made Valve so unique.[1] Including details like these:

- "Nobody has ever been fired at Valve for making a mistake . . . even expensive mistakes, or ones which result in a very public failure, are genuinely looked at as opportunities to learn.
- "Other companies have people allocate a percentage of their time to self-directed projects. At Valve, that percentage is 100."
- "Employees vote on projects [to work on] with their feet (or desk wheels) . . . Think of those wheels as a symbolic reminder that you should always be considering where you could move yourself to be more valuable."
- "Of all the people at this company who aren't your boss, Gabe is the MOST not your boss."

Gabe was Valve cofounder Gabe Newell. Or "Gaben," as he was affectionately called by his legions of fans—many of whom enjoyed sharing memes of Newell, trying to hug him at trade shows, and pledging allegiance to the company he cofounded via the mantra "In Gaben We Trust."[2]

What was it that made Valve's trusty leader such a larger-than-life, meme-worthy figure? Part of it had to be Newell's SantaClausian physique; and part of it, too, had to be his unusual hobbies (like collecting knives and going brony for My Little Pony).[3,4] But beyond the superficial or eccentric, the devotion truly came from this: Newell spoke for PC gamers the same way that the Lorax spoke for the trees. Passionately, protectively, and never mincing words—famously calling Sony's PS3 "a total disaster on so many levels" and declaring Windows 8 "a catastrophe for everyone in the PC space."[5,6]

Eventually, a TSA supervisor showed up to inspect Luckey's headset. It looked fine at first glance, but he wanted to run bomb swabs over it. Once the results came back, Luckey was free to go—but by this point, he had already missed his flight.

Luckey emailed Dan Newell (saying that "TSA held me for more than an hour . . . they did not seem to have a problem with Michael's [HMD], but I was not about to point that out to them") and was able to push the meeting twenty-four hours. This gave Luckey plenty of time to switch his flight and get through security. But it also meant that Antonov had to stay an extra day in Seattle, which he wasn't too thrilled about. Nor was he particularly happy to learn, when Luckey finally arrived, that the laptop with the demo was no longer working.

"Are you serious?" Antonov asked. Their meeting was the next morning and it was already nighttime. "Why didn't you say something earlier? I already went to Best Buy to buy a video camera so that we may record if it goes well!"

"I didn't know," Luckey said. "Sorry."

"Well . . ." Antonov replied, thinking. "I know where there is a Best Buy. Let us go see if they are open."

They made it to the Best Buy (thirty minutes before closing), but the store didn't stock a laptop like the one that Luckey had been using. And unfortunately, even though these were a pair of tech-spec-loving guys, there wasn't any way to know for certain which laptops (specifically which graphics cards) would work with their prototype. So, with a hope and a prayer, they ended up purchasing a silver Samsung laptop.

There's no better time to bond with someone than when you're traveling together and both feeling an equal mix of anticipation and relief. So before heading back to the hotel, the two partners decided to savor the momentary lack of chaos at a Pei Wei restaurant near Best Buy.

"Any progress on the SDK?" Luckey asked, checking in on the software development kit that Antonov would be spearheading.

"When I start on something new, I really like to try and understand it deeply so I can be effective. In some cases that may slow down progress in the beginning, but that makes me feel like I can really move fast down the line and make the right choices."

"That makes sense to me."

"I have not yet worked with gyroscopes before. So the orientation tracking, it is all very new to me. But I am beginning to better understand what needs to be done; I started reading about sensor fusion."

Sensor fusion is, literally, the process of fusing together data from multiple sensors. Specifically, Antonov was referring to the need for a VR headset to track head movements and orientation in a three-dimensional space. Central to accomplishing this is something called an inertial measurement unit (IMU), which is an electronic device that detects the rate of acceleration (via accelerometers), the rate of angular velocity (via gyroscopes), and the intensity of a magnetic field (via magnetometers). Each of these sensors—accelerometer, gyroscope, and magnetometer—yielded three-axis measurements (X, Y, Z) and each of those measurements had to then be fused together. As all that was happening, corrections had to be made for electronics-related blips like drift and tilt. Because if all that didn't happen, the virtual world won't be in the right place when the headset wearer turns his head.

This was a far cry from anything Antonov had ever worked on before. And because a virtual reality industry didn't even really exist at this point in time, it was a far cry from what most anyone had worked on before. It was enormously challenging for Antonov to write code based on concepts and high-level mathematics he hadn't dealt with, though—not unlike John Carmack—he had a talent for teaching himself complex disciplines with relative speed; and, to his point, he was beginning to better understand what needed to be done.

"For example: drift correction," Antonov said. "I don't know what paper Carmack used to write his algorithm for the drift correction, but the one I have been reading is called 'Quaternion-Based Fusion of Gyroscopes and Accelerometers to Improve 3D Angle Measurements'—a 2006 paper by Favre, Jolles, Siegrist, and Aminian[7]—"and let me just say that much of this stuff is very new to me. But let me also say: I find it to be very fascinating."

Mention of Carmack reminded Luckey that he had not yet heard back about getting a short video endorsement for the Kickstarter video. But right now wasn't the time to try and deal with that because, after dinner, Luckey and Antonov had their hands full with something else: laptop issues. The graphics card worked fine, but the software wasn't rendering. There appeared to be a compatibility issue with the GPU. And just when it seemed like the night couldn't get any more off-kilter, Luckey received a disappointing email from John Carmack.

"What did he say?" Antonov asked.

"I don't think I can make an endorsement on your Kickstarter video without upsetting ZeniMax," Luckey read aloud. "There is already a somewhat sour sense that someone is going to make a lot of money in VR, but it probably won't be ZeniMax."

Antonov sighed, though he could not argue with that point. If *Doom 3: BFG Edition* was as good as anticipated, the game would make plenty of money. But the true windfalls typically went to those who controlled the platforms—companies like Sony, Microsoft, Nintendo, and Valve whose expansive distribution networks sold content made by others, for which they took a cut, usually in the neighborhood of 30 percent.

"It is very important," Luckey continued reading, "that you not use anything that could be construed as Zenimax property in the promotion of your product. Showing my R&D testbed with the Rage media would be bad, for instance. The set of links you have been keeping at the web site to all the interviews should cover everything, and you could probably do some kind of montage of them in the video without stepping on anyone's toes, but I am not a lawyer . . ."

"All right," Antonov said. "Did he copy Brendan or Nate?"

"He did not," Luckey answered.

"We should tell Brendan and Nate right away."

"Done," Luckey said, forwarding the response to Iribe and Mitchell.

It was hard for Luckey and Antonov to get back to work after receiving that news. Shortly thereafter, still unable to get their just-purchased computer working, they concluded that this was an unfixable issue.

"There is nothing more we can do," Antonov finally said. "We're just going to have to run it on one of Valve's desktops when we get there."

Doing that would be pretty embarrassing—hey, we're here (after we postponed!) and oh yeah can we borrow a computer!—but, like Antonov said, there was really nothing more they could do at this time. Their best bet was to get some sleep.

"I HAVE A GOOD FEELING," LUCKEY SAID TO ANTONOV FROM A COUCH IN VALVE'S lobby.

Antonov turned to Luckey but chose, instead of words, to deliver a look that said, *Coming from the guy who missed his flight and brought a broken laptop?*

"Don't worry," Luckey said, standing up to stretch his legs. "I've got a *good* feeling about today. I mean: we're at Valve! What could be better than that?!"

Antonov nodded, admittedly feeling a little inspired by Luckey's confidence. He wanted to tell this to Luckey—to share some camaraderie for today!—but by the time he looked up, the kid had already meandered halfway across the lobby. "Hey! Where are you going?"

"This is cool," Luckey said, pointing to the enormous brass valve in the center of the lobby. It was a heavy-as-hell, four-foot-tall pressure controller that had been salvaged from a ship and eventually gifted to founder Gabe Newell by his brother. Toward the top—a couple inches below Luckey's shoulders—was a large wheel wrench that Luckey couldn't resist trying to spin. But the thing wouldn't budge.

"Palmer?" Luckey heard over his shoulder. He swiveled around to see a lanky, long-necked man in a short-sleeved, button-down gray shirt walking toward him. "I'm Michael Abrash. It's great to finally meet you!"

"Oh!" Luckey enthusiastically replied. "Hey! Likewise!"

After a hearty handshake, Luckey noticed Abrash was flanked by a couple of colleagues. The two men—Atman Binstock and Joe Ludwig—introduced themselves and then escorted Luckey and Antonov to a conference room. "We're just waiting on a few other people to come join us."

While they waited, Antonov and Luckey revealed that they didn't actually have a working computer with them.

"Would it be possible for us to borrow one?" Antonov asked.

"Sorry," Luckey added. "I know it's a pain. But it'll be worth it. I promise."

If Abrash was perturbed, he sure didn't show it. "Tell me about the IPD," Abrash said, referring to the headset's interpupillary distance. "Do you plan to address in hardware or software?"

"It'll be adjustable in software," Luckey confidently replied. "Which should be totally fine; unless you're dealing with an outlier. But generally speaking, anything between 58 mm and 75 mm should work perfectly."

Fabian Giesen, Gordon Stoll, and the rest of the Valve crew arrived. So did a computer. But not long after booting it up, Luckey ran into another false start. Although the headset's sensor worked fine, its clock rate ran at 125 Hz. The Rift's software, however, was set to 250 Hz, which mean that the virtual world would only turn half the distance as the headset-wearer's head; as a result, it would almost immediately induce nausea.

And yet . . . Michael Abrash continued to appear delighted.

It took Luckey and Antonov a little while to fully understand the reason for this apparent delight, but it started to make sense when Abrash began detailing what had been going on at Valve since he joined the company in May 2011. Initially, the plan was for him to come and help Valve create a console of their own. But after investigating what steps would need to be taken, and weighing the costs of such an endeavor, it didn't make any sense for Valve to take that kind of risk on hardware. Shortly after coming to this conclusion, Abrash heard about a company called Innovega that had an AR system in development. It was still very much in the prototype phase, but what they were making—augmented reality contact lenses—piqued Abrash's interest. Then at some point shortly thereafter, Abrash asked Atman Binstock— now smiling there beside him—to meet for coffee.

"By the end of that conversation," Binstock said, recounting that meeting for Luckey and Antonov, "I just started thinking through how it would actually work. Like I hadn't really thought about the problems. What it really takes to go from head movement to putting the right photons into your eyeball. That's sort of the core problem—end to end—how to make that work. It sort of seemed exciting and intriguing, but even after that first meeting it took some salesmanship from Mike to get me on board."

"Atman did a *pretty* good job of outlining his objections," Abrash said through laughter. "That's probably what I remember best from that day!"

"What I remember most from that day was what you said near the end," Binstock said. "The part I recounted earlier, about how techno-logical revolutions actually happen."

"It really does make a difference," Abrash said. "Who chooses to work on things. I don't think that's as true with things related to Moore's law. That's about a huge number of interlocking and over-lapping efforts, right? But the sort of thing we're doing—well, both what we started out doing and, now, what we're currently doing—I think it requires finding key people and, when the history is finally written, there will be all these branch points that you can just point to and say: Ah, if not this, then none of this would happen. Or it wouldn't happen this way or as quickly. And of course"—Abrash leaned forward and pointed to Luckey—"you obviously fall into that category. But I'm getting a little ahead of myself."

Abrash rewound his story a bit, returning to the start of Valve's AR experiment. There was a certain irony to this—the way that he and his team nostalgically talked about the "early days"—when, in real-ity, those days were only about six months old. That was when Atman Binstock, along with Fabian Giesen—a young, supersmart German engineer whom Binstock knew and trusted from his time at Rad—decided to take the leap and join Valve to work with Michael Abrash on augmented reality.

Internally, this team became known as the "Vortex Group," and it consisted of only five people: Abrash, Binstock, Giesen, and a pair of seasoned Valve engineers (Aaron Nicholls and Gordon Stoll). The group's original plan was to focus on two soon-to-be-released pieces of wearable hardware. Unfortunately, it turned out not to be soon enough. These two wearables—AR glasses from Lumus and those digi-tal contact lens from Innovega—remained stuck in a prototype phase and Abrash's group was forced to look for other options. So what they started with instead was a flimsy, disappointing pair of AR glasses made by a company called Vuzix. Team members joked that the thing looked like it had been glued together by hand, but the joke stopped being funny when they tried it on. Because the FOV was only 15°.

For a portal into enhanced realities, these glasses were about as lousy as it got. So the first thing this Valve team needed to figure out was what they could even do with this crappy hardware. The idea was to create an app that could display a computer-generated whiteboard on a wall. At some point, as their belief went, this virtual whiteboard would be interactive (so you could write on it), but that wasn't really a concern for now.

The team realized that without any reliable way to track the user wearing the headset (and with that minimal field of view) it was extremely difficult to reliably find, let alone see with any detail, the thing that they were working on. So instead of starting off that way, Binstock built a program, called "The Virtual, Virtual Whiteboard," that would enable them to conduct this experiment on a PC instead of in real life. In other words: they essentially created a first-person-shooter in which the POV character was looking through a virtual Vuzix headset. After a few months of work, they became convinced of two things: this was going to be much harder than they ever imagined and there was nothing more important than having a wide FOV.

Fortunately, around this time (May 2012), Abrash connected with his old friend John Carmack. They commiserated over FOV difficulties—Abrash with AR, Carmack with VR—and then, at some point, Carmack started talking about this kid he had found online, this kid who had created a compelling, low-cost VR headset with an incredibly wide field of view.

As if on cue, Antonov got the prototype working. He handed it to Abrash, who became the first to try out the Rift.

Going into this meeting, Luckey knew that Abrash would be tough to impress. So he was pleased to at least see some flickers of delight on Abrash's face. "You've definitely done a good job with the field of view," Abrash commended, panning his head in the virtual space. "And you've managed to keep it pretty lightweight."

This praise, however, was not offered by Abrash's colleagues. Although none were outright dismissive, Luckey felt that their skepticism extended far enough to border on cynicism. But at least they were vocal about it: Why was this your approach to X? How do you plan to solve Y? When do you intend to address the inevitability of Z? Luckey had anticipated these doubts and did his best to answer them with a positive spin, which led, understandably, to another wave of curiosities. How familiar are you with what happened in the '90s? Do you really think this will be compelling enough to attract the attention of developers? Worst-case scenario—just some devil's advocate here—but couldn't this possibly have the reverse effect and end up poisoning the well for VR?

Midway through the barrage, Luckey came to a conclusion, as good a one as he could find at that moment: this skepticism, he believed, was less about the Rift and more about VR in general. As valid as these questions would prove to be—the folks at Valve just weren't "True Believers." Valve, with its unique structure, was a place where all it took to join a different team was to simply move your desk to the VR component. And in the eight-ish months that Abrash had been working on AR/VR, very few employees had taken the leap. As a result, most of his team were contractors who had been hired to help fill the gaps. This did not mean that contractors like Atman Binstock or Fabian Giesen were any less qualified. But, to Luckey, it meant that most of the people there were only willing to dip their toes in the water.

Luckey's assessment was on point, but the truth was that he didn't even know the half of it. Within Valve, the AR/VR work was a constant source of skepticism and derision. Not a week went by without a clique of Valve employees stopping by to criticize what they perceived to be a "boondoggle." Stop wasting money! Stop wasting time! Stop trying to "make this a thing"!

Beyond whatever psychic toll that comments like those may have caused, there was also a practical cost to the boondoggle perception. Because at Valve, a significant portion of employee compensation came from a peer review system; so working on projects that didn't create a tangible benefit was unlikely to be rewarded; and with less a chance of being rewarded, it made it that much harder for Abrash to recruit colleagues and—as per the employee handbook—get them to "vote on projects with their feet (or desk wheels)."

Neither Luckey nor Antonov knew this at the time, but even if they'd been aware, it likely wouldn't have made a difference. Just the fact that Valve had *anyone* working on this stuff made Luckey's face glow with a big, bright smile. Yet there was still one thing that would have made Luckey's smile grow ever wider. "Do you think Gabe is gonna have time to stop by?"

Initially, Newell had planned to join the meeting, but Abrash explained how pushing it back a day had nixed that plan. "But I think Gabe would really enjoy your demo," Abrash said. "Would you maybe be okay with leaving the Rift here for a few days?"

"Let me check with Brendan," Antonov said, and then texted Iribe: "Can we leave it with Valve? It means that we also have to leave SW [software], which Carmack just warned about . . ."

Iribe joked back that they could leave it there if Valve gave Oculus an endorsement! "[But] seriously," Iribe wrote, "just tell them we need to get approval from Carmack to leave the software, so we'd rather come back and just do the demo when Gabe is here."

Decisions like these were always tough: juggling what you *want* to do with the right thing to do (and then inevitably spending the plane ride back wondering if you should have done something differently). But in this case, those theoretical regrets soon became moot.

"Hi, I'm Gabe," said the bearded, bespectacled, and large man who entered the room. Like a wizard or warlock, Newell exuded a quiet-yet-fantastical aura. Even Luckey, who fancied himself as someone who didn't get starstruck, needed a moment to make sure that this was real life.

"Glad you could make it," Abrash said. "You gotta check this thing out."

From the way they interacted, Luckey could tell that Abrash and Newell were close. But, at the time, he didn't quite realize how far back the relationship went; nor that Abrash was partly responsible for the existence of Valve.

What had happened was, in 1996, Newell and cofounder Mike Harrington flew down to Texas for a meeting with id Software. After several years at Microsoft (thirteen for Newell; nine for Harrington), the two men had decided to start their own game company and were interested in licensing the source code from id's hit game *Quake*. Although id had little interest in arranging such a deal, Abrash—who, eighteen months earlier, had left Microsoft for id—stepped in to help get a deal done.[8]

Over the next two years, Valve used that code to build a puzzle-solving, sci-fi shooter called *Half-Life*. Before the game was released (and would go on to sell over ten million copies), Newell and Harrington asked Abrash if he wanted to become the company's third founder. Tempted as he was, Abrash ultimately declined, opting instead to join the Natural Language Group at Microsoft. But over the years, as Valve grew into a dominant force—after *Half-Life* came *Team Fortress* and *Counter-Strike*—Newell and Harrington periodically asked if he was ready to join up.

For fourteen years, Abrash declined. But in 2011 he decided to take the leap and joined, initially to spearhead a console for Valve. A "platform that Valve could rely on" was how Abrash described it, but shortly into his tenure he concluded that this was not a recipe for success; it would take too much money and too much manpower for something that wouldn't be particularly future-proof. So he ended up setting his sights on the future: augmented reality.

Although AR still remained Abrash's primary focus, he had become increasingly interested in pivoting to VR. It just seemed much more tractable. And it removed five of the biggest challenges facing AR:

- **INPUT:** VR (at least initially) can utilize standard game controllers, whereas the input scheme for AR is currently uncertain.
- **TRACKING:** Since VR is restricted to a single, controlled location, it becomes much easier to develop technologies that can track the user.
- **POWER AND BATTERY LIFE:** Since a VR headset is tethered to a powerful PC, it has access to significantly more computing power than anything going mobile. Also, due to the single location, battery life becomes moot.
- **FORM FACTOR (AND SOCIAL ACCEPTANCE):** Since VR headsets will be used inside the home, they don't need to be particularly stylish; nor does the form factor need to be as lean and socially acceptable as a pair of familiar-looking sunglasses.
- **THE SUN:** As if it wasn't already hard enough to make a phone screen visible from all angles outside, imagine trying to create and sustain an overlaid reality.

In addition to all that, VR presented a hidden benefit as well: it could be used to try and solve several of the problems plaguing AR (including all five above). Similar to how Binstock had used "The Virtual, Virtual Whiteboard" to try and solve AR challenges in a computer simulation, VR could be used as a proving ground for AR ideas; and it'd be much cheaper, too (why bother buying parts and building prototypes when you can create almost anything with a sequence of 1's and 0's?).

"Ready to try it out?" Luckey enthusiastically asked.

Newell smiled, nodded, and strapped on the prototype.

As Newell receded into the virtual world, Luckey and Antonov exchanged a hopeful glance. Here goes . . . everything! And then wonderfully—with each nod from Newell, with each upbeat-sounding *hmmm* from Newell—their nerves began to subside, to the point that by the end of the demo, the Oculus guys were feeling quite optimistic.

Newell thought it was an exciting technology demonstration but was concerned that the cool tech wouldn't be enough to interest developers. "You need to figure out why people should make a VR game," he explained. "You need to figure out what kind of games work best in this. You're not gonna be able to just sell this if it's a bunch of tech demos; you're going to need to be able to figure out how people can make actual games. Not just this one demo that I just saw. It needs to be real stuff."

That's fair, Luckey thought. He knew that his technology had flaws, and he hadn't set out to convince Newell that VR was already here and now. He was merely here to demonstrate what he believed to be the first step in a revolution. And when the time felt right, Luckey asked if Newell and/or Abrash would be willing join that revolution and record a brief testimonial in support of Oculus. Antonov explained a little bit about how their quotes would be used and then, as if to show them how harmless this all would be, he dug through his bag and held up the video camera he had purchased from Best Buy.

After some friendly ripples of laughter, the Valve guys politely declined. At least for now. They needed to talk it over among themselves (and then with their lawyers) before they could provide any sort of endorsement. They apologized for being unable to commit at the moment; but, they explained, if Valve did decide to participate, they would happily record something on their own and then send the video over.

With the Kickstarter going live in only four days, this didn't leave much time. But, at least, there was still a fragment of hope. The same, however, could not be said for the secondary objective of their visit: to see if Valve would consider adding a "VR section" to Steam, so that developers could distribute content for the Rift.

No. Nope. Not gonna happen. "Steam," Newell explicitly explained, "isn't a repository for half-baked tech demos."

That stung. But Luckey and Antonov did their best to remain professional.

"Thank you for meeting with us," Luckey said, as the conversation neared its end.

"We really appreciate all your time," Antonov added, realizing that the meeting had lasted over three hours. But before it reached its last breath, Luckey decided there was still one thing left to do.

"Gabe!" Luckey chirped. "Do you think I could I get a picture with you?"

"Of course," Newell said with a hearty nod.

"And maybe Mike, too?" Luckey hopefully added.

Abrash nodded. "Sure."

As Luckey wedged himself into a pose between Abrash and Newell, he felt a bit embarrassed. "I normally try to be more professional," he said. "But couldn't resist."

"That's okay," Newell replied. "I do the same thing when I meet my favorite Japanese video game developers. And I make the exact same excuse."

After a wave of soft chuckles, Antonov snapped a photo. Although it wasn't quite the trophy Luckey had hoped to leave with that day, it certainly made for one hell of a consolation prize.

CHAPTER 11
KICKSTARTER

July/August 2012

"NATE," IRIBE SAID, "I NEED YOU TO DO ME A FAVOR."

Mitchell—still in China with his stepmom and dad—could tell from Iribe's tone that this was going to be a big favor. And he was right.

"Nate," Iribe said, "I need you to resign from Gaikai. Today."

For a moment, Mitchell felt a strange, upside-down sense of déjà vu—this was just like when Iribe had called him in France and pressed him to join Gaikai; except, of course, this was the complete opposite. Now he wanted Mitchell to resign from Gaikai? Today? Huh?

Iribe apologized for the confusion and how out of the blue this must have seemed, then explained that what had happened was QuakeCon.

QuakeCon was an annual convention dedicated to celebrating the games of id Software.[1] The event had taken place in Dallas every summer since 1996 (when *Quake* was first released) and featured panels, previews of upcoming id content, and an ultracompetitive BYOC (Bring Your Own Computer) gaming tournament. By convention standards, QuakeCon was relatively small; the event typically attracted around ten thousand folks. But of that ten thousand, over 20 percent showed up for the BYOC tournament.[2,3] This was exactly the demographic that Oculus wanted to reach.

That year's QuakeCon was scheduled to take place in a few days—on August 2, at a Hilton Hotel in Dallas. All of which Mitchell already knew, since Luckey would be doing a panel there with John

Carmack and Michael Abrash. But what he didn't know was that Oculus had just been offered a booth to run demos at the show; which was in itself awesome news, except that it wasn't really the kind of thing that Luckey could handle on his own. "So I need you to go," Iribe said.

"Can't we just hire someone to help out?" Mitchell asked.

"Yes," Iribe said. "In fact, I'm setting up a call with that guy Palmer wanted to hire so we'll have additional support, but I mean . . . look, I think Palmer's friend is even younger than he is . . ."

Mitchell laughed. They had co-founded a company with a nineteen-year-old and their first official (non-exec) hire was going to be a teenager? Okay, Mitchell understood why Iribe wanted him at QuakeCon. But he wasn't sure he was comfortable with resigning to make that happen.

"I know it's a lot to ask," Iribe told him. "But you can't go on behalf of Oculus while you're still employed at Gaikai. So we have to decide what to do here about QuakeCon . . ."

"WANT TO COME TO QUAKECON?" LUCKEY MESSAGED HIS FRIEND CHRIS DYCUS over Skype.

> **CHRIS DYCUS:** What's happening there?
> **PALMER LUCKEY:** We have a booth. And I am on a panel.
> **CHRIS DYCUS:** Ooh, sounds sweet. But wouldn't it be sold out by now?
> **PALMER LUCKEY:** We don't need no stinking passes! By the way: Brendan emailed you. Wants to have a phone chat. I should brief you really fast
> **CHRIS DYCUS:** All right. Whatcha wanna tell me?
> **PALMER LUCKEY:** So, Brendan was the co-founder of Scaleform, a middleware company. Their software was in over 1000 games. Call of duty, Gears of War, God of War, blah blah blah. Sold the company for $38 million
> **CHRIS DYCUS:** Dang!
> **PALMER LUCKEY:** Now he's an executive at Gaikai. Gaikai just sold for $400 million. And he does not want to stay, he does not like being in a big company. So he is going to be our main business guy and investor

CHRIS DYCUS: Is that it?

PALMER LUCKEY: As far as you go . . .

"WHO IS CHRIS DYCUS?" NATE MITCHELL ASKED OVER THE PHONE, AS HE walked through the streets of Beijing with his family.

"Palmer's friend," Brendan Iribe replied. "He's gonna fly down to QuakeCon with Palmer and help us out."

"Oh, good. That's huge."

"Yeah, no, it is. But look: I don't know the guy and even if I did . . . I mean, I think this Chris guy is even younger than Palmer."

Mitchell laughed, spurring Iribe to do the same. "Man," Mitchell said, "when I was nineteen, I was definitely not launching a company."

"Well, I was," Iribe reminded him, "and it sure as shit wasn't easy. I think it's really important that you get back here."

Mitchell glanced across the street—at the smiling face of his mom, the stoic one of his dad—and knew they'd be upset about this. For most the trip, he'd been stowed away from the family and working on his laptop, finalizing the messaging and going back and forth with Win Bates about the Kickstarter video. It was clear that his mind was elsewhere, and Iribe's request gave him the excuse he needed to go where he belonged.

"Thanks so much, Nate," Iribe said.

"Of course," Mitchell replied.

"And you don't need to be there until the first, so you can probably sneak in a couple more days of Mitchell family bonding."

With the Kickstarter video finally locked, that was exactly how Mitchell pitched it to his parents: Sorry, guys, I need to go back to the United States, but at least we can now maximize the next two days! Except Mitchell ended up needing to abandon that when, hours later, the folks at Valve sent over a video.

"I got to meet Palmer Luckey and try out the Oculus Rift," Michael Abrash said, speaking to the camera. "And I have to say it was a very exciting moment. It could be the beginning of a whole new industry that leads us eventually to having true augmentation all the time. Every place. And I'm really looking forward to getting a chance to program with it. And to see what we can do."

PALMER LUCKEY: How did it go?

CHRIS DYCUS: Went very well! We settled on a $40k salary for now. Did a lot of discussion about what I'd do, where the company's going, etc.

PALMER LUCKEY: Awesome!

PALMER LUCKEY: Did you talk about QuakeCon?

CHRIS DYCUS: Yeah, that I'd basically be slave to you and Nate. :P

CHRIS DYCUS: And manning the booth, I suppose. So, what time on Monday were you planning on getting me?

PALMER LUCKEY: No idea at all yet

CHRIS DYCUS: Man. You have like no sense of planning.:P

PALMER LUCKEY: Thank god for Nate

PALMER LUCKEY: He is our creative director

PALMER LUCKEY: And manages all kinds of things

CHRIS DYCUS: Awesomesauce.

PALMER LUCKEY: $40k is not bad considering the benefits

PALMER LUCKEY: Free top of the line computer, free food, etc.

CHRIS DYCUS: Sounds very good to me!

PALMER LUCKEY: I don't want much of a salary from Oculus myself

PALMER LUCKEY: Would rather the money be spent making things happen

FROM THE DIMLY LIT CODING CAVE IN HIS APARTMENT, MICHAEL ANTONOV watched the video again. He cheered quietly to himself as the Abrash clip segued into an endorsement from Gabe Newell. "It looks incredibly exciting," Newell explained in the clip. "If anyone's going to tackle this set of hard problems, we think that Palmer's gonna do it. So we strongly encourage you to support this Kickstarter."

"Brendan," Antonov said over the phone. "This is *very* good."

"Right?" Iribe excitedly replied. "You and Palmer must have killed it."

"It is not too late?"

"We should be okay. I spoke to Nate and he said, basically, everything is already timed out, but he and Win think they can just lay it on top of the video that's there. So we'll have to cut some stuff. But definitely worth it."

"Oh yes."

WHILE WAITING FOR HIS FLIGHT TO BOARD AT BEIJING CAPITAL INTERNATIONAL Airport, Mitchell was busy editing the Kickstarter video on his laptop when he got a call from Iribe. "Mike is still feeling good about $500,000. So we're basically all on the same page, except for Palmer."

"I'll talk to him," Mitchell said.

"Make sure he feels heard," Iribe replied. "We've all raised money before, so he's coming at this from a different place. I want him to feel comfortable."

"Understood. It's not worth going to battle over."

"Yeah. We'll be fine either way. Just nothing embarrassing, please."

"WHAT'S WRONG WITH A GOAL OF $100,000?" LUCKEY ASKED.

"It's just . . ." Mitchell replied, still in the airport. "We aren't just selling a product. We're selling a version of the future. We want to exude a sense of confidence."

"But if we don't hit our goal, we get nothing."

"Hence the *confidence*."

To try and build Luckey's confidence, Mitchell ran through a list of recent Kickstarter success stories. "Ouya has already raised over $7 million. Pebble raised over $10 million. That Double Fine game even raised more than $3 million."

"I know all that," Luckey replied. "And by the way, Pebble's goal was $100,000. But here's what I'm saying: even if we only sell a thousand of these things, or even maybe high hundreds, we're still kind of a success. If Microsoft, say, launched a new console and they got a thousand developers to get a development kit, that would be a massive success for them. They don't get anywhere near that number."

"But—"

"—Now, they have a different strategy, of course. They try to target people they've worked with in the past. Reward them for their loyalty, et cetera. Work with people who have a proven track record of success. But the fact remains that if we sold a thousand of these things, we'd really be in a very good place. And we could easily go to investors and say: hey, we got a thousand game developers to buy these things. There's already hundreds who are concretely working on stuff. That's better than any game console in the past few years. That's pretty good."

"That *is* pretty good," Mitchell replied, packing away his computer and readying himself for the flight. "But what you created—and what we're doing—is so much better than pretty good. And I have a feeling that no matter what number we settle on, at some point tomorrow you will come up to me and say: Nate, we should have gone higher."

"Well, if you really think so . . ."

"How about we compromise? $250,000?"

"Okay."

"Good. Get some sleep and I'll see you tomorrow in Texas. And Chris, too. I'm excited to meet our first employee!"

"WHICH ONE IS NATE?" DYCUS ASKED THE FOLLOWING MORNING AS HE AND Luckey scrambled to pack everything they might need for QuakeCon.

"What do you mean: which one?" Luckey asked, disconnecting some wires.

"Is Nate the suit?" Dycus asked. "Or is he the vest?"

"Ohhhhhhhhhh . . ." Luckey chuckled. Since Luckey had started calling Iribe "the Suit," his partner in crime—Nate—became Mitchell "the Suit in Training." But that sounded clunky, so Luckey borrowed from a different formal garment. "Nate's the Vest," Luckey answered. "You'll like him. He's good people."

Luckey and Dycus finished packing and headed to the airport. And at some point along the way, two things happened:

- Chris Dycus confided that he had never been on a plane before
- Oculus's Kickstarter campaign went live.[4]

RARELY DOES ONE'S MOST EMBARRASSING MOMENT COINCIDE WITH THEIR most thrilling. But as Brendan Iribe rose and slowly scanned the Gaikai office, he couldn't help but feel this strange convergence had occurred.

With the exception of Michael Antonov, who was doing his best to act like business as usual, everyone in the office was blasting the Oculus video. As everyone watched it over and over, it was quickly becoming clear that Iribe's "little side project" might have been significantly larger than he let on.

. . . My name is Palmer Luckey and I'm a virtual reality enthusiast . . .

Iribe was less concerned with what his colleagues might think of him than he was about the possibility that a similar scene was playing out in a Sony office somewhere; and that with plans to acquire Gaikai for $380 million, the Sony people might not like seeing one of Gaikai's key stakeholders shilling for a different venture.

Keep it together, Iribe thought. Just focus on the good.

"DANG!" DYCUS EXCLAIMED, UPON CHECKING HIS PHONE AFTER HE AND LUCKEY arrived at the airport. "It's already passed $100,000!"

Luckey's eyes bulged with disbelief.

This was actually happening. This was real! And slim as the possibility might be, it looked like Oculus would at least get a chance to bring back virtual reality.

Then, in an instant, Luckey's eyes debulged and his mood stabilized. "Good stuff," he calmly said, and then he segued the conversation to a technical issue impacting one of the three headsets they'd packed for the trip. "I'll show you how to fix it when we get to the hotel. It's on the fritz, but it shouldn't be that bad."

At first, Dycus wondered if this was perhaps just an act, an attempt to mask the happy-happy-joy-joy that Luckey must have been feeling. But as Luckey talked about gluing driver boards onto fragile display screens, Dycus realized it was no facade at all. His friend just had an incredible talent for compartmentalizing his emotions, and with so much still in flux for QuakeCon, problem solving took precedence over fist pumping. So when it was time to turn off his phone for the first leg of their flight, Luckey didn't compulsively panic about disconnecting from the growing buzz. Nor did he particularly worry that the phone's battery was dying (and he was without a charger). He took it all in stride—knowing it was out of his hands, literally, anyway—and thought through various permutations of how the next twenty-four hours might go.

AFTER TOUCHING DOWN AT LAX—BEFORE MITCHELL COULD EVEN CHECK THE Kickstarter—his phone was pummeled by a barrage of text messages: Is this your thing? Duuuuuuuuude! Don't forget me when you're famous!!!!

Mitchell slid aside the well-wishes and connected to the Kickstarter page. It had been less than two hours since going live, and they'd nearly surpassed $250,000.

Refresh: $250,126

Refresh: $251,325

As Mitchell walked through LAX, he felt true unbridled joy. In addition to the money they were quickly raising (Refresh: $253,446), the messaging had resonated with gamers and technology enthusiasts. Oculus was getting love from everywhere: GameSpot, Ars Technica, and even the BBC![5,6,7]

Jet lag be damned, this was going to be a good day. Even though somehow, in the next couple of hours, he needed to go home, shower, unpack (repack), and make it back to LAX in time for his flight to Dallas. Because at least, throughout it all, he could just lift his spirits with a quick refresh.

DURING A LONG LAYOVER IN PHOENIX, LUCKEY BEGAN TEACHING DYCUS—NOW a proud, newly minted air traveler—how to build VR headsets. He also used this time to start going through emails in his now-overflowing inbox. Most, he could ignore. But two piqued his interest.

One was from Jon Malkemus, the artist who had done the graphics for Oculus's Kickstarter video: "You guys better hold on, cause you're about to become VERY, VERY popular!" And the other came from Carmack: "How are you hitting that price point, considering that Hillcrest wanted $90 just for the trackers, and you were originally looking at $500 for the kits?"

Luckey explained how Oculus would essentially be selling the devkits at cost, and how—instead of using one of Hillcrest's trackers—

they planned to design their own in-house. In fact, he already had someone in mind for this task (a hacker who went by "nrp" on MTBS3D).

"Three hundred is a really awesome value," Carmack agreed. "[But] I still think you are making a mistake cutting your margins so thin—you would have been able to move plenty of units at $500, and being able to sock away some cash would have been a good thing."

Some cash would definitely be a good thing. But even better would be to get these the devkits out to as many content-creators as possible. Even if that meant taking a loss on each headset they sold. That obviously wasn't ideal (and could be avoided if things broke right), but taking this gamble was something that Luckey felt strongly about. Because, in his mind, VR only had one chance to take off: if someone was willing to bite the bullet and play the long game.

IT WAS ONLY A MATTER OF TIME BEFORE DAVE PERRY, THE CEO OF GAIKAI, called Iribe into his office. Perry—a longtime friend and mentor—was as perplexed and disappointed as Iribe had feared. "Brendan. The whole office is watching your Kickstarter. What is going on?"

"DP, I'm sorry. This thing is turning out way bigger than any of us expected. We did this thing on the side and we didn't know where it would go. But clearly it's going to go somewhere and we should probably talk about how a few of us—myself, Nate, and Mike—are going to transition down that path."

The rest of that conversation was not easy, nor did it end that day, but ultimately Perry (and Gaikai) were as understanding as Iribe could have hoped. Three days later, he, Antonov, and Mitchell would officially exit to focus on Oculus.

AFTER A LONG CONNECTION, A FLIGHT DELAY, AND A LONG WAIT FOR THE AIR-port shuttle, Luckey and Dycus finally arrived at their hotel around 3 AM. And though it was very late, there were a couple dozen people milling around the lobby. And just about all of them looked angry. As did Luckey and Dycus when they learned the reason: the hotel's computer system was down, so they were unable to check people into their rooms.

"Do you know what room the Vest is in?" Dycus asked Luckey.

"Yeah," he replied. "But I'd prefer not to wake him. He came straight from China."

For about an hour, Luckey and Dycus sat on a couch in the lobby. Then, still without an end in sight, they decided to go to Mitchell's room and knock on his door.

"Palmer?" Mitchell asked, still half-asleep and in his underwear.

"Hey Nate, this is Chris," Luckey said. "He's our new employee! Can we stay in your room tonight?"

"Of course," Mitchell groggily replied, welcoming the guys into his room. He then woke up enough to formally introduce himself to Dycus and grip Luckey with a celebratory hug.

"I'm just over the moon!" Mitchell said about their first day on Kickstarter.

"I don't regret where we set our fundraising goal," Luckey said. "But man . . . I severely underestimated this thing, didn't I?"

After a few minutes of jubilant small talk, the guys decided to call it a night. So they pushed together couple of chairs to form a little nest-like crevice for Dycus to sleep in, and then Mitchell and Luckey used some pillows to divide the bed in half and then climbed onto opposite sides.

"Thanks for being right," Luckey said.

"Thanks for inventing the coolest thing I've ever seen," Mitchell replied. "And don't forget: this is only the beginning."

And finally, Mitchell was going to be there from the beginning. There—from the very start of a start-up. This was it, this was his life. Wow, he thought, reality was pretty damn good. But, even so, virtual reality was going to be even better.

Lying in bed, Luckey felt that same boundless optimism. But he also felt something else. What was it? The pressure of seeing a project through to the end? The realization that he now actually had to deliver on all the promises that Oculus had made? Or was it just the weight of a question he had meant to ask earlier in the day?

"Hey Chris," Luckey whispered. "Are you still awake?"

"Yeah, what's up?"

"I forgot to ask you: were you scared? On the plane?"

"Nope," Dycus said. "I was a little startled by the acceleration—because the only thing I really had to compare it to was a roller coaster, and I didn't realize that planes accelerated that quickly. But I wasn't scared. I was like: oh, this is cool."

HOW TO BUILD A COMPANY

CHAPTER 12

THE KING OF POP SOFTWARE

August 1, 2012

THIS IS A PORTRAIT OF PAUL BETTNER:

:-)

Bettner, thirty-four, the cocreator of *Words with Friends*, always seemed to have a smile on his face. Since founding his first game studio that emoticon also summed up exactly the type of content he wanted to create: playful, accessible, and iconic.

Unfortunately, for the past two years, he hadn't been able to do much of that. His flow of vision and creativity had been stunted by circumstance. But on the afternoon of August 1—logging on to Kickstarter from his office in McKinney, Texas—Paul Bettner was feeling inspired as he emailed his wife:

FROM: Paul Bettner
TO: Katy Bettner
SUBJECT: The future of virtual reality

The Oculus Rift guys (the team that Carmack has been working with) have FINALLY announced their kickstarter!!

Their top backer level is $5k, and includes a trip to their offices and a day with their development team. I'd really like to do this, both to

support an effort that I think is going to define the future of computing,
and also to get the opportunity to meet and network with this team.

Can I back them for $5k?:-)

I love you

While waiting to hear back, Bettner recalled 2008, the last time
he had felt like there was an imminent technological shift on the horizon. Bettner was still at Ensemble Studios then, the Dallas-based developer famous for games like *Age of Empires* that he had been working at
since1997; and which his younger brother (David Bettner) had joined
in 2003.

At some point during their time at Ensemble, Paul Bettner set up
a chess board in his office so that he and David could play what, essentially, amounted to the opposite of speed chess. The way their game
worked was one brother would make a move and then, the next time
the other brother passed the board, he would make his move (whether
or not the other opponent was present). The game would continue in
this fashion—toggling back and forth, each at his own pace—until
there was a winner. Sometimes it would take weeks.

For years, this ongoing chess game was nothing more than a form
of fraternal, workplace bonding. But in 2008, when David Bettner suggested the brothers break away to start their own company, it became
the backbone of a vision that the Bettners had about gaming, one that
broke away from the bigger-stronger-faster mentality of traditional
gaming (on console or PC) and, instead, leveraged the growing ubiquity of smartphones.

It was around this time—two years after BlackBerry introduced
their first camera-included phones; one year after Apple introduced
their first iteration of the iPhone—that smartphones began evolving
from luxury items for professionals to mainstream devices for everybody. Foreseeing this future coming wasn't what set the Bettners apart;
what did was trying to figure out what kinds of games people would
want to play.

Based on the iPhone's initial lineup of games, the answer seemed to
be shorter, lower-quality versions of titles that had appeared elsewhere
before; like *Bejeweled*, which had been popular on browsers since 2001,

or *Super Monkey Ball*, which had first appeared on consoles that same year. To the Bettners, there was nothing wrong or even surprising about this—handheld devices, classically, have been home to watered-down versions of "real games"—but the brothers felt that developers were missing a real opportunity to take advantage of the unique features of a device like the iPhone. And so, in August 2008, the Bettners left Ensemble to try and do just that.

Now, four years later, as he waited for his wife to reply about Oculus's Kickstarter, Paul Bettner couldn't help but recall how crazy she must have thought him to be when he first floated the idea of leaving Ensemble. Hey, honey! I know you're pregnant with our first child, but would it be cool if I quit my stable job so that David and I can start making games for this device without any proven business model for developers?

Just thinking back on the awkwardness of that conversation, Bettner's stomach sunk. But those nerves just as soon disappeared—just as they did back then—upon remembering his wife's response: absolutely.

The idea that had earned her blessing was still being fleshed out, but the general concept went something like this: in terms of graphics and power, the iPhone paled in comparison to traditional handheld gaming devices (like the Game Boy or PSP). But the iPhone did have one advantage: it was always connected to the internet, which made it a great fit for play-at-your-own-pace, turn-based gaming. Or, put another way, for a digital version of something like the Bettners' asynchronous, at-the-office chess game.

To bring this idea to life the Bettners formed a company (Newtoy) and started meeting every morning at the McKinney Public Library. "We're building a game for the iPhone called *Chess with Friends*," the Bettners would explain to curious librarygoers. "It's like the gaming equivalent of text messaging; you make a move—which sends a little package of information to your friend, and then they see it when they check their phone and send you a move right back."

After months of work, the Bettners completed *Chess with Friends* and released it on Apple's just-four-months-old App Store.[1] The game was by no means a runaway hit (ranking far below the likes of *Bejeweled* and *Super Monkey Ball*), but *Chess with Friends* succeeded in a way games like those did not: retention.

Unlike most of the games that populated the mobile market—whose retention rates hovered around 1–5 percent after thirty days—*Chess with Friends* had managed to retain over 50 percent of its players. In light of this development, the Bettners concluded that their problem wasn't the gameplay, but rather the game itself. They needed something more fun than chess. Something that kids and mothers (and grandmothers!) would be tempted to pick up and play. Something like . . . Scrabble.[2]

Again, from the McKinney Public Library, the Bettners got to work. But this time they brought company: Shawn Lohstroh, a veteran Dallas-based programmer. The guys knew Lohstroh from Ensemble (where he'd most recently worked on *Age of Empires III*) and brought him on to build the game engine with what remaining funds they had. That gamble—to go all in on *Words with Friends*—would eventually pay off. But in the months between launching the game (July 2009) and watching it become a worldwide phenomenon (October 2009), it became obvious that, unless something changed, they'd be out of money by year-end.

Words with Friends was doing pretty well, about the same as *Chess with Friends* had done. But when you're a no-name publisher without any marketing budget, anything short of excellent often isn't good enough, especially in the mobile space, where the content was being sold so cheaply that the Bettners had to price their games at $2.99 just to stay competitive.

Paul Bettner felt like he'd let down his little brother. Logically, he knew it was about time to throw in the towel. But how could the fate of these games—these games that were now nearing a 60 percent retention rate!—just burn out into oblivion? Years ago, Bettner had shut down his BBS business because he knew it had no place in the world of tomorrow. But this was the opposite; this was him at the start of something big to come. It's almost fitting that what eventually pushed *Words with Friends* over the top would come from a bona fide pop star: John Mayer.

Seven words from him was all it took: "*Words with Friends* is the new Twitter."

On October 5, 2009, John Mayer tweeted those words, and the

Bettners' lives changed forever. Within hours, the install rate for *Words with Friends* multiplied by fifty times the norm. That initial spark, coupled with the game's uncanny retention rate, made it a mainstay atop Apple's top-grossing list for the next several years. By 2010, *Words with Friends* had reached 1.6 million daily active users (each of whom played the game for an average of one hour per day).

Soon enough, Disney, Electronic Arts, and other top-tier entertainment companies began reaching out to the Bettners to see if their promising young studio might be for sale. With lofty numbers being tossed around, the brothers found themselves facing the question that all successful start-ups inevitably face: to sell or not to sell?

Prior to this, the brothers had always agreed on critical business decisions. They'd argue, sure, oftentimes for hours but, eventually, they'd move forward with a united vision. In this case, however, they couldn't agree. Paul vigorously wanted to remain independent, whereas David was keen on selling, because—assuming they stayed on to run the acquired studio—they'd have a new tier of resources at their disposal. Although they were unable to reach a consensus, the Bettners both agreed that there was no reason not to at least hear out their suitors.

The most aggressive of the bunch was Zynga, which created games for Facebook, for which they had been able to leverage a trio of early hits (*FarmVille*, *Café World*, and *Texas HoldEm Poker*) into a dominant foothold; they had boasted over 250 million monthly active users by the time they welcomed the Bettners to their office. "What's it going to take to make this happen?" someone from Zynga asked the Bettners.

If only I knew, David Bettner wanted to say! But he remained silent and let Paul do the talking. "You guys flew us out here. Surely you have something in mind," Paul said.

They did: $20 million.

"Twenty million?" Paul Better repeated with skepticism. He knew that was a lot of money, enough to provide his family with all sorts of incredible things; but as he looked around the room—hungry faces everywhere—he realized this didn't matter. He already felt like the happiest person in the entire universe. Why would he ever want to give that up? He thanked the Zynga folks for their time and abruptly ended the meeting.

The what-was-that conversations that followed—between the Bettners, in the elevator, and the Zynga folks, still befuddled in the boardroom—were awkward. But not nearly as awkward as the one that took place on the street when Paul and David Bettner were chased down by the executives they had just left.

"You can't just do that!" one of the executives said.

"Well, what you just offered was insulting," Paul Bettner said. "And you know it's insulting. So what's the point of sticking around?"

"That wasn't our intention, okay? Why don't we start over? Just tell me what would it take?"

For the first time, Paul Bettner allowed himself to imagine. "Two hundred million."

The executive was so stunned he didn't reply.

"Look," Bettner elaborated, "I know that two hundred million is ridiculous. But I'm not here to bluff. The fact is that I believe so much in what we're building that the only way you could possibly have us now if you want us is to pay something as ridiculous as that. Don't take it personally. If you don't want to talk anymore, that's fine. But that's kind of in the ballpark of what it would take."

"Um, okay. Thanks. We'll get back to you then."

Zynga did get back to the Bettners, and though they weren't willing to hit that figure from the street, they did up offering something very close: $180 million.

Still, Paul Bettner was reluctant to sell. He didn't want to give up this special thing that he and his brother had built. Not for $180 million, not for $200 million, not even for $500 million—to him, what they had built was literally priceless. That said, Bettner knew that if he blocked this deal from going through, then he'd be sacrificing his brother's dreams (to be financially independent) at the expense of his own (to own a game company). And so, on December 2, 2010, Zynga officially acquired Newtoy.[3,4]

Over the next two years—between selling his company and sitting at his computer now, curious about this young company Oculus—Paul Bettner would tell portions of this story many times. And just about every time, no matter the audience, he'd eventually be asked something like "What the hell?" What the hell made you walk away from $20 million? Didn't you have a wife and baby at home? What was it, exactly, you were trying to accomplish?

Bettner didn't have a good response. He knew that to some he'd appear naive, to others greedy, and, perhaps to most, as furiously idealistic. But the philosophy he always carried with him, the one that illogically guided his actions, was that life is short and you don't have a single moment to waste doing something just because you're told. You just need to do the things that your heart tells you to do and make the most of your time that you have here on the earth.

After selling Newtoy to Zynga, he and his brother had to stay on for at least four years per the terms of their deal, to run the studio. While there, they hoped for the best; all the magic of before, but now with better resources at their disposal. Except, unfortunately, it didn't work like that. Zynga cared too much about the bottom line, extracting every last drop of profit from all their products. They had every right (and incentive) to do so, especially after paying such a hefty sum for the *Words with Friends* franchise.

For two years with Zynga, Paul Bettner had been trying his best to navigate the reality of the situation: he had sold his employees and all the games that they loved. And even though he had two more years to go, he was nearing the end of his rope. So was his brother. Neither necessarily regretted the decision they had made, but they certainly regretted coming in to work each morning, forced to face these employees they loved and constantly have to tell them no, sorry, that's just the way things are now.

It sucked. It really sucked. Paul had started to think about what would happen if he left early; what kind of fallout there would be and how much money he'd be leaving behind. Too much of both to not just wait it out—until eventually he could bear it no longer. He began to seriously consider leaving Zynga and starting a new game studio, the timing of which seemed almost perfect, with an exciting new, lost-cost gaming platform on the horizon: Ouya.

Like many game developers during the summer of 2012, Paul Bettner was intrigued by Ouya. With their $99 microconsole, and record-breaking success on Kickstarter, they appeared to be in a great position to leverage the ubiquity of game engines and the rise of independent game makers to disrupt a $50+ billion industry that, for the past decade, had been dominated almost exclusively by two companies (Sony and Microsoft). As much as Bettner loved those two companies, their bigger-stronger-faster business model had somehow lost what he loved so much about games: the fun.

Where was this generation's *Super Mario*? Or *Sonic the Hedgehog*? Or even *Bonk* (with his crazy adventures)? Maybe Bettner was nostalgically yearning for a return to the days when story, character, and mechanics were more important than speed, graphics, and explosions. *But* there was a chance that he was not alone in these desires, a chance that the Goliaths of today had forgotten about the average gamer and left the door open for someone who could capture today's audience with the wonder of yesteryear and the technology of tomorrow.

Up until today, Bettner had thought that someone might be Ouya. And given all the momentum they had, they still seemed like a strong candidate to take that mantle. But as he watched Oculus's Kickstarter video for a fifth time, then a sixth, he couldn't help but think of a different sort of future, and an email he'd sent to John Carmack in early 2012.

"When we were creating *Words with Friends* I kept thinking: We're screwed. This is such an obvious idea," Bettner had written. "VR/AR feels the same way to me right now. It must be completely obvious to the Apples and Sonys of the world that the tech is here today (has been here for months), right? For sure they've got ultra convincing VR prototypes running in their labs right now and killer products destined for market next Christmas—or maybe not; maybe it's more likely that nobody has made the connections and figured out the problems that need to be solved to make this really work . . . I don't know where all this goes but I know I want to be a part of it. It's world-changing stuff and I've been daydreaming about it since I was a kid. Sure it's still probably 5+ years before my wife buys her first set of consumer-grade VR gear, but I'm pretty sure that the 'Jobs/Wozniak garage edition' of *real*, working VR tech can be built now . . ."

Oculus's Kickstarter reiterated to Bettner that he had not been crazy to think that the time for VR was coming; and that the "Jobs/Wozniak garage edition" he had mused about to Carmack was already here—made by Palmer Luckey. Wow, Bettner thought, VR really *could* be the future. But whether or not it *would* be the future depended largely on the success of this nineteen-year-old kid and his ambitious, just-formed company. Because what Oculus needed to build wasn't just a compelling piece of hardware (or software); they needed to build an entire industry. They'd need to bet on the right people and the right content and form alliances with the right companies.

Bettner wanted to support Oculus's Kickstarter, and the only thing standing between him and taking that leap was a reply from his wife to his email from earlier in the day. He wasn't looking for permission to spend some money, but rather her approval, that same "Go for it!" that she had given him when he had wanted to leave his job and start a company that made mobile games. He got exactly what he wanted:

FROM: Katy Bettner
SUBJECT: Re: The future of virtual reality

absolutely

CHAPTER 13
SHOWTIME

August/September 2012

"I'M ALMOST A LITTLE SCARED THAT THINGS ARE MOVING TOO FAST RIGHT now," Carmack said the afternoon of August 2, delivering what would end up being a three-and-a-half-hour keynote speech. "The original goal that Palmer was doing with Oculus was 'I'm going to make one hundred kits and sell them in the VR community . . .' and in many ways, I thought that was going to be a really great thing. Because there *are* things that need to be figured out about this now."

When it comes to new technologies, the line between hype and hope is often very thin. So it seemed wise for Carmack to preach caution—to remind those watching him speak that Oculus's Kickstarter was for an experimenter's kit and that, right now, there was little to no software for the Rift. And yet, Carmack also said he thought "people are going to look back and say the tail-end of 2012 is when things took off," noting that those in the crowd now had the chance to "really be at the forefront of something that I think is going to be a very big deal in the coming years here."

Comments like this stoked the enthusiasm of those who wanted to believe. As did news that hours later, Oculus' Kickstarter eclipsed one million dollars. And by the following night, when Carmack, Luckey and Valve's Michael Abrash all took the same stage for a panel called "Virtual Insanity," it felt like virtual reality had been resurrected in some small-but-tangible way.

"I initially set out to look into AR specifically," Abrash said. "Because AR, when it gets here in its full glory, is really going to change everything . . . [but] something that I came to realize a little later in that process—and that John and Palmer kind of helped me think about a whole lot harder in a hurry—is that virtual reality is here now . . . This is, this is really it. This is the seminal point where there's a reasonable chance that out of this will explode all the wearable computing gaming stuff that we've been treating as science fiction for so many years."

If true, then why now? "VR has been talked about since the '60s," the panel moderator noted. "Why now are the hardware and the software there to open up the opportunities?"

"Like John has said," Luckey answered, "the design is pretty simple. And some people are saying, 'Oh, this is amazing! Imagine the advances that you must have made in this field to arrive at this. How come such-and-such big company hasn't done this?' And the answer is that it isn't really even me, or Oculus, or any other VR company that's making this happen, it's much bigger, more massive markets like the cell phone industry and motion controllers in gaming . . .

"This could have happened five years ago," Carmack added. "The displays were there, the MEMS disk controllers were there. It takes someone deciding to put it all together . . ."

The following day, Oculus started doing demos at QuakeCon and almost immediately a line began to form around the Oculus booth.

"Hey!" Luckey boomed, glad-handing the first person in line. Having never demoed the Rift at a trade show before, this was all new to him, as well as Mitchell and Dycus. So when it came to talking about the product, they mostly planned to learn what worked best as they went. But before they even got started, they had agreed on one sacred objective: don't oversell the experience. Especially to this crowd, which consisted of hard-core PC gamers who lived on computing's bleeding edge.

With this in mind, the Oculus crew held true to their objective. Along the way, they noticed something funny happen time and time again: all the hype and hyperbole that Luckey, Mitchell, and Dycus refrained from saying would instead be said *to them* by those who had received demos. Everything from "That was incredible!" and "This is going to be HUGE!" to one person, almost trembling, who said, "Mark my words: VR is going to change how our species lives."

"REALLY REEEEEEALLY NEED YOU TOTALLY AND COMPLETELY FOCUSED ON OCU- lus SDK . . ." Iribe texted Antonov late at night on August 4. "Dillon's resigning from Autodesk on Monday and joining Oculus!"

"You don't think it's a bit early?" Antonov replied. "What will he be doing a month from now, when the Kickstarter is over and we don't yet have SDK & Kits to give out? I guess organizing forums and Korean community?"

"Localizing Kickstarter, press release, first press event, etc.," Iribe said. "We need Korean devs buying Kickstarter kits."

"Well that is obvious . . . just to me it sounds like part a part time job until kits actually ship. Right now we are in a hype-spin marketing wave; but then we'll need to buckle down and actually get it all ready . . . I just feel you are not allowing ANY time for R&D."

"Carmack's demo allows us to do so. Dillon will be visiting tons of studios, demoing, getting devs interested, announcing their support for each game."

"Yes," Antonov replied. "Then you'll be coming in every day asking - where's SDK? Where's SDK? And we really need few months for that."

"You have 3 months," Iribe told him. "Focus on the SDK."

"Or four months," Antonov said, "with preview in three."

"Yep . . ." Iribe replied. "We have to move quickly. And scale aggressively. Trust me, selling 1m consumer units next Christmas will require incredible ramp."

"IT WAS CRAZY," LUCKEY TOLD EDELMANN, AFTER RETURNING FROM QUAKE- Con. "People came up to me and, like, they wanted my autograph."

Luckey assumed that this was just a quirky ask from a hardcore gaming crowd. But then it happened again two days later, at SIGGRAPH in Los Angeles. Then again at Gamescom in Germany (August 15) and at Unity's Unite event Amsterdam (August 22). Throughout the month, the excitement for Oculus continued—eventually surpassing $2 million on Kickstarter with pledges from over nine thousand supporters (including heavy hitters like *Minecraft* creator Markus "Notch" Persson, who backed the guys to the tune of $10,000). On the heels of that momentum, Meteor Entertainment and Adhesive Games—whose almost complete mech shooter *Hawken* was headed to PCs in December—announced that they'd be porting an iteration of the game to work with the Rift.[1] Now, in addition to *Doom 3 BFG*, there'd be another game ready to go when the headsets shipped in December. And it looked like there might even be a third: a new game from Notch—a space-themed follow-up to *Minecraft*—who the guys had managed to track down at Seattle's Penny Arcade Expo. Persson didn't have much time, but enough to be suitably blown away. "I'm 100 percent impressed," Notch proclaimed, before vowing to stay in touch and then running off.

That was the perfect end to a magical month, which Iribe and Mitchell celebrated that evening at a party thrown by the *Hawken* guys, sneaking off to not so subtly check their phones—refreshing, refreshing for last-minute Kickstarter pledges—until finally, with a 5–4–3–2–1 countdown to midnight, the guys clinked together their champagne glasses: $2,437,429.

That was the final tally, nearly ten times their fund-raising goal. And into September the momentum kept building until Iribe set up his first VC meeting—with Mitch Lasky at Benchmark Capital—and after what seemed like a good pitch found out that Benchmark wouldn't be making an investment; largely because they were concerned about "motion sickness." Because as it turned out, several of the partners who Iribe had given demos to wound up getting sick later that day.

To educate his colleagues, Luckey put together an overview of what they were dealing with. "On a basic level," he wrote, "there are two kinds of sickness."

The first was "Motion Sickness," which was caused by a disparity between what you see, what your brain expects you to see, and what your inner ear feels. This is why some people get sick in boats, cars, or even just playing fast-paced FPS games on a normal 2D monitors. And the second was "Cybersickness," which was caused by additional VR factors such as wide field of view, increased latency, head pressure/weight, and visual problems (i.e., bad stereoscopy).

"Our job is to fix the second category," Luckey explained, "by making our hardware as good as possible. Positional tracking is going to make a huge difference and that is the best we are going to be able to [address the nausea issue]."

Positional tracking was, in his opinion, a MUST for the eventual consumer version of the Rift. And if Oculus ended up pushing the consumer launch to first launch a second development kit—a "DK2"; which was an idea they had already begun to discuss internally—then positional tracking would probably be a must for that too.

"For the average gamer, though, it is not going to be a problem, and they will be able to adapt to it over time. Fighter pilots need training to overcome motion sickness, the same goes for astronauts. If we market our product as being on the same level of intensity as those things, people will hopefully understand that this is not something that 100% of people can handle with no problems . . . I am confident we can get to the point where 95% of gamers are fine."

This was comforting for Iribe to hear as it had become increasingly clear that compared to Luckey, Mitchell, Antonov and Reisse, he was way more likely to feel queasy when demoing content in VR. So if they could really hit that 95 percent estimate, then he would probably be fine.

"I ran an installation at Sundance film festival with one of my HMDs last year," Luckey said, before linking to some suggested reading on the subject. "And out of 170 or so people I saw go through the 10 minute experience, only about 15% of the people reported any kind of nausea on the survey . . ."

Although Iribe was hopeful that Oculus could solve motion sickness, he acknowledged that Benchmark's concern was incredibly valid. As were the other concerns that they raised: dangers of the hardware business, VR's previous failures, etc. Everyone who had joined Oculus to this point—Luckey, Iribe, Mitchell, Antonov, Reisse, McCauley, Seo, and Dycus—all knew that these risks were serious. But they didn't care. They *couldn't* care. They needed to believe that they were destined for a different ending. Because if they didn't believe that—if they weren't able to trick themselves into that delusion—then virtual reality would never happen.

But even if these eight dreamers were willing to take that risk, the challenge was simply too big for them to do it on their own. As exemplified by Paul Pedriana, Oculus was too small, too risky, too underfunded to bring in traditional A-players. What they needed was a crew of anomalies.

CHAPTER 14

THE ANOMALIES

September/October 2012

NIRAV PATEL

Growing up, Nirav Patel developed an obsession for taking things apart.

"Might be a remote control car, might be a Game Boy, might be a toaster," Patel—now 25—recounted to Luckey and Iribe during a mid-September meeting at McCauley's workshop. "Whatever it was, it was always the same: instead of appreciating the thing for what it was, I had to know what actually made it work. Or how someone actually got it to do the thing that it does."

Luckey nodded. He knew exactly what Patel was talking about.

"So you go and get the screwdriver," Patel continued, "take it apart, and put it back together. Sometimes it works, sometimes it doesn't, but it's the process and the technology that are more interesting than the thing itself."

It was this type of curiosity that led Patel to Carnegie Mellon, and then Apple and then to create side-projects like the Adjacent Reality tracker.[1] This was a small, wireless head and hand tracker—weighing only 9.2 grams—that, if added to the Rift, could capture the three degrees of freedom necessary to deliver an immersive experience.[2] In essence, it could replace the Hillcrest tracker that Luckey & Co. had previously been using. This meeting was to explore that possibility; or, even better, explore the possibility of Patel leaving Apple to come join this crazy little start-up.

Patel, however, had no idea about these possibilities; he simply thought Luckey had invited him over to finally check out his VR prototype. But when Iribe and Luckey shared conspiratorial glances over whether or not they should discuss what was going on, Patel grew suspicious.

"We're just having a screen issue," Luckey blurted.

What had happened was that in the lead-up to their Kickstarter, McCauley had worked with Chimei Innolux (a Taiwanese LCD supplier) to get Oculus a sweetheart deal on a line of discontinued five-inch panel displays. By acquiring these so cheaply, Oculus was able to price their devkit at $300. But the problem—the "good problem to have"—was that Oculus ended up selling seventy-four hundred devkits, which was significantly greater than the number of five-inch panel plays that Chimei Innolux had remaining (around five thousand). And because this model was now discontinued, Oculus couldn't just up the order.

The guys had been able to scramble and obtain a comparably priced display of relatively equal quality, but it came with a caveat: these were seven inches long. And as with microchips and ulcers, bigger was not actually better when it came to HMD displays. Even more especially in this case because integrating a panel that was 40 percent bigger would require Oculus to drastically redesign their product.

"But problems like these," Iribe waved off, "are just par for the course. We'll get everything squared away and, when we do, it's going to be absolutely incredible. You'll see—let's get you in a demo—and you'll see that we're going to *change* the world."

After nearly four years in Silicon Valley, Patel had become immune to this sort of hyperbole. Or so he thought. Because when Iribe made that claim, Patel found himself believing this guy whom he'd only just met. Perhaps it was because, in Iribe, Patel recognized a version of himself, a casual cocksureness that might have been off-putting if it weren't buoyed by supreme confidence. In other words: Iribe wasn't posturing. He legitimately believed exactly what he said. And while there was certainly a chance that Iribe could be wrong, that underlying certitude primed Patel toward believing.

That was one theory, but there also was another: Patel believed Iribe because, deep down, he so badly wanted to believe. For years, as an avid science fiction reader, Patel had fantasized about futures like those in William Gibson's *Neuromancer* or Neal Stephenson's *Snow Crash*. But up until about six months, all that just seemed like pure fantasy. That all changed, however, when he read Ernest Cline's *Ready Player One*.[3]

Set in the year 2044, *Ready Player One* depicts a world in which people jump in and out of virtual reality with the ease of using the internet. In this fiction world, the government has migrated all schooling from the physical world into the virtual one. This may sound sinister, maybe even antisocial, but it's quite the opposite: in this virtual universe, the OASIS as it's called, human interaction is just as beautiful and horrible as it is in the real world. Except here the resources are infinite. So say good-bye to budget cuts! And au revoir to those old socioeconomic divides! In the OASIS, everybody has access to the best and a chance to reach their true potential.

As with most things in life, new solutions bring about new problems. But just as significantly, those unanticipated issues don't turn *Ready Player One* into a cautionary tale. In this respect, by avoiding either a utopian or dystopian fate, *Ready Player One* felt uniquely authentic to Patel. It felt like how technology happens in real life: dazzlingly, haphazardly, and with an agnostic sense of inevitability.

"Have you guys read it?" Patel asked.

"Not yet," Iribe replied. "But people keep telling us we need to, ASAP."

"I finally read it a few weeks ago," Luckey excitedly said. "It's totally awesome. And the tech in RPO is all near future tech: no major breakthroughs required. Which is kind of the opposite of SAO. Have you seen SAO yet?"

"What's SAO?" Patel asked.

SAO was short for Sword Art Online, a series of Japanese light novels by Reki Kawahara that had recently been adapted into a sci-fi anime series. The show takes place in a not-so-distant future where 10,000 gamers battle one another to reach the top floor of a treacherous floating castle in a mysterious new virtual reality game. To get into the game, players link to a "NerveGear" VR headset, whose high-density microwave transceivers are able to hack the brain and induce artificial senses.

"So it's basically the coolest game ever," Luckey explained. "But as the players soon learn, this is more than just a game! Because if you die in the game, then you die in real life."

By the time of this first meeting with Patel, only ten SAO episodes had been released. But already, Luckey could tell that this was going to be one of his all-time favorites. Between this hot new anime, and the best-selling heat of *Ready Player One,* it was starting to feel like virtual reality was having a cultural moment.

All of this fed into Patel's desire to see science fiction morph into science fact. But one obstacle still stood in the way of him fully allowing himself to believe: Palmer Luckey's prototype.

If Luckey's prototype wasn't convincing—if it was unable to convince you, at least minimally, that you had been transported to somewhere else—then none of the science or engineering really mattered. As Luckey handed him the HMD, Patel felt a pinch of anxiety. This was it! The glimpse into what kind of a future might one day be. Anxious, Patel entered virtual reality.

"This is *amazing!*" Patel declared.

Luckey and Iribe traded glances. Amazing enough to leave Apple?

"What are you thinking about for inputs?" Patel asked.

"Oh," Luckey replied, "we're looking at a lot of things. Definitely want to have wands ready by the time we launch DK2. Hands are gonna be key."

"Have you guys looked into what Leap Motion is doing?"

Leap Motion was a San Francisco start-up that had burst onto the scene back in May with a dazzling demo of their hands-free motion controller.[4,5,6] They believed that the days of navigating computers with a mouse were coming to an end, soon to be replaced by touch-free gestures like scroll, swipe, and pinch-to-zoom. Where this was relevant to Oculus was the sensor-based technology that made finger-tracking possible.

"It's funny you mention that," Iribe said with a grin. "Because we're headed over there to meet with them tomorrow. So . . . you know . . . if you were to sign with us before then, we could take you along!"

Patel wanted to say yes, but Oculus couldn't come close to matching his salary at Apple; nor could they match any semblance of that stability. And creatively, Patel was worried that wearing many hats might derail him from his specific passions, or that market forces would change the scope of his project. These were valid concerns but ones that Luckey and Iribe believed they could overcome. So they tried appealing to his ego ("You're too good for Apple!"), his cynicism ("You'll get lost at Apple"), and his sense of idealism ("Think different . . . for real!"). Given that Patel's résumé literally opened with "Objective: to make Science Fiction fact," it was no surprise that this last tactic was the one that offered the best shot.

Patel still needed some time to weigh his options. Even if he said yes, he'd still need a few weeks to give Apple proper notice and get his affairs in order before relocating to Irvine. Luckey and Iribe needed someone now, though. They needed to start solving their tracker issues ASAP. Under no circumstances was Iribe willing to lift his foot off the pedal. "It just so happens we're turning in our cap table tomorrow," he said, referring to the company's breakdown of ownership percentages. "If you can commit by then, we can get you on there and give you founders' stock."

"Can I have a night to think about it?" Patel asked.

"Of course," Iribe replied. The following morning, he sent Patel an email.

FROM: Brendan Iribe
SUBJECT: Oculus follow up

Ready to change the world?;-)
    We're headed to see the latest Leap demo around 5:30 pm today. If you're on board and free this evening, we can meet before to discuss your position and you could join us at Leap. Or if you need more time, which is certainly understandable, we can talk in the near future.

FROM: Nirav Patel
SUBJECT: RE: Oculus follow up

Brendan,
    I am ready . . .

Later that evening, after returning from Leap Motion, Patel received another text from Iribe. This one had a copy of the aforementioned (and newly updated) cap table:

STEVE LAVALLE

STOCKHOLDER	CLASS A	CLASS B	% OWNED
Palmer Luckey	160,000		16%
Brendan Iribe	160,000		16%
Michael Antonov	110,000		11%
Reserved for Seed Funding	300,000		30%
Nate Mitchell		30,000	3%
Jack McCauley		30,000	3%
Andrew Reisse		10,000	1%
Greg Deutsch		10,000	1%
Dillon Seo		10,000	1%
Nirav Patel		10,000	1%
Reserved for Advisory Board		50,000	5%
Reserved for Employee Options		120,000	12%

"Hmmm . . ." hummed Steve LaValle, a world-renowned roboticist who had flown in from Finland. He was wearing the duct-tape prototype, standing almost exactly where Patel had demoed it just days earlier; though oddly—like a dog trying to sneak up on his tail—LaValle kept quietly looking forward and then suddenly jerking his head up (then down). "Sorry," LaValle said, lifting the headset above his eyes. "I think I broke it."

"What?" Iribe asked. "No, that's impossible."

"Not impossible," LaValle explained. "It just means you have a math problem."

"Really?" Luckey enthusiastically asked. "What kind of a math problem?"

LaValle's eyes lit up upon hearing the question. He noticed this and tried to dim the excitement a bit. He didn't want to seem *too* geeky. But the professor in him couldn't help it. Or maybe, now in retrospect, that was why he became a professor.

"So there's this thing called a kinematic singularity," LaValle explained, speaking in a pedantic-very-unpretentious Jeff Goldblum–esque manner. "It's related to gimbal lock, which you might remember in relation to Apollo 11—No takers? That's okay, interesting story for another day!—the long story short is that when you're dealing with 3-D transformation, if you don't really, really know what you're doing, then quick movements straight up and straight down will misalign the tracking system."

Iribe opened his mouth but didn't know how to reply. He didn't understand what in the world LaValle had just said. And yet, the more LaValle spoke the more convinced Iribe became that Oculus *needed* this guy.

As Iribe processed this odd feeling, and strategized the best way to convince LaValle this need was mutual, Luckey playfully chimed in. "Wait. So basically what you're saying is that you broke my headset on purpose!"

"I *tested* your headset on purpose!" LaValle replied, laughing. "But, yes, then it broke. Don't worry though! It should reset when you reboot the system. And if it's any consolation: I do feel a little bad."

LaValle really did feel a little bad, but he also felt a little relieved. Over the course of his past seventeen years spent in academia—getting his postdoc at Stanford, teaching computer science at Iowa State, and then becoming a professor at University of Illinois—he had never been tempted to press his luck in the business sector and the prospect of doing so seemed daunting. It seemed even more daunting after viewing Oculus's slick Kickstarter video and realizing that these guys might actually be onto something here. And so, after weeks of worrying that he'd be out of his element—that these tech guys, these twentysomething dudes from the video game industry—might see him solely as the outsider that he was, LaValle felt relieved to now know that at least he had something to contribute.

Although LaValle had been feeling tremendously jet-lagged, he noticed that the more he spoke—the more he proved himself capable of providing value to these kids with their crazy VR dreams—the more comfortable he felt. And yet he noticed something else as well: the more comfortable *he* felt, the less comfortable *they* felt.

"We really appreciate you flying out to meet with us," Iribe said later in the day, sharing a round of beers with LaValle and Antonov upstairs. "All the way from Finland, that's quite a distance! What is it you're up to over there?"

"Well," LaValle began, "in academia it's very common to take a sabbatical every seven years. So I've actually taken the year off to write a book."

"Cool," Antonov said. "What about?"

"Ah, good question," LaValle replied. "That still remains to be seen. But essentially it summarizes some of the stuff that I've discovered since the last book I wrote during my sabbatical in 2004. Back then, I had gone to Poland and written a book about robot motion. A big, fat, thousand-page book."

Planning Algorithms, the big, fat, book LaValle had written in 2004, explored the crossroads between robotics, artificial intelligence, computer graphics, and control theory (an area of engineering that deals with behavior in complex systems: achieving certain outcomes, incorporating feedback, etc.).[7] LaValle's book boldly tackled these disparate areas of study and garnered universal recognition for presenting a unified treatment that tied together numerous different types of planning algorithms (i.e., motion planning, sensor-based planning, kinodynamic planning).

"Now an obvious question seems to be," LaValle continued, "what the hell does any of this have to do with virtual reality? I mean, that's what I was wondering when out of the blue I receive an email from some guy named Jack McCauley."

"So what made you respond?" Iribe asked, glad to see the conversation loosening.

"Let's see," LaValle began. "Hmmm . . . well, the Kickstarter video was pretty slick. That's not it, though, but it was pretty impressive."

"That was Nate's baby," Iribe said. "Nate Mitchell, one of the founders. You'll meet him the next time you're out here."

The next time I'm out here? That seemed presumptuous, LaValle thought. Or maybe that was a tactic: build a relationship by making it seem certain.

"Sounds good!" LaValle said, playing along. "So yeah, there was the video. And then there was also Jack McCauley. He had impressive credentials and he wrote a very nice email. At least enough to warrant that I do the same. But there was one other thing, though, and it was this: in robotics, I noticed that a lot of times we often run into trouble with a technology because we don't have the necessary component. It doesn't exist yet. Like the right sensor hasn't been invented. It happened with robotics in the '90s, until SICK (a German company) invented these game-changing sensors. And all of a sudden all these computer scientists started looking smart. But it was really all because of this $8,000 component. So you really need that commodity, that component. That jibed with what I knew from robotics and so I was willing to believe you guys might be on to something. And I was willing to believe that I might have something to add."

"Well," Iribe said, "we'd love to have you—"

"Oh!" LaValle blurted. "Going back for just a moment, there's someone else I need to credit before I forget: my wife. Anna Yershova. She's also an engineer. And I remember, I was sitting in a coffee shop in Rotuaari Square in Oulu, Finland—I was all ready to just refer Jack to a couple of my colleagues—but then she texted me that maybe this would be interesting. Maybe we could work on it together. It's also worth noting that she is bored out of her mind in Finland, so that was maybe part of her reaction."

"To Steve's wife's boredom!" Antonov toasted, raising his beer.

JOE CHEN

"Joseph Chen!" Luckey exclaimed, when they finally made it to Chen's office.

"Dude: no," replied Chen, a thirtysomething, bald-by-choice product development manager who had been working in the tech sector since 2000 (and at Epson since 2007). Yet despite all those years of experience, there was something about Chen—a VinDieselly, get-shit-done fire his in eyes—that suggested his destiny was not behind a desk, but on the ground getting his hands dirty. "Call me Joe."

"Joe!" Luckey cheerfully parroted.

"Palmer!" Chen shot back, matching the sarcasm of Luckey's tone. He introduced himself to Dycus and thanked the guys for coming by. "Especially since I know things have gotten *super*busy for you since SIGGRAPH."

SIGGRAPH was North America's largest computer graphics conference, the most recent of which had been held in early August at the Los Angeles Convention Center. It was there, at Epson's mostly deserted booth, that Luckey and Chen first met. After a demo for Epson's crude augmented reality glasses, Luckey determined Chen was exactly the kind of guy that Oculus needed to hire. Now, in Luckey's pocket was an employee agreement he planned to whip out and get Chen to sign before leaving his office.

Chen had no idea about any of this. Like Luckey, he had come to this meeting with his own surreptitious agenda. "Before I take you around," Chen said, "I'm dying to check out the Rift. Can you give me a quick demo?"

"Well," Luckey replied, "that's gonna be a bit of a problem. Because there's really only one that works at the moment, and Brendan has it."

"Dude!" Chen barked with disappointment. "Like half the reason I invited you here was to see this thing. To know if you're legit or if it's vapor."

"It's totally legit," Luckey assured him. "And don't worry, we are literally in the middle of building a whole bunch of new prototypes."

To prove this point, Luckey and Dycus explained how, for the past few weeks, they had been sharing a room at a two-star hotel in Long Beach (the SeaPort Marina Hotel) and were basically just building prototypes around the clock.

"It's kind of ridiculous," Dycus explained. "Like we have all sorts of hot glue guns, soldering irons, wires, and foam core everywhere around this hotel room."

"It's pretty awesome!" Luckey added. "The only thing that sucks is we can't really do room service because, you know, they'd probably think we're making a bomb or something. So, yeah, the room's kind of this giant mess and my girlfriend says it reeks of Greek fries, but other than that it's really conducive to getting stuff done."

"So like Palmer said," Dycus explained, "a bunch of these will be finished soon. But we did, actually, bring something." What they'd brought was a Nintendo 64 that Dycus had turned into a battery-powered handheld. "Palmer said you had made a custom PCB with a composite video input. So we brought this try out on your AR headset."

This was not how Chen had wanted to spend this meeting, but he obliged—partly because this bizarre little demo would be a good time to bring up the reason he had arranged this meeting: to pitch Palmer on the idea of appearing with him in a short promo video that explored the differences between AR and VR. "The point would be to educate people," Chen explained.

Luckey couldn't help but feel an ironic sense of déjà vu. It was only a couple months ago that Oculus was begging industry's players to appear in their video, and now the reverse was happening (albeit on a much smaller scale).

The big difference, however, was that Valve, Unity, Epic, and all the others had already earned their reputations and, as far as Luckey was concerned, Oculus had not yet accomplished a thing. They were months away from shipping a devkit, years away from a consumer product, and a culture shift away from convincing the generic public that this time the hype was real.

That said, Luckey understood why Chen was asking. Because even though the number of people Oculus had already convinced was relatively small, the audience they had cultivated—a mix of hard-core gamers, sci-fi lovers, and early adopters—believed in what they were doing with an almost religion fervor. And so, by extension, they were devoted to Palmer Luckey and his plans to change the world.

"You've attracted quite the following," Chen said, earning a giggle from Dycus and sarcastic eye roll from Luckey. "What? You have. I read the AMA."

Chen was referring to the barrage of unsolicited affection that Luckey received during a late August "Ask Me Anything" Q&A session on Reddit.[8]

"So what do you think?" Chen asked. "It would just be a short little video and I'd be happy to work around your schedule."

"Honestly," Luckey replied, "I think the video thing is a pretty solid idea. But I actually have a *much* better idea: you should quit this job and come to Oculus."

Chen laughed, assuming this was a joke.

"I'm serious!" Luckey replied, presenting Chen with the employment agreement. "Joe, come work with me at Oculus. All you have to do is sign this paper."

"Dude," Chen replied, "you're in *my* office! I'm pretty sure that's not how this is supposed to work."

Luckey leaned forward, his excitement exceeding its threshold. "Come and make a difference. Get out of this place. They don't care about you. They don't care about your ideas. They're not going to let you make a difference."

"I'm not gonna sign this, I have no idea what you guys even do."

"You know enough."

"I can't just sign this and leave Epson," Chen said.

"Of course you can!" Luckey replied. "But don't just do it for me. Do it for the people, Joe. The people!"

"I mean," Chen replied, trying hard not to sound condescending. "Has anyone else you've asked said yes?"

"HAVE YOU EVER DREAMT ABOUT BEING PART OF SOMETHING LARGER THAN yourself?" Iribe asked a few weeks later from a private room in the back of Capital Grille. With Luckey by his side and a wineglass in his hand, Iribe was amping up his A game to reel in one last recruit for the initial wave: Joe Chen. "Have you ever thought: I really want to do something special, like *really* special, with my life?"

Although Chen figured that Iribe probably made a pitch like this several times a week—a pitch that, he couldn't help but notice, *did* kind of make him feel like he was being recruited to join a cult—there was a darelike quality to Iribe's recruitment that resonated with something deep inside him. Even so, that wasn't something he liked to vocalize. "I don't mean to interrupt your spiel," Chen said, "but what kind of person hears those questions and is just, like, flat out: nah, I'm good!"

"Good point," Luckey said. "Like," he continued, adopting an overly dramatic tone, "actually, I try to avoid doing interesting things as often as possible. And don't even get me started on the dangers of doing things larger than myself!"

Iribe loved the vibe between these two. It was now starting to make sense why Luckey had wanted to hire this guy so badly.

At the time, Iribe had thought Luckey was just being a bit head-strong. But the longer this meeting lasted, the more easily Iribe could see what Luckey saw: Chen was a glue guy. He was that go-with-the-flow, no-job-too-small guy that every successful start-up needed.

"So a couple years ago," Chen began, "I started feeling . . . I don't know. I was bored at my job. I was doing random marketing things. I mean, I started off in product development, and that was pretty awesome, but then we got to the point where we were actually making money, so our parent Japanese company was, like, 'Oh, you don't do product develop-ment anymore, we'll take over that business. You just market the stuff to the US again.' So I thought that was stupid. And I realized . . . okay, so my background is in Texas, but I had now lived in California for a while and I was just kind of tired of this consumer culture. So I was, like, you know what? I'm gonna do something with my life. I'm gonna join the army."

Chen told the guys about how he got in shape, quit his job, and signed up right before his thirty-fifth birthday. Being in his midthirties made him a senior citizen among the recruits—"old man" they called him or "grandpa"—but he excelled at his newfound career. After Chen finished Airborne School in March, he was put in an aggressive pipeline to join Special Forces. But while running through a drill with seventy pounds on his back, he injured his left foot. As a result, he was dropped from the course.

"It's the weirdest thing ever," Chen said, a little somber now. "When you're there, it's hell. It's your body being broken over the course of weeks and months. But once you're out, and you've failed, all you care about is getting back."

"Can you go back?" Iribe asked. "Is that allowed?"

"There's probably a lift," Luckey suggested.

"Yeah," Chen said. "There's a one-year break from the time you're discharged. So I have a year—well, minus a few months—until I can try to get back in."

"That's crazy!" Iribe said.

"That's amazing!" Luckey said.

"So yeah," Chen said, scoffing at the memory. "I busted my foot, got cycled out, went back to my old job, and got told to market these stupid transparent TV glasses and that's what led me to Palmer."

Iribe didn't go as far as to call the sequence of events a blessing in disguise, but he tried to skate the line without sounding like an asshole. Whatever it was about Special Forces that had appealed to Chen, Iribe was certain that Oculus could offer more: a chance to make history.

"Look," Chen said, "you don't need to give me the hard sell. I wouldn't be here if I wasn't interested in what you guys were doing. But what it really comes down to is . . . the hardware business is brutal. So why the hell do you guys think you have a chance?"

Having recently put together a pitch deck, Iribe pulled out his laptop and walked Chen through his plans for the business. It was what Chen expected—"We're the best!" "We know John Carmack" "We know how to use exclamation points!"—until Iribe reached the slide about software.

"So you've worked in hardware before," Iribe said not long into the presentation. "You know that going around just selling hardware isn't really a sustainable business. Maybe if you're Samsung, but that's not our goal. We don't want to be those guys, we want to own the platform. We want to be Google, we want to be Apple, we want to be Microsoft."

Iribe's emphasis on building a platform began to convince Chen that this could be a real business. Oculus wasn't just some teenage kid trying to throw together a video game goggles company. It was a tech company focused as much on software as hardware. Highlighted by objectives like the following:

- Build a digital store where devs can sell VR-ready content ("like the App Store!")
- Develop casual first-party games ("like *Oculus Chess!*")
- Create and support additional peripheral devices ("like motion controllers, body monitors, or even shoes!")

"So . . ." Iribe said, turning to Chen. "What do you say? I promise you: it'll be the most fun you've ever had."

The glow in Chen's eyes told Iribe everything he needed to know. Chen was in, ready to join the cult.

"YOU GUYS READY TO PULL THE TRIGGER?" IRIBE ASKED, WALKING WITH Luckey, Mitchell, and Antonov through a potential office location in Irvine. "This place is *in*credible!"

Situated on the second floor of a fourteen-story black granite tower, this place—19800 MacArthur Blvd, Suite 450—was seven thousand square feet of sleek, uncongested Class A office space. In its current state, the beige carpeting and vanilla walls seemed a bit too stale and corporate for the Oculus brand; but in time that could all be remedied. Either way, the succession of windows that wrapped around the floor gave the office a bright, airy feel.

"So what do you think?" Iribe asked the group as they inspected a little reception area by the entrance. "I really think this is the one."

"Well," Antonov replied, holding back a chuckle, "it is certainly 'the one' that is closest to your apartment!"

This was true. Though amid the ensuing laughter, the guys acknowledged that it *was* also the nicest space they had seen. It was also the most expensive, by far, which concerned Luckey—telling Iribe that he thought they money should probably be used to hire additional engineers.

"This office will help us recruit those engineers," Iribe said, also noting that a space like this would also help with attracting investments. Luckey couldn't argue with that. And so, heading into November, Oculus got their first office. "Now," Iribe said, "we just need to find someone to run the place!"

HEIDI WESTRUM

"Being a good office manager," Heidi Westrum said, while interviewing with Iribe for the position, "means, basically, being everyone's best friend; you need to be that friendly voice they can always count on and always one step ahead, anticipating their desires."

For the past fifteen years, Westrum had been with a relocation company called Alexander's Mobility Services; so leaving that behind and becoming the Office Manager at Oculus would be a significant transition. But after so many years of doing the same thing, she felt like she was ready to take a leap; and though leaping to a few-month old startup like Oculus seemed like a pretty risk move to make, Westrum—by the end of her interview—decided that she was up for the challenge. Largely as a result of two things: Iribe's track record building companies; and how frequently he used the word "family" to refer to members of the Oculus team.

This idea—of one's work family becoming their second family—was something that Westrum had always felt and very much believed in. And in Iribe—as he raved about Luckey like he were a little brother; or about the importance of celebrating successes as a company—Westrum saw a leader who appeared to truly share her sensibility.

Okay, Westrum thought in early November, I'm in.

"Great!" Iribe said with a smile that made her feel like she'd already made the right decision. And then, before she left, he added one more thing she'd never forget: "Make sure to always have champagne on ice; because, with this company, there's gonna be a lot to celebrate."

LYLE BAINBRIDGE

It would be many months before Lyle Bainbridge—a New-Zealand-born, Minnesota-based PC board designer—would first meet the Oculus team in person, but in the meantime he'd play a key role in helping to build Oculus' first devkit.

"Lyle's awesome," Patel—who worked closest with Bainbridge—would tell Luckey not long after they brought him up, "This is very specific, but a tell for great electrical engineering on the kinds of products we're building is passing FCC radiated emissions testing without sticking a bunch of shielding on the mainboard. And, well, I can already tell that this guy is a truly elegant architect."

THE GHOSTBUSTER, THE REDIRECTED WALKER AND THE KING OF MTBS3D

In look, livelihood and location, Peter Giokaris—a Canadian living in San Diego—Brant Lewis—a proud lifelong Texan—and Andres Hernandez—a born-and-bred New Yorker—had little in common. Except for two things: over the past few years, each had devoted most of their free time to work on virtual reality projects; and, in the fall of 2012, each were amongst the first to interview for jobs at Oculus' new Irvine office.

"So this is what I've been working on lately," Giokaris told Luckey, Iribe and Mitchell, showing off a backpack-based VR system.

"This is awesome," Luckey said, impressed with the craftsmanship of Giokaris' backpack. "And as a bonus: it kind of looks like a Ghost-buster proton pack!"

Giokaris laughed, and then told the guys about how he had built it, about how—in 2010—he had left the games industry after 16 years to join Qualcomm and work on their augmented reality platform Vuforia; and about how—ever since seeing *TRON* in 1982—he had been convinced that his life's mission was to build "synthesized worlds." Or, Giokaris clarified, at least to help others build virtual worlds. Which is why, after failing to convince the Vuforia team that they should start looking into VR, he had decided to build a VR SDK of his own.

"Something agnostic to headset and inertial device," he explained. "The idea being that as VR becomes more popular there will be enough headsets and inertial units to start building more and more content. But as a guy who—literally, back in 1995—was building games for Nintendo's Virtual Boy . . . I know that VR 'becoming popular' is not, uh, exactly a foregone conclusion."

Andres Hernandez and Brant Lewis shared that same cautious optimism. They had known each other for years from MTBS3D, but it wasn't until they both flew out to Irvine for mid-November job interviews—with Hernandez, the top moderator at MTBS3D, interviewing for a community manager role; Lewis for software engineer—that they both finally met each other for the first time. Not only that, but together they shared a glimpse of the future.

"This should work," Lewis said, after convincing Hernandez, Luckey, Iribe, Mitchell and Antonov to join him at a football field on some local community college campus. It felt like an odd choice, especially as the sky got dark. But Lewis needed a big open space to show off his "redirected walking" demo, which delivered a full six-degrees-of-freedom game experience so that you could actually walk *anywhere* in a game environment.

Naturally, the guys who Lewis had led to the football field were skeptical. Probably Luckey and Hernandez most of all, since they each knew more than almost anybody in the world what a *bad* virtual experience feels like. And yet . . .

And yet . . .

Hernandez was utterly speechless. It was like he'd been swimming with weights for all these years, and stuck with only three degrees of freedom at the bottom of a pool. But this . . . kicking free . . . this let him move with a freedom and invincibility that he had never felt before.

CHAPTER 15
THE HONG KONG SHUFFLE

November 2012

"MARLBORO CIGARETTES?" MITCHELL ASKED, AS HE AND LUCKEY APPROACHED a duty-free shop at LAX.

"Marlboro *Red* cigarettes," Luckey replied, reading off an email list of things McCauley wanted them to bring to China. "He needs three cartons."

"Does it really say that?" Mitchell asked, half distracted.

"Cone eyepieces, cylindrical eyepieces," Luckey recited from the list of things McCauley needed them to pick up before meeting him in China. "Plexiglas control box and three cartons of Marlboro Red smokes."

"Oh," Mitchell remarked offhandedly. "Wow."

Luckey could tell Mitchell was preoccupied, and he had a pretty good idea what with: ZeniMax.

Over the past couple of months, the relationship between Oculus and ZeniMax (the parent company of id Software, where John Carmack worked) had begun to sour. Following mutual enthusiasm at E3—ZeniMax grateful that Luckey had freely lent them his prototype, Luckey appreciative that ZeniMax's *Doom 3: BFG Edition* demo had helped put Oculus on the map—a lot had changed in a short amount of time. Ultimately, the wedge was a result of Oculus and ZeniMax being unable to reach an agreement about how best to work together going forward. Up to this point, there had been informal collaboration between the companies—the type of casual back-and-forth that typically exists between hardware makers and software providers; though slightly enhanced in this case due to Carmack's passion for VR and his unique contractual freedom to go above-and-beyond with hardware vendors. But despite their loose collaboration, Oculus and ZeniMax had not yet hammered out any sort of formal agreement. In fact, save for an NDA that ZeniMax had Luckey sign before E3 (to receive the testbed that Carmack had offered him), there wasn't a single piece of documentation at all between the two sides.

By early August, both companies knew this ought to be resolved. Formal discussions between the two sides began on August 7, with a call between Iribe and Todd Hollenshead, the CEO of id Software, and the negotiations got off to an optimistic start. No numbers were discussed, but it was clear from the mutual respect and exuberant enthusiasm that both sides saw this partnership as a when-not-if inevitability.

The problems began a month later, when Iribe sent over an initial proposal to "kick off" negotiations. There were many facets to the offer but the crux of it was this:

- ZeniMax would receive 2 percent equity in exchange for allowing John Carmack to serve as a technical adviser to Oculus.
- If interested in a larger stake, ZeniMax could obtain an additional 3 percent equity by investing $1.2 million (at either a seed round or Series A)

From Carmack's perspective, accepting the technical adviser offer seemed like a no-brainer. "I already do that for free," he told ZeniMax management—so, to him, 2 percent equity for merely formalizing that arrangement seemed like "free money." As for the investment opportunity—that wasn't really up to him. Though, technically speaking, none of it was really up to him. Ultimately everything went up the chain to ZeniMax CEO Robert Altman, who thought Oculus's initial offer was an insult. And he felt even more insulted when, later that month, Oculus sent over a newer "investor prospectus," which listed

Carmack as an adviser/endorser. This had not been finalized! Nor had ZeniMax given Oculus approval to use the DOOM logo (nor, for that matter, had ZeniMax even officially agreed to produce a VR version of *Doom 3: BFG Edition* for the Rift). Finding all this unacceptable, Zeni-Max sent over a counterproposal on October 19 that they felt better reflected their "past and continuing contributions" to Oculus.

As with Oculus's initial proposal, there were many facets to Zeni-Max's counter. But what it boiled down to was that with*out* making any financial investment ZeniMax wanted 15 percent nondilutable eq-uity in Oculus.

To Iribe, this was a ludicrous offer. Forget whether or not 15 per-cent was fair, or even a percentage that Oculus would be willing to part with, but "nondilutable" equity stakes—meaning those where a stake-holder's ownership percentage remains the same even as new money is raised—were pretty much unheard of in the investment community.

For the first time, the exec team at Oculus began to consider the possibility that a deal might not be struck and they'd be unable to ship *Doom 3* with DK1. Especially after Iribe's latest reply to ZeniMax, which described their counterproposal as "so far out of the ballpark, we're left wondering if there's any hope." That was sent November 13, the same day that McCauley requested those Marlboro Reds. And though there had been no word from ZeniMax, there would be plenty of chatter from McCauley now that the guys had finally arrived in China.

"Wercome . . . to the Puhr-maan Hoterr!" McCauley exclaimed, speaking in an over-the-top, stereotypical Asian accent that he hoped would make Luckey and Mitchell shake their heads as their cab arrived at the Pullman Hotel. Located in Dongguan, the Pullman was the only Western-style hotel within about twenty miles of Berway (the factory where the Rift would be manufactured).

"Oh man," Luckey whispered. "It's 2012: you can't say shit like that!"

"Especially not in front of the valets!" Mitchell said, embarrassed. As they would soon discover, McCauley recycled that same stupid off-color impersonation *every single time* they pulled up to the Pullman.

Given McCauley's age, the guys just rolled with it and chalked it up to Jack fulfilling his dual role as Oculus' wacky uncle. And what choice did they have? Until the day all ten thousand headsets eventually shipped (or whatever the number now was), the fate of Oculus rested in Jack McCauley's hands.

To be fair, McCauley was no tyrant. He could get persnickety, sure, and he was a bit set in his ways, but it was obvious to anyone who spent five minutes working with the guy that what motivated him most wasn't money or power; it was an unquenchable quest for product excellence. Pure as those motives appeared to be, that didn't change the fact that McCauley wielded an incredible amount of power. He was the linchpin of Oculus's manufacturing operation, single-handedly responsible for contracting every vendor needed to produce the devkit—all of whom were based in China and who would have been invisible to any American who lacked McCauley's expertise; the guys had to trust him. Not only in regard to the price of driver circuits and other technicalities like that, but in other areas, too.

Like when McCauley decided that he *didn't* want to move down to Irvine, and instead wanted to keep working out of his lab in Livermore. And when McCauley said that he needed to hire (and sometimes travel with) an office assistant whose previous job was with the Taiwanese American Junior Chamber of Commerce Northern California. And when McCauley explained that he needed the company to pay for near-weekly trips to China? Sure. Why the hell not! Of course, there was still the issue of McCauley being a loose cannon on email. Typically, Iribe would be the one to try and rein him in. But since that didn't seem to be working, Luckey had recently tried his hand at peacemaker. "We want to be careful about putting Carmack on emails," Luckey wrote McCauley. "He wants to help, but it's probably best to pool our knowledge at Oculus before contacting him." Luckey, hoping that his loose cannon might respond better to humor, ended his email by explaining that "even if we don't really know what we're doing, we should make an effort to appear as if we do!"

"I'm just gonna throw this out there . . ." Mitchell started, before glancing over his shoulder to make sure McCauley was out of earshot as he and Luckey got off the elevator at their floor one morning. "*Maybe* there's some higher level to Jack's joke that we're just not getting. Is that possible? Like, is it a reference to something?"

"Nah," Luckey replied. "Jack's just a crazy old man!"

The guys harrumphed in laughter, and then synced up about what time to meet tomorrow (way too early), if there had been any word from ZeniMax (nope), and whether Jack's particular brand of crazy more closely resembled that of a loon or of a fox (inconclusive). Then before disbanding to their rooms for the night, Luckey asked Mitchell if Iribe (who had visited Berway in August) had told him much about what to expect at the factory.

"Okay," Mitchell said, with a grin growing on his face, "you know those super-high-tech factories? Like that one Amazon has, with the robots? Well, Brendan said that Berway is basically the total *opposite* of that."

"What does that mean?" Luckey asked.

For a moment, Mitchell considered the best way to relay what Iribe had told him about Berway. Should he talk about Elaine Chan, the enigmatic managing director, who had worked with McCauley for over ten years? Or maybe he should warn them about the smell— the ferocious scent of hot plastic—that dominated not only the factory but the entire city as well? Or perhaps the best preface was to describe just the commute, to tell the guys about how a factory employee would arrive tomorrow in a van, and then zip them through all these crazy off-the-beaten-path roads that were lined with little shops that sold tools, components, and assorted electronics. With each of these thoughts, Mitchell chuckled to himself, which only made Luckey even more curious. This is why it felt a little cruel when Mitchell decided to forgo these little details and respond with this:

"Like virtual reality, Berway is something you'll just need to see for yourself."

"SHE SINGLE-HANDEDLY BUILT THE COMPANY," MCCAULEY EXPLAINED THE next morning, during their bumpy van ride over to Berway. "She's just amazing."

McCauley was talking about Elaine Chan, relaying her bio before they soon met her at the factory. "She's from a family of nine people," McCauley continued. "Ten, actually, but one of them died. The father died, too, before Elaine could remember him. The mother raised those kids. No welfare, they didn't have that back then. So she had to make it work. They all lived in apartments stacked up in bunk beds. They didn't have the money to send her to college, so Elaine got this job on an assembly line. Then she got a job over in an office and she excelled at it. She's very hardworking and clever, in a way. She's not technical, but she's very smart and she's very shrewd. And she's made of iron. Once she makes up her mind on something, you cannot change her mind. And if you get on her bad side, she won't forgive you for a long time."

"Have you ever gotten on her bad side?" Luckey asked.

"Oh yeah," McCauley replied. "Yelling matches. Most over money."

With Oculus, it hadn't quite escalated to that level, but it came close. Mostly because Chan had little interest in fulfilling the order. Even for a friend, an order this small just didn't seem worth the headache. To sweeten the deal, McCauley suggested that Iribe also use Berway to manufacture the controllers for the other venture he had recently invested in (Wikipad). Chan, however, still had doubts; but after all she'd been through with McCauley, how could she resist?

"So don't forget: she's doing us a favor," McCauley reminded them as the van dropped the guys off outside a beige factory in downtown Dongguan: Berway Technology Ltd.

It was neither like the robot-roving factories of the future nor the *I Love Lucy* factories of the past. It was something in between, but so unlike the visuals we associate with automakers (gritty, pristine), food processors (orderly, sanitary), or even sweatshops (oppressive, dangerous). To Americans experiencing it for the first time, it felt like a place of contradictions.

Starting with Elaine Chan, whose relaxed gait, cheerful welcome, and perpetually smiling heart-shaped face seemed to betray McCauley's description. This was no Iron Lady, it seemed, as she began walking them through the factory. But she soon broke away, ever so politely, to bark at an employee with scalding ferocity. Then just as politely as she had left, Elaine Chan returned with that unbreakable smile. "We go now!" she chirped, and then continued with the factory tour.

Technically speaking, Berway was big, occupying two entire city blocks, over five hundred thousand square feet in size. Yet almost everywhere inside felt kind of claustrophobic—whether from the tooling machines or the two dozen on-site engineers. Upon further observation from Luckey, the oppressiveness was probably a result of drab coloring (the floors were forest green, the rest faded whites and grays), relatively low ceilings, and a suffocating smell of plastic.

Another idiosyncrasy was the level of noise at Berway. The place had 32 assembly lines, 128 set injection machines, and over a thousand workers on the floor; yet walking through it all, there was hardly any sense of cacophony. Where was the clatter? The moans and groans? The bloated grumble of machines? Those were all there, if you listened hard enough, but something about the layout seemed to deter such moments of observation. At first, Luckey was tempted to chalk this oddity up to some sort of high-level feng shui. But by about the third time Mitchell whispered what they were all thinking—"this place is so messy"—it was too hard to believe that any harmony here was a by-product of design. And that's when Luckey realized that the opposite was true; this place was such a bastion of chaos that no singular instance of disarray really had any chance of standing out.

After the tour, the guys sat down with Chan to chart out a production schedule for manufacturing DK1. McCauley did most of the talking, making no bones about the fact that they desperately needed this done ASAP.

"How much ASAP?" Chan asked.

"We need to ship by March," McCauley said.

"*This* March? Of the 2013?"

"Yes, Elaine. Remember? We talked about this already."

"I don't know this talk," Chan said (somewhat unconvincingly). "March in five months. What talk is this?"

"Come on," Mitchell interjected. "There's gotta be a way. I mean, before the screen issue, you were going to be able to do December. Right?"

Something about Mitchell's comment rubbed Chan the wrong way. And though the tone of her reply lacked bark, there was a glint of ferocity in her eyes. "Quality very important," Chan said. "My company—not very big company—but our R&D team very strong. We can work on famous product with our customer. And very quick to complete the product. Normally it take one year. But we can do it six months."

"Is there *any* way to accelerate that?" Mitchell asked.

"We're down to do anything on our end," Luckey added. "I know you work on a lot of products. A lot of great products. It's amazing! But the Rift . . . it's our morning, noon, and night. There's not a thing in the world I wouldn't do (I don't think)."

"It is special," Chan said, looking deeply at Luckey. "But you are very young. You make the idea, but you do not know how to make the product."

Luckey smiled. "That's why we have Jack!"

Chan giggled. "Okay! We work! But: first I want to say a thing." She then paused for a moment to think, searching not so much for the words, but to momentarily strip away any sense of facade. "For me, I'm just manufacturer. I make the product. But personal, I think many company will want to do similar product. So Oculus, you will need to maintain the special. More features? Yes, but also other. Okay then! Now we go through schedule and planning. Then you meet engineering. We try March."

As much as the guys were tempted to probe Chan about maintaining "the special" (and what she meant by "the other"), they couldn't ignore the chance to finally work out a road map that shot for March. Ultimately, they decided upon this:

PHASE	DESCRIPTION	# OF DAYS
Design Verification	Selection and validation of components	21
Design Finalization	Final specification locked	1
Tooling 1 (T1)	Manufacturing of injection molds	56
Tooling 2 (T2)	Fine-tuning for part function and fit	10
Tooling 3 (T3)	Final fitting, matching, and polishing	6
Preproduction (PP)	Small verification build (aka "pilot run")	5
Mass production (MP)	Full-scale build (500 pieces per day)	12
Chinese New Year	Annual factory shutdown	28
Sea Freight	China to US (including cus-toms)	28
Estimated First Arrival	Approximately 5,000 developer kits	n/a
Shipping	From California to final destination	7–28

There was just one problem: the schedule was nothing more than words on a page. To hit this timeline, they needed Berway's engineers to make it happen, and at first glance, the engineers seemed to think this timetable was hilarious.

"Oh, Jack!" laughed HK, Berway's in-house electronics guru.

Unraveled and confused, Luckey and Mitchell huddled together to try and make sense of what had just happened. What was the point of even drafting the schedule if these dates weren't realistic? And if the timetable was so far off base, why hadn't McCauley spoken up? Wasn't he supposed to be some kind of expert? McCauley proved then that's exactly what he was; though what he demonstrated here was a unique and unusual brand of expertise.

"So *that's* what the cigarettes are for," Mitchell noted, as McCauley broke out a carton of Marlboro Reds and started divvying up packs to the engineers. And little by little, the scheduling estimates started to improve. That mold they said would take a week? Maybe it would really only take three days.

"I feel like I'm watching a prison movie," Mitchell said, as the guys watched McCauley wheel and deal.

"That's just how it is," Luckey replied. "When you open your heart and mind to the wonders of the Puhr-maan Hoterr."

As the weight of the day finally fell off Luckey's and Mitchell's shoulders, they felt a little guilty about giving McCauley such a hard time. Granted, it was all behind his back (and granted, he *was* a little nutty), but, seriously, where would they be without this guy?

He confidently handled the "Hong Kong shuffle," which was a phrase that McCauley used to describe the tricky dance of transporting components between China and Hong Kong in a way that minimized costs while adhering to tax, custom, and certification laws. To be perfectly honest, the Oculus guys didn't *totally* understand how it all worked—why the tariffs were ultimately cheaper if components passed through Hong Kong or why certain items from abroad weren't required to meet certain Chinese regulations—and, even more honestly, they didn't really want to know. Because whether or not it was kosher to swoop through all these loopholes, doing so was necessary to keep their scrappy start-up afloat.

"How did you even learn all this stuff?" Mitchell later asked Mc-Cauley, as they chugged through rural China on a rickety little train.

"Who could have even taught you all this stuff?" Luckey asked, kicking his foot against a bag of cheap Chinese luggage. Inside were hundreds of resistors that McCauley had acquired no more than an hour ago, picking them up at some small podunk resistor factory in exchange for a pile of cash. Now they were on their way to some other obscure factory so they could be traded for something else they apparently needed. "You're like some kind of evil import-export genius!"

"Oh no, no, no," McCauley said, his face nearly blushing. "You just pick up these things, you know? Because here's the truth of the matter: I'm not the smartest guy. Really, I'm not. But I love designing and working on things and I've done it my whole life. It's the most important thing to me. More important than my loved ones, believe it or not. I hate to say that, but it is."

Like many times before, he had failed to answer the question. But that was okay; McCauley had earned the right to ramble. What his words lacked in specificity he more than made up for in sincerity.

"I'm getting pretty fed up with ZeniMax," Mitchell said a little later, speaking in a firm whisper so as not to wake a half-dozing McCauley. "I mean: as big as ZeniMax is, they're not Epic. Or Valve. Or a dozen other companies that wouldn't be such dicks. Okay, I get it, we're a start-up. But come on. We're not, like, some dudes-in-their-dorm-room start-up. But"—Mitchell continued, now playing devil's advocate to himself—"the press will *kill* us if we lose *Doom*."

"Two things," Luckey said. "One: we can't lose *Doom*. We just can't. And for that matter, we can't lose John. And two: while I agree with you, obviously—we're not some little rinky-dink start-up—it *is* interesting that *now* is when we're having this conversation. Just in the sense that in almost *every* way, start-ups are at a distinct advantage from big companies. Even middle-sized companies! But in this specific instance— shuttling between weirdo factories to circumvent . . . whatever—this is like the *one* case where being small actually helps us. Like, no company moves that fast on a consumer electronics product. And people *imagine* that big companies can move fast because they have huge resources, but because we're under the radar, too small to worry about any red tape, we're actually able to get shit done. We're actually able to move *faster* than any legitimate company could ever move."

"I like that," Mitchell said. "I'll take that silver lining. And you know what? This is Carmack we're talking about. He wants to do this . . ."

"Yup."

"So," Mitchell finished, "there's gotta be a way to figure it out."

In the silence that followed, as their train wrapped through a slew of smog-filled cities, it was hard for Luckey and Mitchell not to stare out the window and wonder: But what if there's not?

CHAPTER 16

JUST KIDS

November/December 2012

"VLATKO IS, UH, A *COLORFUL* CHARACTER," IRIBE WARNED ANTONOV SHORTLY before they arrived in Maryland on November 21 to meet with Bethesda Softworks president Vlatko Andonov, who was now the point person on id Software's negotiations with Oculus. It didn't take long for Vlatko to live up to his, uh, colorful reputation—striking a hostile tone with Iribe and Antonov, and telling them, "You guys are just kids, you should be working with us."[1]

Given the leverage that ZeniMax had (and the fact that Oculus had already promised *Doom 3: BFG* to thousands of backers), Iribe expected that Vlatko would demand some obscene number for each copy of the game. But instead, he said that if Oculus didn't sign the deal that ZeniMax wanted, they would no longer let Carmack work on VR.

Iribe was shocked by Vlatko's threat, though he probably would have been less surprised if he had known that, internally, Vlatko frequently referred to the Oculus guys as "clowns"; or that just a few weeks earlier Vlatko had complained to a contact at Sony PlayStation about how Carmack was wasting his time on "VR support for *Doom 3* and the (stupid) Oculus Rift."

"THAT'S CRAZY!" DYCUS DECLARED.

Luckey shrugged, dabbing a tiny screw with isopropyl alcohol.

"*You*'re crazy!" Dycus chided.

"Eh, maybe," Luckey replied. "But do you *know* of a more efficient solution?"

Luckey and Dycus were at the office trying to determine the ideal eye-relief settings—the ideal distance between lens and eyeball—that ought to be implemented into the eyecups of DK1 before manufacturing began in China.

At a normal company, this would all be calibrated with some sort of fancy computer vision system. But this being a start-up, always short on time and money, they needed a faster and cheaper solution.

"Chris! I got it!" Luckey had shouted minutes earlier. "We're going to drill a hole in the center of the lens and then run a flat-top screw through the opening so it's jutting out, you know? We'll start it out at a safe distance and then just keep rotating the screw until, basically, it ever-so-slightly pokes you in the eye!"

"What? No," Dycus had replied. "You can't do that. It'll give you an infection."

This was a valid point, which is why Luckey was now coating the head of a screw with isopropyl alcohol. As he did that, Dycus—ever skeptical—drilled a hole in the center of a lens; when they both finished, the disinfected screw was threaded through that opening.

"Are you sure you want to do this?" Dycus asked. Before he finished the question, Luckey had the headset against his face.

As per the parameters of their rugged experiment, they had started with the screw placed at a safe distance from the eye. Even so, Luckey instinctively flinched.

"There's no shame in backing out!" Dycus offered. "Seriously, Palmer: you're going to scratch your eye."

Luckey removed the headset to adjust the screw. "One: I've scratched my eye before and it's not that bad. Two: it's a perfectly smooth metal dome that is nice and polished. It'll probably be the nicest thing that's ever touched my eye! And three: let's say—*worst-case scenario*—I *do* scratch my eye real bad and end up needing to wear an eye patch for a while? Boom: modern-day pirate!"

As a handful of employees came to gather around Dycus, they were surprised to see their beloved founder subject himself to this Clockwork-

Orange-esque experiment; and even more surprised to see that it actually worked. After several rotations, the screw was flush against his eye. And—score!—they now had a way to measure the immeasurable.

"Got it!" Luckey cheered, the screw now barely grazing his eye.

"Aren't you uncomfortable?" Dycus asked, surprised by Luckey's cheer.

"I mean, it's superspooky," Luckey replied. "But not uncomfortable. You'll see."

"Me?"

"Yeah," Luckey said. "You and the rest of those lookie-loos. We need to see how it differs by facial structure, eyeball shape, and everything else. So . . . who's up next?"

"YOU!" MCCAULEY CALLED, POINTING TO THE FACTORY WORKER WHO WAS NOW at the front of a sixty-employee-long line at Berway. "You, it's your turn."

Like Luckey, McCauley was conducting an impromptu ocular experiment of his own, though what he wanted to determine was the ideal focal length—the distance between a person's eye and the screen—that should be used for the next batch of prototypes. Since this was a somewhat subjective measurement setting, McCauley wanted to collect as much data as possible.

A week or so later, McCauley would visit the office in Irvine and bring with him the very first models of DK1. Although he would be satisfied with Berway's initial pass, the rest of the team would not be. They would have issues with just about everything: the shape, the weight, the optics, and so on. They wouldn't say anything *too* harsh, but it was clear that these kids were not pleased. And so, when they got to the focal length, McCauley wouldn't hesitate to push back.

"Look," McCauley would explain. "There's a subjective aspect to all this. Especially the settings. But I tested them on sixty people and that's what we came up with."

Despite this explanation, the Oculus guys would remain displeased. Especially that kid Nirav Patel. But what does he know? Who cares what *he* thinks? McCauley would ask himself these questions many times during that visit because—to his surprise, to his horror—

Brendan, Palmer, and Nate seemed to actually value his opinion. Why? How did that happen? As McCauley's mind would begin to swirl with questions like these, he'd force himself to cut it out. Politics was the part he hated most about companies.

But that was a week or so away. Right now, running factory workers through his focal length experiment, he was at his finest; commanding, resourceful, and relentlessly focused. This, McCauley thought, is how it should be.

"NO," ANNA YERSHOVA SAID, SHAKING HER HEAD. "IT SHOULD NOT BE LIKE this."

Her husband, Steve LaValle, was moving through the apartment they had rented in Finland, conducting a last-minute electronics sweep while also layering up to brave the weather outside. "It's sweet of you to worry, honey. But I'll be fine. Besides, you get to stay home and play with the new robot!"

Snowy and hushed with only four hours of sunlight, this time of year was brutal on almost everybody. But, as Yershova liked to tease, her husband "loooooooved the weather." What could he say? The weather *was* brutal—cold, dark, and numb—but he greatly enjoyed its by-product tranquility. Intellectually, he found it conducive to clearer thought; and physically, he'd never felt better. For whatever reason, living in Finland had freed him from the chronic cases of asthma and allergies that had plagued him his entire life. Which is why, as he pedaled through the snow—searching for someone who might be able to repair the custom circuit boards McCauley had sent him from China—he sported a frosty, boyish grin.

For Yershova, however, living in Finland was brutal. Having grown up in Ukraine, it wasn't so much the cold weather as it was the lack of sunlight. That was brutal. She also missed Chicago, where she received her Ph.D. and had been lecturing since 2011. The sense of purpose she had felt at the university was part of the reason she had encouraged LaValle to explore working with Oculus in the first place.

When her husband came back from that first trip and said, "They're just kids, but I think they've got something special and they might be onto something," it was enough to pique her interest; enough that she (and her husband) were willing to help Oculus solve their tracking issues. But as much as she relished the challenge (and enjoyed working closely with her husband), she still thought VR was kind of silly. Until after LaValle returned from his second trip (to Irvine this time) and he brought back one of Luckey's prototypes.

"This is mind-blowing," she said, after trying it out. "Yup, this is it."

Yet despite now truly believing in this technology, both she and her husband were still reluctant to accept Iribe's recurring offers to officially join the company, to go from consulting in Finland to working full-time in California. Because while the opportunity itself was incredible—a chance to be on the forefront of a new consumer technology; to solve the unsolved and shape the evolution of VR—there were some very real concerns. LaValle was a tenured professor, one currently on sabbatical to write an important book. Was he really willing to give that up? Were both of them willing to trade academia for life in the private sector? What were the chances that Oculus even survived its first year?

Then, of course, there was the financial component. If they were to join Oculus full-time, they'd basically be giving up their university salaries in exchange for a lottery ticket; they'd primarily be paid in company equity, which could very well end up worth nothing. Could they even take that risk? LaValle had two children from his previous marriage, and he and Yershova had plans to start a family of their own. And speaking of kids . . . Oculus had been founded by a teenager and was being run (mostly) by guys who were in their early thirties. Sure, several of those guys—the "Scaleform Mafia," LaValle affectionately called them—had created a successful middleware business, but virtual reality was an entirely different beast. And last, when it came to age and experience, the generational difference could not be ignored. "For whatever it's worth," LaValle had said after his most recent trip to Irvine, "they're all perfectly nice guys." To Yershova, however, that wasn't worth much. "Not because I don't trust your opinion," she had said. "But because it's not about nice. *Anybody* can be nice. My worry is about the type of relationship this will be. I think it will be an abusive relationship."

Yershova, of course, didn't mean physically abusive. She didn't exactly mean emotional abuse, either. Because even though she was a little worried that (based on what she'd seen over email) these guys didn't quite appreciate how renowned LaValle really was, that wasn't her concern. What concerned her was how transactional everything appeared to be: a question is asked, a request is made, a result is provided. If things go well: Awesome job, dude! If they go poorly: Can you just fix it? Maybe it just felt that way due to the asynchronous nature of emails. Or maybe this was just how things worked in the business world (and therefore had nothing to do with age). Either way, Yershova sensed something that felt antithetical to the (generally) collaborative nature of academia, in which ideas are presented, argued about, and built upon to construct a merit-driven solution. This felt dangerous; this felt abusive.

"And they always need it right away," Yershova had said, before LaValle left their apartment for his bike ride. McCauley had sent some custom circuit boards that broke; LaValle tried to repair them on his own (he even got out his soldering kit!) but that didn't work and instead of new ones being sent, her husband now had to ride through town and search for someone—if such a person even existed—who could solder these pieces back together. "I am hopeful," she had said before he left. "But my worry: it is less about best and more about fast. And it should not be like this. No, it should not be like this."

"DID PALMER TELL YOU WHAT VLATKO CALLED US?" IRIBE ASKED JOE CHEN, AS they drove through Orlando on the evening of December 2, searching for their hotel to get some rest before exhibiting the following day at I/ITSEC (the military's annual Interservice/Industry Training, Simulation and Education conference).

"Nah," Chen replied. "What'd he say?"

"Kids!" Iribe said. "He said we were 'just a bunch of kids.'"

Chen laughed. His new colleagues were young, sure, and maybe had a juvenile sense of humor, but they worked 'round the clock (sometimes to the point of forgetting to sleep or shower); and the groundbreaking work they were doing was anything but child's play. Working at Oculus felt a lot like his time in the army, which was a comparison

Chen felt guilty making because the stakes were so different. One was life and death, the other success or failure. But fair or not, Chen felt the two experiences shared a lot in common: the strong sense of mission, the us-against-the-world mentality, and the trial-by-fire bond that quickly formed with those who were there by your side.

For Chen and Iribe, that bond grew during their trip to Orlando. Starting with them arriving at the hotel and then hustling over to Kinko's to print up flyers, business cards, and other assorted materials to try and look professional-on-a-budget for the next day's show. At some point during their late-night Kinko's excursion, Chen realized how much Iribe really loved this stuff. Not the actual grunt work, but the never-ending, under-the-gun feel of building up a little start-up. And he was good at it, too, Chen noted. Really good. The kind of guy who probably played his best *Mario Bros.* when the time started running out and that *da-da-da-da-duh* music kicked in.

"How's the demo working?" Iribe asked later that night, back at the hotel.

"Good," Chen said, his face pressed against one of the prototypes. "Probably not as impressive as *Doom 3*, if I'm being honest. Not as flashy. But this will definitely do the trick."

To start distancing themselves from ZeniMax, the Oculus founders thought it would be best to demo something other than *Doom 3: BFG Edition*. The problem was that there was literally no other Oculus-ready software out there. Nothing even close, really. So they'd need to put together something in-house. There wasn't enough time to build anything from scratch, but Andrew Reisse's integration for the Unreal Engine was far enough along that (with some help from Epic, especially Nick Whiting) Oculus was able to sample some stock models from *Unreal Tournament 3*. While it worked, it didn't necessarily flash the Rift's top potential, which was what they needed to do at the January 2013 CES (Consumer Electronics Show).

"What do you think we'll show at CES?" Chen asked.

"That's definitely something we'll need to figure out," Iribe said. "For now, I just hope what we've got holds up."

"It will," Chen said, head still against prototype. "Andrew did a great job on the integration. Seriously: he crushed it. If there's anything we should be worried about, it's the PC. Just from wear and tear."

"Good call," Iribe said. With Oculus's round-the-clock focus on building a headset (and building the software to drive that headset) it was easy to forget that all of that was moot without a PC powerful enough to make it all run smoothly and compatibly. This was a lesson that Luckey and Antonov had (kind of) learned the hard way during their trip up to Valve. It was something they would need to be conscious of going forward. In the meantime, they would just need to hope for the best at I/ITSEC.

Compared to the big annual game conventions (like CES, GDC, and E3), I/ITSEC was a small show. It attracted roughly fifteen thousand attendees each year, almost all of whom worked for militaries or governments.[2] In this regard, it was kind of an odd show for Oculus to attend, but it was important to see how Oculus stacked up against the competition. "Remember," Luckey had said to Chen a week earlier, "the military is essentially the only industry that has continued to invest in VR since the '90s. They are essentially VR's only actual 'customer.' *But* while that might make you think you're about to see some crazy awesome shit, you're probably not. Because that whole industry has been in a deep freeze for the past decade."

After a keynote from Major General Glenn M. Walters (Commanding General, Second Marine Aircraft Wing) at the conference, Chen got his first glimpse of exactly what Luckey had meant, a sense of how truly deep that freeze had been.

MILITARY OFFICER [PREDEMO]: I suppose I'll give this a try.

MILITARY OFFICER [POSTDEMO]: That was incredible! How much does it cost?

JOE CHEN: 300.

MILITARY OFFICER: Any chance you can do any better? Maybe closer to $250,000?

JOE CHEN: Sorry. I didn't mean $300,000. I meant just $300.

MILITARY OFFICER: No way! Are you serious?

JOE CHEN: Absolutely, sir.

MILITARY OFFICER: [head explodes]

That's US dollars? And the SDK is free? Wait, am I on *Candid Camera*? Over and over this pleasant surprise occurred. All these people—many of whom had been hardened by a lifetime devoted to combat—couldn't help but beam like a giddy child who'd just seen his very first magic trick. Which left many with one question: How is it, exactly, that you guys plan to make any money?

CHAPTER 17
OCULUS VS. OUYA

December 18, 2012

"THE MAGIC'S IN THE SOFTWARE!" IRIBE EXCLAIMED, RUNNING THROUGH A RE-cent pitch deck for four guests who had earned a visit to Oculus HQ for backing their Kickstarter at the highest level.

Although this was technically true—one of these guests *had* backed Oculus for $5,000 and *was* collecting his reward—Iribe had forgotten the full context. So when, toward the end of their meeting, the guest in question suggested that they work together, Iribe was a little confused.

"Work together?" Iribe asked, not yet realizing that the guest in question was *Words with Friends* cocreator Paul Bettner.

"Yeah," Bettner replied, smiling his best to keep things upbeat. "Yeah!"

"*Work* together?" Iribe repeated, now squinting at Bettner and the devs he had brought along. They had flown in from Texas for a couple of meetings in Southern California—meetings that Bettner hoped would help determine the next phase of his career.

"I'm sorry if that seems too forward," Bettner said. "I'm not trying to step on any toes here. But we thought we could maybe help create, like, the *Words with Friends* of VR. I don't mean that game specifically, just something equally catchy."

"Wait, hold on," Iribe said. He quickly glanced to his side—to Luckey, to Mitchell—but they seemed equally confused. Or were they just confused that Iribe was confused? Dammit, Iribe thought, I'm so burnt the fuck out.

"We're the guys who did *Words with Friends*," Bettner offered.

"Ohhhhh," Iribe replied, trying to mask his embarrassment with a casual cool. "I thought that meeting was tomorrow. I thought you guys were here for a Kickstarter reward, just to visit."

"Well, in fairness," Bettner said, "we *are* here because of the Kickstarter reward; but, yeah, we were hoping that today's meeting would actually just be the first of a much longer—and mutually rewarding!—conversation."

As soon as the Oculus guys realized who their guests were, that was exactly what they wanted as well. This was almost a dream scenario, after all: a talented developer (Bettner) with the street cred of a hit game (*Words with Friends*) who wasn't beholden to the costly bureaucratic whims of a big studio (aka not ZeniMax). It wasn't yet clear exactly what this relationship should look like; but, yes, let's figure out a way to work together; yes, absolutely, let's make this the first of a much longer conversation. Some of which continued in the conference room—Bettner talked about his vacation to San Diego and his love of being "early to new platforms"—before the conversation moved into Luckey's office.

"Are those from *Chrono Trigger*?!" Bettner asked, pointing to the war-ready action figures on Luckey's desk. As Luckey nodded proudly, a bond between the two began to form—mutual admiration, maybe friendship—with Bettner impressed that Luckey shared a love for Japanese RPGs (especially one from before the young founder's time); and Luckey impressed by how quickly Bettner had identified the characters (and the fact that this *Words with Friends* guy was more than just some superficial casual gamer). "Wait," Bettner said. "Were you even alive when *Chrono Trigger* first came out?"

Luckey laughed. "Let's see . . . *Chrono Trigger* came out in 1995. In May I think. So yeah, yeah I was alive! Two and a half years old! Come on, Paul. I'm not *that* young."

"But how—" Bettner stopped himself, unsure how to ask what he wanted without sounding condescending. What he really wanted to know was why Luckey, unlike most of those his age, actually had any interest (let alone respect) for something that was created longer than, like, six months ago. "I mean, why is it—"

This time it was Luckey who cut Bettner off. "Because," Luckey explained, "a great game's a great game. And if there's a great game out there, I'm gonna find it."

This led to a discussion about Nintendo, a company both Luckey and Bettner held in singular admiration when it came to making great games. "I am legitimately awed by what they're constantly able to do," Bettner said, shaking his head. "I mean, I really fell in love with video games thanks to Nintendo. With the NES. And that was what—twenty-four years ago? Twenty-five?—either way, that was *before* you were alive. That's a long, long time. And yet, to this day, despite *everything* that has changed—the hardware, the software, the culture of gaming, whatever—they still put out the highest-quality games."

Luckey nodded, no argument there. Even with Nintendo's latest console, the Wii U, which had come out one month earlier to mostly mild reviews, there was near unanimous agreement that the games themselves were nothing short of incredible.

"I know this is going to sound cliché," Bettner said, "or maybe arrogant, but my goal with this new studio is I want to create the 'Nintendo for the Next Generation of Gaming.' I know, I know: what a cliché!"

"I will admit," Luckey said, "you are not the first person to express this desire."

"I know it," Bettner said. "To think that this studio I'm just now starting—a studio that, thus far, consists of me and the three guys you met today; oh, and we don't even have a name!—I mean, I know it's an insane goal; but that *is* the goal. And I'll tell you something people often forget when it comes to Nintendo: part of the reason (and it's admittedly just a small part), but part of the reason their games feel so perfect is because the teams they have working on the software get a head start; they know what new hardware is coming—they're kept in the loop—and because of that they are able to create games that optimize every aspect of the hardware. That's one of the reasons why I'm so glad that we were able to connect. I don't yet know what level of commitment we'll be able to make to VR, but we want to get in early."

"I'm curious," Luckey said. "You don't need to tell me, like, dollar amounts, but what sort of commitment do you have in mind? And how aggressive do you want to be? Like, are you looking to build a studio that makes VR games? Or a studio that makes games . . . including some for VR?"

Bettner pursed his lips. Having had his first VR demo no more than an hour earlier, this was a tough question to answer. He was still trying to process what he had just experienced, to balance the awe of his emotional reaction—holy shit, I literally feel like I'm somewhere else—with the warning bells of his intellectual reaction—holy shit, building a viable VR ecosystem is gonna be nearly impossible.

"I'll be honest with you," Bettner said. "The first thing I thought when I tried on your headset was: wow, this is the future! I've only ever felt that way once before, when I got my first computer. So that's amazing, like truly amazing—I honestly never thought I'd feel that way ever again—and that's how I know I want to do *something* with you guys. But is it *the* future or *a* future? Like cards on the table here: we've also been talking to Ouya. And they have a vision for the future that, let's be honest, is much less exciting than Oculus, but there's a much more obvious path to get there."

"It's really interesting, isn't it?" Luckey asked rhetorically. "Like nobody would ever say that Oculus and Ouya are in direct competition because what we're doing is so different. And I actually *like* what they're doing; like, I backed their Kickstarter and—you can't tell this to anyone—but we've actually even been talking to them about the possibility of doing something together. But at the same time, if you really think about it, we are competing with them in this really strange way. Like, we both have these really different ideas about what the future of gaming looks like and, in the end, those futures are basically mutually exclusive."

"There can only be one!" Bettner joked.

"Exactly! But it's kinda true. Because we're both trying to build these ecosystems and there just aren't enough devs to support more than one. At least as long as consoles are also still around. And so, like, where it gets *really* interesting is: both of us—both Oculus and Ouya, that is—we're trying to sell these totally different versions of 'inevitability.'" Luckey paused momentarily, worried he might have revealed too much; but Bettner was right there with him, nodding along.

"You nailed it," Bettner said, awed by the minimonologue that Luckey had just delivered. It became obvious to Bettner that Luckey was a natural showman. What surprised him was that he hadn't noticed it before.

"Oh!" Luckey said. "One other obvious difference between us and Ouya is that we're *only* shipping devkits, and they're sending out devkits *only* a few months before their launch."

"Ha," Bettner said. "I had been thinking about that."

The following morning, Bettner met Ouya's energetic and hyper-focused CEO Julie Uhrman for breakfast date at a posh brunch spot in Brentwood.[1,2,3]

Like most disruptive ideas, Ouya sounded great in theory. To Bettner—who aspired to create fun-first, ultra-accessible "pop" software—the idea of a low-cost device that brought casual gaming into the living room sounded downright magical. But jazzed as he was, he was curious if gamers would be willing to forgo the bigger-stronger-faster mentality that drove console innovation for the freedom and ingenuity of something smaller-weaker-cheaper. And, well, if the past six months were any indication, then the answer was a resounding yes.

In addition to dominating Kickstarter, Uhrman had already managed to recruit a handful of notable developers, luring the likes of Kim Swift (best known for designing *Portal*), Tim Schafer (*Psychonauts*), nWay (*ChronoBlade*), Tripwire Interactive (*Killing Floor*), and several others to create games exclusively for Ouya.[4,5] This lineup of known talent was made even more promising by Ouya's commitment to empowering the next generation of Kim Swifts and Tim Schafers. In the spirit of shattering that barrier to entry, Ouya was also helping to fund titles by unproven talents, betting on young developers like Matt Thorson (whose arena-based archery project, *TowerFall*, had dazzled at a recent Game Jam in Vancouver) and new studios like Robotoki (founded by Robert Bowling, the former creative strategist for Infinity Ward).[6,7]

There was one other major reason to be feeling bullish about Ouya: their software development kit (nicknamed "ODK") had been released earlier that month (and already downloaded over ten thousand times). This meant that it was possible for Bettner to immediately get to work. Meanwhile, Oculus's SDK wasn't even done yet.

Bettner left his breakfast with Uhrman feeling inspired—inspired like he hadn't been in years, though also very much torn. And shortly after processing these emotions, he emailed a few close friends with what he was thinking.

FROM: Paul Bettner
DATE: December 26, 2012
SUBJECT: Hopes and Dreams

I need to find my footing. I've been restless since we saw all this stuff in CA. I want to get to work. I'm ready to bring together some truly brilliant people and build something. But where do we start? It's so big, nebulous . . .

I want to build a profitable company. Not an R&D lab . . . My passion is to create. I aim to ship insanely great products that change *millions* of people's lives. We did it with *Words with Friends*. It's time to do it again. My favorite place to do this is on a frontier. I love to try and see what's coming next and then to build that thing people don't even know they want yet. Who could have possibly predicted that the whole world wanted to play Scrabble on their mobile phones? We found this because we built something that *we* wanted and something we thought others would want too, even though they didn't know it yet.

The frontier is awesome and powerful, but it's also dangerous and deadly. Companies die quickly when they choose the wrong frontier, or they move too far out ahead. Surfing the wave of the "next big thing" is a tricky sport.:)

So I have asked myself two simple questions:

#1 Will there be a breakout VR hit in the next few years? *No.* It is beginning to finally show awesome potential, but it's not ready yet. Maybe 3–4 years from now. Between now and then there will be some powerful signs that this tech is coming, but no breakout/mass hits. Not yet.

#2: Will there be a breakout hit for a $99 dollar hi-def console in the next few years? *Absolutely yes.* Multiple breakout hits. It is an *open, democratized* $99 dollar home console platform. This will redefine the industry again. It will do to TV gaming what the iPhone has done to handheld gaming. *Everyone* will want one and everyone will be developing games for one (eventually.)

These answers come readily to mind and feel correct to me . . . I now believe that the ***next*** great opportunity to do that—to tap into yet another huge new wave of consumers—is with this revolutionary $99 dollar console.

VR will be that opportunity too, probably sooner than anyone expects. But until we can create real VR products that change millions of people's lives, it won't be where my real passion is. I think that's maybe the wave that comes *after* this next one.

What do you guys think?

CHAPTER 18

GOOD, BETTER, AND BEST AT THE 2013 CES

January 2013

LAIRD MALAMED, FORTY-FOUR, OCULUS'S JUST-HIRED CHIEF OPERATING OFFI-
cer, walked into a conference room full of a twentysomething engineers.[1] To him, this babyfaced group did *not* look like a collection of innovators poised to change the world, but rather like a crew of dudes that he might find playing *Halo* with his stepson. What the heck did I get myself into? he thought.

"Basically," Iribe explained to the crew, "he's here to organize everything that's going on. Laird, why don't you introduce yourself to the guys? Tell them how you got into the industry and what you've been up to lately."

"Sure," Malamed replied, waving a hearty hello to his new colleagues. "I suppose I should start by answering the question you're probably all wondering: yes, I *do* realize that I'm the only one wearing a button-down shirt."

The uncool appearance of his group, while putting him at ease if only because he didn't have to try *too* hard, also reminded him of how he would've appeared while working at Activision twenty years earlier, the game publisher responsible for powerhouse franchises like *Call of Duty*, *Guitar Hero*, and *Skylanders*.

Activision was an exciting place to be when Malamed joined in 1994, at a time when the biggest buzzword in the industry, *multimedia*—which was why Activision had targeted him in the first place—offered endless upward mobility, despite the toil it sometimes took to get work done. It was the struggle of that experience that made him fall in love with making games; and, along the way, Activision fell in love with him making them. From there, Malamed quickly ascended to directing and/or executive-producing titles like *Zork: Grand Inquisitor* (1997), *Star Trek: Voyager-Elite Force* (2000), and the first-ever *Call of Duty* (2003).[2]

"Flash-forward several years," Malamed told the group, "and this will connect to why I'm here with you today: we bought Red Octane, the creators (along with Harmonix) of *Guitar Hero*. The notion at the time was that we wouldn't merge them into Activision; they'd stay their own business unit, their own supply chain and hardware team. So Mike Griffith, our CEO at the time, asked me to be the SVP of production on that." Malamed got lost in the memory, a smile dangling from his face. "I remember saying to Mike: But I know nothing about hardware! I've been working on software my entire career. He told me: you'll learn it. And he was right. I had a really good team, so that was helpful. But we did it. We did it together. And so I was there to take *Guitar Hero* all the way up"—referring to the runaway success of *Guitar Hero* and its many sequels, and then with a self-deprecating grin Malamed finished by saying, "and then, yup, all the way down"—referring to the bomb of *Guitar Hero 5* and all the iterations that came after.[3]

Tiny moments like that—where Malamed felt the beginnings of a bond with the small group in this room due to his self-deprecation—served as a great reminder that his difference in age was irrelevant in the face of their shared love of games. It had been over a year now since he had left Activision to pursue some the nongaming interests he had missed out on over the years; teaching part-time (at USC), consulting for a charity (the Bill and Melinda Gates Foundation), and training for an upcoming slate of marathons (because he loved running almost as much as making games).[4] But as he recounted some of the greatest hits from his Activision days, he realized how much he craved achieving that kind of success again.

"*Guitar Hero* just blew up," Malamed explained. "I was a part of that cultural phenomenon. And my little piece of it—making sure that it go out on time and on budget and had good marketing—was just *one* piece of everything needed to make that happen; but it was a piece that . . . I don't exactly know how to describe it. But it feels like everything."

To some in the VR community, Oculus' hiring of Malamed was a cause for concern. As summed up by one user on MTBS3D: "I hate to jump to conclusions because I don't know the guy, but Activision is one of my least favorite companies, and their business strategy . . . [has] made for some really unethical business-over-game-design choices . . . Maybe a little compromise needs to be made to get to big markets with big hardware, but I'm still concerned about the 'open-sourceness' spirit going forward." These sorts of gone-corporate concerns were enhanced when, in an introducing-our-new-VP interview with *GIBiz*, Malamed talked about competing with console-makers and suggested that the consumer version of Rift might cost more than originally thought.[5]

As someone who typically operated behind the scenes, Malamed wasn't used to being in the crosshairs of internet drama. Which is why it meant so much to him when, unprompted, Luckey addressed these concerns head-on.

"Laird is a fantastic guy," Luckey wrote on MTBS3D, "and I would not have hired him if he were not a perfect fit for Oculus . . . he has a skill that, frankly, we lacked as a company: getting a consumer product into the hands of as many people as possible. If this guy were in it for the money, he had plenty of options that don't involve a scrappy startup! We would not have hired him if we thought his vision was different than ours, and his position as COO still reports me to me at the end of the day. If I don't want to go in a particular direction, then we don't. We want to make a great headset that gamers can afford, not a luxury item. We know what happens when you launch a consumer product at $599, no need to repeat history."

As the new COO, Malamed wanted to prove his usefulness right away, and he figured that the best place to begin was by taking all the finance stuff off Iribe's plate. To ask and answer questions like: What do we have in the bank? What bills do we have to pay? And ultimately, how are we doing?

"This is a logistical nightmare," Malamed told Iribe, by the end of his first week.

"Is there anything, specifically, that you're referring to?" Iribe asked.

"One of our Kickstarter backers lives in Angola," he said. "I have no idea how we are going to get a unit to this person. And that's just one country. We have Kickstarters in 109 more! I started talking to the freight-forwarding company that Nate had been in touch with and they basically said, 'Oh, we're going to bring all the units from China to San Francisco, then we'll send them by small parcel out to the rest of the world.' I mean . . . that's crazy! Then you throw in the fact that 35 percent of our sales are outside the US, and also—correct me if I'm wrong—but our Kickstarter doesn't say we're going to be charging anyone extra shipping anywhere."

"No, you're right," Iribe replied. "We've just been so focused on designing this thing, and trying to get it made on time, that we just . . . you know."

"Sure," Malamed said. "Switching gears, what's the latest on Zeni-Max? I remember you said you were still negotiating with them, but is anything drawn up already?"

"Not yet," Iribe said, shaking his head. "Hang on a sec, I'll forward you the most recent thing."

The most recent thing was unfortunately not all that recent. It was from December 10, when Iribe sent over the company's latest proposal. In it, he offered ZeniMax a 15 percent ownership stake in exchange for a $6 million investment. If that was more than they were comfortable investing, which Iribe knew was a possibility, ZeniMax could still purchase an equity stake at that same rate ($400,000 per 1 percent). And as long as ZeniMax purchased more than a 10 percent stake, they'd receive a seat on Oculus's board of directors.

In addition to that, and regardless of whatever amount ZeniMax invested—or even if they didn't invest at all—the proposal also offered a 2 percent stake in exchange for John Carmack serving as a technical adviser to Oculus.

"That's the most recent?" Malamed asked. "They haven't countered yet?"

"They haven't even responded yet," Iribe replied.

Malamed nearly did a double take. Oculus *needed* to finalize this deal. Soon. The relationship with Carmack was a pivotal asset. But Malamed tabled the issue for now, moving on to the issue of cash flow.

At this point in time, the entirety of Oculus's funds came from either the seed investment that Iribe and Antonov had made ($2.5 million), or the money raised on Kickstarter ($2,437,429 minus about 10 percent in fees). Since the hardware was being made at cost, all the Kickstarter money (and a portion of the seed) was being used to pay for the initial batch of ten thousand headsets. This left a little under $2 million, which may sound respectable for a start-up (where the payrolls are generally lean and low due to the true incentive being equity). But even so, the cost of operations quickly added up—especially for a hardware start-up—in which rent, travel, and insurance were compounded by the likes of testing, tooling, prototyping, and paying for wildcard items (like the $30,000 robot that Steve LaValle requested). Plus, of course, there *were* still software expenses, not only related to building the SDK, integrations, and plug-ins, and so on, but content development as well. From the get-go, Oculus knew that content was king. And while the whole point of shipping DK1 was to empower and entice outside developers, there was still a vital need for in-house content creation. In a few years, the guys at Oculus hoped they would be in a position to fund numerous projects—not just internal ones, but external ones too, funding them from scratch, or postdemo, or during that critical home stretch. That was the true goal, the hopefully-one-day dream. But for now all Oculus could afford to do was contract Paul Bettner's small studio. And as Malamed was explaining to Iribe, they couldn't even really do that. Not at the rate that Bettner and Oculus eventually worked out: $100,000/month.

"That's a sizable expense," Malamed commented. "It's one of many sizable expenses. And we're a small start-up."

Iribe nodded, appreciative. As much as all this sucked, he needed to hear it.

"I mean," Malamed continued, "I did a quick burndown of our

cash chart. And by my rough calculations—and keep in mind I put this together quickly, and obviously don't yet have all the necessary information—but roughly speaking: even if we continue to sell devkits online, and even if we can avoid bringing on any new hires, we're going to be completely tapped out by midsummer. I'd say we've got six months, tops."

"Six months?" Iribe asked.

"Something like that," Malamed replied.

"Okay . . ."

"Yeah . . ."

"Well," Iribe said, almost chuckling, "that's a little tighter than I was expecting. But not totally surprising. I guess we'll just need have to be more aggressive with our Series A."

"Exactly," Malamed replied. "I'm looking forward to that. And forgive me for stating the obvious here, but given the timing, a good showing at CES would really go a *long* way."

DAYS LATER, ON JANUARY 6, THE WORDS *LAS VEGAS* SPARKLED IN NEON PINK above the entrance to Central Hall—the crown jewel of CES, where the latest in audio, video, and entertainment were all on display—as a seemingly endless flood of showgoers elbowed their way toward a panoply of soon-to-be-released smartphones, computer processors, and glorious 4K televisions.

New and old, big and small, hundreds of companies demoed upcoming products from booths on the trade show floor. Oculus, however, was nowhere to be found. Because all those booths had been rented many months before. As were all the conference rooms on hand, as well as at nearby hotels. In fact, miles away from the hustle and bustle, all Oculus could get was a suite on the thirty-sixth floor of the Venetian.

Although this location would make it difficult to lure media coverage, the Oculus guys on hand—Luckey, Iribe, Mitchell, Patel, and Chen (plus their PR duo: Jim Redner and Eric Schumacher)—scrambled to make their suite look as professional as possible. All things considered, the guys had done a pretty good job—hanging a few banners, scattering some swag, and setting up a photo-friendly demo station beneath a chandelier in the center of the room. The room itself was opulent—

with a piano, antique chairs, and so much gold trim that King Midas jokes felt obligatory—but compared to a booth at the Convention Center, or even just a typical hotel conference room, Oculus's ostentatious setup felt like an odd place to try and launch a revolution. "That is: *if* we can actually get *The Verge* to make the trek."

"Yes," Mitchell said, laughing. "If we can somehow get *The Verge* to the Liberace Suite—and *especially* if we can get [Editor in Chief] Josh Topolsky to stop by—then we absolutely need to bring our A game."

"But, Nate," Luckey playfully objected, "whatever do you mean? Here at Oculus we always bring our A game. We don't even have a B game. Who do you think we are: Ouya?"

"Hey!" Iribe chided half jokingly. "None of that, no negativity! We are shiny, happy people and we just want to share the Oculaid."

"But of course!" Luckey cheered in response. "Oculaid for all!"

Over the next few hours, Redner and Schumacher were able to get some reporters to trek to the Venetian. Including a small crew from *The Verge*, led by journalist Nathan Ingraham.

"Thanks so much for making it out here!" Mitchell chirped, ushering them inside.

As the crew set up to film a "hands-on" video that would serve as a visual complement to Ingraham's article, Mitchell casually asked if *The Verge* cofounder and editor in chief Josh Topolsky might be joining them.

Unfortunately, Topolsky was back at *The Verge*'s trailer, where he'd likely be holed up most of the week, overseeing the multitude of podcasts, video streams, and articles that they'd be running throughout the week.

For Mitchell, not getting to demo the Rift for *The Verge*'s key tastemaker was a bummer, but hardly any reason for disappointment. Most important, Oculus had managed to drag a high-profile outlet out to the Liberace Suite; and if this went well, then perhaps word would spread.

The resulting article—published later that day—wasn't especially glowing.[6] But there were enough pluckable quotes in there—like "the immersion trumps all" and "the story behind Oculus is almost as interesting as the Rift itself"—that Redner and Schumacher could dangle to other outlets and possibly parlay into additional coverage. And while the PR guys focused on that, Luckey and Chen carved out time to walk the show floor. With thousands of vendors and booths as far as the eye could see, it was hard not to feel a little bitter about Oculus being so far away from the melee.

"If we were down here," Luckey said, "we'd win every Best in Show. Can you imagine what our line would be like? I bet we'd cause a fire hazard!"

"Dude," Chen said. "Did you just take pride in causing a theoretical fire hazard?"

Luckey laughed. "You know what I mean."

"Yeah, I also think that if we were down here we'd crush all these jokers. But whatever coverage we could get by being on the show floor here, we could probably still at GDC. And we'll actually have a finished devkit to show there. So, yeah, getting some love from WIRED or IGN would be make us feel good. But tangibly speaking: What's the gain?"

"Brendan says that VCs love that stuff," Luckey responded.

"He's probably right," Chen said. "But since he's the one who has to deal with VCs, that's what we call a Brendan Problem! Have a little faith in Eric and Jim. The fact that they were able to line up *any* interviews on short notice is impressive."

Luckey's attention shifted. "Oooooh!" he said, looking at his phone. "HipHopGamer is here!"

HipHopGamer (real name: Gerard Williams) was a passionate, do-rag-wearing YouTube personality whom Luckey had clicked with during an interview at PAX Prime. He loved HipHopGamer's carnival-barker-like enthusiasm (declaring the Rift "the best device ever created—ever created!—in the video game industry") and his eccentric style (i.e., he traveled around with a WWF-sized golden championship belt).

"Tell him to visit us later at the Liberace Suite," Chen suggested. "Or tell him to meet us at the Steam Box. We should probably check that thing out."

The talk of the show thus far (at least among gamers) was the first iteration of Valve's long-rumored "Steam Box." Those rumors started in March 2012 when—you guessed it: *The Verge*'s Josh Topolsky—

reported that Valve had been "secretly working on gaming hardware for the living room."[7] Dubbed the "Steam Box," this gaming machine would be powered by Valve's still-in-development, Unix-like operating system (SteamOS) and mark Valve's first foray into the hardware business. Unlike traditional hardware makers, however, Valve would actively be seeking third-party manufacturers to partner with. This was because, at the end of the day, Valve's interest in hardware had almost nothing to do with hardware at all. Their hardware play, really, was about protecting their software. Specifically it was about protecting Steam.

Steam was Valve's online store, the place PC gamers logged into each day to play their favorite games. With an install base of over 125 million users and an annual revenue of over $3 billion, Steam had a near monopoly on the PC gaming market. And like all wise companies' market leaders, Valve was fixated on protecting that sacred turf.

The biggest threat to that turf were other companies' operating systems since—as an application—Steam was largely beholden to the whims of that OS, such as that of Microsoft's, which was the most used by PC gamers. Thus far, Microsoft's updates to Windows hadn't managed to slow down Steam's spigot, but this possibility was always a concern.

To avoid being at the mercy of Microsoft and other OS makers, Valve decided to create an OS of their own (SteamOS). The problem was that most people were generally content with the current operating system—at least content enough to avoid the hassle of finding and installing a new one—and there was little incentive for anyone to migrate to SteamOS. Instead, Valve felt their best bet was to create hardware that came with their OS preloaded . . . the Steam Box.

The first company to help Valve build their box was a Utah-based computer maker named Xi3. And earlier in the day, Xi3 unveiled that fruition publicly for the very first time: a modular computer, based on their X7A model that was tentatively dubbed "Piston." Naturally, Palmer Luckey and Joe Chen were dying to get a look at this thing.

"What can you tell us about Piston?" Luckey had asked Xi3's chief marketing officer, David Politis, as he welcomed the guys into his booth.

"So Piston is a development stage product," Politis said proudly, "that has been optimized for gameplay within Steam in Big Picture Mode. You guys are familiar with Steam, right?"

"We are," Chen replied.

"Okay, great. So I don't need to go through that whole rigmarole. Instead, let me tell you about how this computer packs a punch." Politis picked up a nearby Piston unit and handed to Chen. "Small, isn't it? No bigger than a grapefruit! But don't let that fool you . . ."

"Don't worry," Luckey joked. "We'd never let that happen."

Shortly after visiting Xi3, Luckey checked his phone and found an in-box overflowing with media requests: outlets hoping for a demo, or an interview, and now willing to make the trip over to the Venetian. Among those queries was another request from *The Verge*, this one for Palmer Luckey to appear on *Top Shelf*, their popular, live-streaming show that highlighted the best of what they'd seen at CES. This, Luckey thought, must have been why Mitchell had made such a big deal out of "crushing" that demo with *The Verge*.

ALTHOUGH THE SIZE OF *TOP SHELF'S* AUDIENCE WAS MODEST (ABOUT A HUN-dred thousand) and this episode was being filmed in a small, makeshift studio (since *The Verge* was on location), the interview was still an important one for Luckey; after all, Iribe, Mitchell, and Chen had persuaded Luckey that, for the love of God, he needed to wear a blazer for this one.

Watching Luckey explain his vision from backstage, Mitchell and Chen glanced at each other, as if to celebrate with a telepathic high five. They were pleased by how much more polished Luckey had become in the four months since Oculus's Kickstarter; he seemed to finally be finding his voice.

At first, Mitchell and Chen tried to class him up—at least to the extent that they could get him to dress like a normal human (shoes, no Hawaiian shirts, willing to wear the occasional blazer). They also tried to temper his energy and enthusiasm and the velocity with which he thought and spoke, which came off as impulsive. Luckey trusted Mitchell and Chen and was willing to attempt something more re-fined. But dialed-back Luckey was much worse than amped up be-cause, despite the risks and occasional rawness, he lacked his usual and necessary spark to light this revolution. They needed Palmer Luckey: unleashed . . . but with a few exceptions.

They needed him to be conscious about how long he talked; not *how* he talked (he had carte blanche, sound bites be damned) but *how long* he went for. And most crucially, they needed Luckey to keep things positive—not just when talking about Oculus, but when it came to partners, competitors, and people online. Palmer Luckey was a child of the internet, which implied that the congruity—between real life and online—was, to him, the epitome of authenticity. It was the authenticity Oculus needed, but as a face of the company he needed to be careful with what Mitchell described as "that flame-war mentality." Luckey understood this, and agreed it was for the best. He would stay positive, stay engaged, and focus on spreading his love for this technology he loved so much.

"One cool thing about VR," Luckey said in the interview, "is you actually feel like you're inside a space. You actually feel like you're inside the space and that's huge for immersion . . . and just being there might be powerful enough on its own."

"There are nerds across America who want nothing more than to just be lost inside the *Marathon* spaceship," joked cohost Nilay Patel, referring to the spaceship from Bungie's popular sci-fi shooter *Marathon*. "I want to put it on my face really bad!"

Luckey handed Patel the headset for Patel to put on.

"OH GOD" Patel blurted as soon as it was on his head.

"You doing pretty good?" Luckey asked.

"OH GOD! OH GOD I'M IN A SPACESHIP!"

As Luckey explained what was going on inside the headset, he accidentally kicked the headset's cable and, suddenly, the screen went black. "Uh-oh." For a few seconds, Luckey's mind was blitzed with thoughts of how embarrassing this must have looked. But as he held his breath, the image miraculously returned. Luckey raised his hands in celebration and sighed with tremendous relief. "Phew," he said, putting his hand on his chest. "I swear I had a heart attack."

There he is, Chen thought. That's our guy! That authenticity—to be vulnerable, to be sarcastic, to be a kid on Christmas morning—that was Palmer Luckey.

"This is the best day of my life," Patel raved. "This is ridiculous. Like, I don't like 3-D things—I usually get a headache and hate myself—but this is legitimately awesome 3-D."

Upping the ante, Luckey toggled the control and transported Patel to a beach. "So now," Luckey exclaimed, "he's in a completely entirely different place in just a few seconds."

As Patel got the lay of the land, he started leaning his body forward and back to see if this movement was tracked (like his head). "So I can't, like, peer at things?"

"No," Luckey answered. "So right now all you have is rotational tracking. Our tracker's currently an accelerometer, a gyrometer, and a magnetometer. So . . . if you roll your head, we have it on a neck model . . . but if I lean forward and lean backward, you don't get anything. That's something that we're working on for the consumer version. It's basically a must-have kind of feature to keep players from feeling disoriented, to connect that feel of immersion."

Shortly thereafter, Patel removed the headset and Pierce asked him how he felt. "I will say," he explained, "that taking that off and coming back to this space actually felt like *returning* to this space. As opposed to, like, not looking at a screen anymore."

"That's what it's all about, being inside the game."

Boom! Chen and Mitchell high-fived each other as the segment came to an end. It had gone so well that Mitchell was able to parlay this success into a visit with *The Verge*'s editorial staff at the trailer where they were working. And there, finally, he was able to demo the Rift for Josh Topolsky.

"CAN WE TALK ABOUT THE OCULUS RIFT?" TOPOLSKY ASKED A FEW HOURS later on *The Vergecast*, his outlet's flagship show. "It's a virtual reality headset—it's VR goggles—and it is the exact actual promise of VR that I grew up on hoping and praying and wishing would happen. I put them on tonight and I had the most incredible, inspiring . . . it blew me away on almost every level that I could be blown away on."

To add some color for the viewers at home, Topolsky's colleague Chris Grant chimed in: "So Josh and I," Grant began, "at times in the past have partaken in hallucinogenic substances. And I'm just gonna straight-up-on-ya-here: when you take it off, it is a little bit like when you come down from an acid high."

"Yes," Topolsky agreed. "It's like acid. It is like the experience of being out of your head in a way . . . I'm not paid to endorse this product, I've just never been more excited."

Between Luckey's appearance on *Top Shelf* and Josh Topolsky's unshakable enthusiasm for the Rift—"I used this today," he tweeted, "and it seriously changed my life"—Oculus became the unforeseen darling of the Consumer Electronics Show. Every journalist wanted to try the Rift, to experience the device that supposedly delivered the promise of VR they remembered from their childhoods. By the third day of CES, Oculus's schedule was so packed that the team had to start double-booking visits to the Liberace Suite, kicking off a carousel of demos and interviews (with either Luckey, Iribe, or Mitchell) that blurred together over the next seventy-two hours. . . .

TOM'S HARDWARE
Holy Shit!
MACHINIMA
(turning head)
The first thing you notice is that you can look at everything.
WIRED
You feel as if you are within another world in a way that's never been possible before—at least not in a consumer product.
IGN
It's immersion on an entirely different level.
CNET
As someone who has always been interested in the idea of VR headsets, and always heartbreakingly let down by the reality of VR headsets . . . I figured it would just be more of the same. [But] I was wrong. Not sure I've ever been more wrong.
TECHCRUNCH
So it seems like something that you see in science fiction

movies and things like that—as an idea that's come around—but you're saying that no one had really figured out a way to do it affordably yet?

PALMER LUCKEY

I don't know if people hadn't figured it out. There were people who were going along the same lines [as me] . . . but they weren't trying to make consumer virtual reality.

HAK5

This has been the dream for quite a while now, where did you guys come from and how did you get into this space?

NATE MITCHELL

The company was actually founded by Palmer Luckey . . . he actually designed and invented the Rift in his parents' garage over the course of two and a half years. He was always superpassionate about head-mounted displays and virtual reality. And he wanted something that actually allowed him to jack into the matrix for video games.

PALMER LUCKEY

The biggest change is that we've developed our own motion tracker sensor chips . . . [which] gives us better data, more samples to work with when we're doing our sensor fusion, so we can get better tracking overall; and most importantly, because it's running at 1000 Hz (instead of 250 Hz) and it's four times faster, we can actually have less latency. Less time between when you make a motion and when it shows up on-screen.

TESTED

So right now you have, with head tracking, roll . . . but you don't have depth yet. Is that something you guys are looking forward to doing?

NATE MITCHELL

Absolutely. We're definitely interested in adding positional tracking; both for the consumer version and potentially as an add-on for the developer kit.

POPULAR MECHANICS

When will there be a consumer product for gamers to buy?

PALMER LUCKEY

It would be irresponsible for me to say when we'll have consumer products. Because I have no idea what feedback we'll get from developers. And if they say the device needs some new functionality, we're not going to release a product until it has that functionality!

Hour after hour, these interviews were watched by gamers around the world, piquing the interest of thousands, then hundreds of thousands, then even millions worldwide. But even with all this fanfare, the gaming community's interest in Oculus still had nothing on their love for Valve, which is why those interviews with Luckey and Mitchell couldn't hold a candle to an unexpected one that had been granted by Valve's very own Willy Wonka: Gabe Newell.

Newell rarely spoke to the press—because, as described by Valve engineer Nat Brown, "[We] prefer to speak to customers with products or directly when we have something specific to say"—but, from Valve's private booth on the CES show floor, an exception was made to speak with *The Verge* about Steam, their plans for the Steam Box, and even the future of gaming.

THE VERGE

So you're working on your own Steam Box hardware. Why work with so many partners when you have your own ideal design in mind?

GABE NEWELL

What we see is you've got this sort of struggle going on between closed proprietary systems and open systems. We think that there are pluses and minuses to open systems that could make things a little messier, it's much more like herding cats, so we try to take the pieces where we're going to add the best value and then encourage other people to do it. So it tends to mean that a lot of people get involved. We're not imposing a lot of restrictions on people on how they're getting involved.

THE VERGE

We've heard lots of rumors about the Steam Box, including

that Valve's own hardware would be "tightly controlled." Can you tell us more about Valve's own hardware effort?

GABE NEWELL

The way we sort of think of it is sort of Good, Better, or Best. So, "Good" are like these very low-cost streaming solutions that you're going to see that are using Miracast or Grid. I think we're talking about in-home solutions where you've got low latency. "Better" is to have a dedicated CPU and GPU and that's the one that's going to be controlled. Not because our goal is to control it; it's been surprisingly difficult when we say to people "don't put an optical media drive in there" and they put an optical media drive in there and you're like "that makes it hotter, that makes it more expensive, and it makes the box bigger." Go ahead. You can always sell the Best box, and those are just whatever those guys want to manufacture.

"Wait," Luckey said, as Chen read aloud some interview excerpts before they fell asleep. "What did Gabe say were the characteristics of 'Best'? Did I miss something?"

Chen quickly scanned through what he had read. "Uh, nope. Don't think so."

"That Gabe is a wily gangster," Luckey said with admiration. "What else did he say?"

"Let's see," Chen responded. "Less laggy . . . some stuff about Apple TV . . . Oh! Here's a good one. They ask if he really thinks that Valve will 'disrupt the home entertainment space and compete with Microsoft and Sony'? And he says, 'The internet is super smart. If you do something that is cool, that's actually worth people's time, then they'll adopt it. If you do something that's not cool and sucks, you can spend as many marketing dollars as you want, [they] just won't.'"

"Did he say anything about us?" Luckey asked.

Newell had not mentioned Oculus during the interview, but behind the scenes virtual reality was indeed on Valve's radar. The Oculus guys knew this because they had remained in contact with Michael Abrash and the dozen or Valve employees investigating AR/VR since meeting with them days before the Kickstarter.

There was, however, one potentially significant development. It had started in October when a couple of guys from that AR/VR Team—Tom Forsyth and Joe Ludwig—said to themselves: You know what? It looks like this Oculus thing is really gonna ship! We should try and make a game for virtual reality! Not a toy, not a demo, but a completely immersive, fully fleshed out game.

Given this objective (and given that it often takes years to make a "fully fleshed out game"), Forsyth and Ludwig weren't thinking about creating something from scratch. Instead, what they wanted to do was port over an existing Valve game to work in VR. Not only would this be a cool thing to do, but it would be a great learning experience in what works, and what does not, in current VR.

So, doing exactly what's described in Valve's famous *Handbook for New Employees*—the one that explains why the desks at Valve have wheels—they rolled their workstations over to the group that worked on *Team Fortress 2*. The reason they selected *TF2* and not, say, a game from the more popular *Half-Life* or *Portal* franchises was because they wanted to port a first-person shooter (believing this genre would translate nicely into VR) and because it had an active developer base at Valve (consisting of about twenty employees who worked to ship out updates every two weeks).

Over the next few months, Forsyth and Ludwig tried to figure out dozens of things. And while the challenges were plentiful, they generally fell into one of five areas:

- **LATENCY:** Figuring out how to minimize the lag between when a player enters a command and that command is executed in the virtual world
- **STEREOSCOPIC RENDERING:** Figuring out what, literally, needs to be drawn on-screen to create a 360-degree image that works in VR
- **USER INTERFACE:** Figuring out where, and how, to best depict game menus (i.e., welcome, pause), status bars (i.e., maps, strength), and assistance features (i.e., aiming crosshairs)
- **INPUT:** Figuring out how to implement head tracking in a

way that feels both natural (to humans) and familiar (with similar FPS games)

- **VR MOTION SICKNESS:** Figuring out how to reduce movements and game cues that induced nausea, headaches, and cold sweats

This last challenge accounted for many of the biggest surprises that Forsyth and Ludwig encountered. For example: decapitation. There are times in *TF2* when an opponent will defeat you by chopping off your head. On an ordinary game screen, this isn't a problem; you (standing behind the screen) just watch your avatar's head get lopped. But when you're *in* the game (and your POV is the head of your avatar), this presents a challenge. Because the default would be for your view to stay as your virtual head, but the resulting mismatch—between your virtual head (now rolling on the ground) and your actual head (upright, wearing a VR headset) will make you sick.

Another mismatch had to do with the weapons. In the normal, non-VR version of *TF2*, your weapons—when in use—ended at your elbows. Which made sense with the typical FPS point of view since your head, without tracking, could only swivel in a few preordained directions. But in VR, with that level of immersion, it was jarring to use a weapon without seeing your upper arm and shoulder. Fortunately, in this instance, there was already some semblance of a fix available; Forsyth and Ludwig could borrow from the third-person perspective—containing the upper arm and shoulder that other players would see when looking at you—and, from there, they could build out a solution that worked in virtual reality.

Even with that fix, there were no shortage of questions that the guys at Valve faced. Why do some players feel sick when their avatar goes up and down stairs? What should we do about the "Scout," a game character who can run twenty-plus miles per hour, and whose inhuman speed can sometimes be unsettling? And, since everybody's perceptual system is slightly different, how do we come up with some sort of a standardized approach?

"Do you think they'll have a version of *TF2* ready by GDC?" Chen asked. "Not something perfect, obviously, but just something for the launch?"

"I honestly don't know," Luckey replied. "But I hope so, because ZeniMax is dragging their feet [with *Doom 3: BFG Edition*], so we can really use it."

Hoping to end the night on a more uplifting note, Chen reflected a bit on how far Oculus had come in such a short amount of time. "I mean, I figured we'd grow fast but this is pretty nuts."

Luckey wasn't one to wax nostalgic, but the whirlwind of positive press had nudged him toward reflection. "It *is* nuts," Luckey agreed. "When I was younger, my grandpa used to drop off a huge bag every week, completely full of magazines. They weren't *all* tech magazines, but he's the one who introduced me to *WIRED* and *Popular Mechanics* and all that other stuff. My grandpa would say that reading that stuff was a 'small investment to be an informed man.' I didn't read them all every week, but a lot of them I did read through."

"That's awesome," Chen said.

"Yeah. But it's also kind of sad."

"Sad? Why sad?"

"Well," Luckey said, now sounding a little frustrated. "Look at all the stuff that's been written about us this week. There are so many errors, factual mistakes."

"Dude. You can't seriously be complaining about the press we've been getting."

"Don't get me wrong: it's awesome. I love that the press loves us. But I just never truly realized how many things they get wrong, until they started writing about me and Oculus."

"You're such a stickler!" Chen said, playfully rolling his eyes. Chen hadn't really noticed that trait of Luckey's until this week. The kid would read a review of the Rift and care more about whether they got the specs right, or quoted him accurately, than the actual review about him and the Rift. Chen wondered where this came from and guessed that it was either a vestige of Luckey having previously wanting to be a tech writer, or because—being so obsessive about the details—Luckey always expected the same from others.

"I know that sounds petty," Luckey said, "but it adds up. And what happens when they mess up the not-so-small stuff?"

"They wouldn't do that."

Luckey shot Chen a quizzical look. "Anyway," he continued, "it's sad because either journalists have gotten worse at their jobs, or these articles have always been filled with little mistakes and I just never knew it."

"Whatever," Chen replied. "Point is it's been a hell of a ride so far. I mean, a year ago Oculus didn't even exist, and I was doing army stuff. But hey: that sleep deprivation training is totally paying off!"

"One year ago, I was here, at CES. Did I ever tell you about my meeting with Sensics?"

"Nah, I don't think so."

"They were giving demos of their newest headset—the one that costs seventeen grand—and I applied for a demo. They confirmed it and said they'd do it. I showed up, but someone else was in my slot. They said: don't worry, come back in fifteen minutes." So I came back exactly fifteen minutes later and their meeting room was empty and they had gone off to get lunch. Then I managed to harass them the next day into giving me a demo. But I felt pretty sour about it."

"Wait, how was their demo?"

"Oh," Luckey said, smiling. "It sucked. I actually told them they should make a headset that returns to their roots and focus on VR, not AR. I actually talked about various optical designs. And they were like, 'Oh no, that's what consumers want. They want glasses. Nobody will take it seriously if it's a helmet.' Then we parted ways . . ."

"THIS IS MY FRIEND PALMER LUCKEY," SAID JOE CHEN THE NEXT MORNING, speaking to a hotel employee in the lobby of the Venetian, "and he is, uh, special. You know . . . he's 'on the spectrum.'"

This was not true, but after realizing they were locked out of the suite (and that the suite was registered under Mitchell's name), this was the ploy that Chen and Luckey had chosen to try and get another key.

Drawing on his days as a childhood actor, Luckey did his best to channel Dustin Hoffman in *Rain Man*. He opened his mouth to introduce himself, but quickly lost the courage to speak; instead, with a flustered tic, he sort of hid himself behind Chen.

"So . . . could you help us out?" Chen asked, nearly batting his eyelashes.

The hotel employee glanced again at Luckey. She appeared skeptical about the situation—but couldn't deny that Luckey did look rather troubled. "Sure."

Play it cool. Stay in character. Don't break . . . don't break . . . until . . . BING! The elevator doors opened on the thirty-sixth floor and Luckey and Chen exited triumphantly. "I told you it'd work," Luckey said. "I told you!"

"Well done!" Chen replied. "You were quite the thespian."

"Did I ever show you *BB Rex and the Cowpunk Gang*?" Luckey asked.

"No. What's that?"

"It's a student film I was in, when I was twelve years old. I did a lot of student films, but that is by far my favorite. Remind me when we get back to Irvine, I'll send you a link."

After arriving at their suite, the guys scrambled to get it ready for a day full of demos. And after conducting about a dozen or so, the guys were greeted by somebody from *The Verge*, who came to present Oculus with an award for their VR demo. Not just any award, but the crème de la crème: Best in Show. Meaning that out of every company at CES—many with thousands upon thousands of employees—this little start-up, with no more than a dozen employees, would be taking home the ultimate honor. It wasn't the only honor that Oculus won. During the final days of CES, the Rift took home "Best Prototype" (from *IGN*), "Best Gaming Gear" (*PC Magazine*), and "The Coolest Thing We've Ever Put on Our Face" (in *WIRED*'s annual "Best of CES" roundup).

As the awards piled up, Chen couldn't help but wonder if these accolades would go to Luckey's head. After all, Chen reasoned, a compliment from a girl would have probably been enough to blow up *his* ego when he was only twenty years old. And while the final afternoon of CES was way too soon to determine if (and how) Luckey might change,

Chen found himself encouraged by their final meeting of the week. A demo for the one and only HipHopGamer.

"Yo, Palmer, my boy!" HipHopGamer said, entering the Liberace Suite.

"HipHopGamer!" Luckey shouted, greeting his old pal. "You made it!"

Since HipHopGamer had already interviewed Luckey before, he took the opportunity to chat with Mitchell. "It's your boy here, Hip-HopGamer. I'm with Nate. It's the Oculus Rift. Shut up, listen . . ."

Looking on, Chen's heart pumped with pride. He and his boys had done real good this week and they were only going to get better. Competition would inevitably arrive, of course, but Chen believed that no one—not even Sony (if that rumor was true)—no one could match the magic that Oculus had; because they were the ones who were going to do it first, do it best, and do it with the most passion. They were the true believers, the guys who cared the most; and they were also the only ones who had a Palmer Luckey.

Say what you will about the kid—and people were starting to say things: He's a genius! He's a visionary! Nah, he was just in the right place at the right time!—say whatever you want, but Chen knew this kid was special. Not special in the way that had led them back into this suite, but special in the way that would take Oculus to the promised land. And, fingers crossed, it wouldn't go to his head. At least that's what Chen thought after HipHopGamer made a final request.

"Your shirt," HipHopGamer said, pointing to Luckey's T-shirt. It was gray and said I LOVE VR (except the "I" was actually Oculus's logo: an eye). "Where can I find one?"

"You'll get one in a couple months!" Luckey said. "We're having them made for all our Kickstarter backers."

"Come on, man!" HipHopGamer plead. "I can't wait. Can I buy that one off you?"

"It's the only shirt I brought with me," Luckey confided (meaning he'd been wearing the same shirt all week).

"Please, man! You know I love Oculus."

Luckey looked to Chen for advice and received a shrug. This was their last demo of the week, so sure, why not? "But I'm not going to make you pay," Luckey said, removing his shirt. "You can just have it. I want you to have it."

"Thank you! You're such a good dude. You literally gave me the shirt off your back."

Chen and Mitchell laughed, and they continued to feel good about Luckey's generosity until HipHopGamer left and they discovered a small problem: there didn't appear to be anything in the suite for him to throw on. Which meant that the founder of this newly minted Best in Show company would need to wander through the hotel topless in order to get back to the room where he and Chen were staying. And, well, that did not seem like a good look.

"You can wear my shirt," Chen volunteered.

"Then you'll be in the same position as me!" Luckey replied.

"No," Chen clarified. "I'll keep the blazer and wear that. You can have the T-shirt."

Okay. This seemed like a reasonable solution. Except that Chen's shirt was smaller than what Luckey normally wore so it was pretty snug.

"You look ridiculous!" Mitchell said, cracking up.

"You look like you just robbed a Baby Gap," Chen added.

Whatever. Good enough. Final day of CES, no one will notice. So Luckey—in flip-flops, cargo shorts, and his Baby Gap shirt—along with Chen—in jeans, no shirt, and a blazer—exited the suite and slithered through the hotel. Fortunately, they managed to avoid running into anyone they knew. Except, for a split second, they thought they saw the hotel employee from earlier. The one they had coaxed into giving them the keycard. And even though, on second thought, they determined it probably wasn't her, they couldn't help but imagine what would have been going through her mind: Yup, yes-indeedee, that's one special kid right there!

CHAPTER 19
PROVERBS 29, VERSE 18

January 2013

COMING OUT OF CES, OCULUS WAS RIDING HIGH. AND JOHN CARMACK, WHO HAD believed in Oculus from the start, wanted his company to be riding by their side. But as negotiations between Iribe and Vlatko at Zeni-Max frayed, Carmack wanted clarity from his supervisors on how they would like him to proceed. So he contacted the chief decision makers at ZeniMax, asking for some direction on the VR work.

As Carmack awaited a reply, he could not help but note the parallels between this current situation and how ZeniMax had approached mobile gaming. Back in 2009, right after ZeniMax's acquisition of id, Carmack had been given leeway to explore the nascent mobile space. At the time, he believed that mobile gaming—then just a small market—was about to get a whole lot bigger; so he set out to explore innovative techniques (that could optimize games for the mobile space) and experiment with alternative pricing strategies (like giving away games for free and then generating revenue through ads or microtransactions). Initially, ZeniMax permitted Carmack to do this. At times, they even encouraged his exploration. But when no imminent windfall followed this pursuit, Carmack was instructed to return to doing what ZeniMax did best: producing AAA titles for consoles and PC.

There was nothing wrong with this; ZeniMax was running a business after all. But Carmack believed that if they had just stuck with mobile, they would have wound up with not only more than just a windfall but a reliable new revenue stream.

In other words: Carmack believed that ZeniMax had made a mistake with mobile—they had demonstrated a lack of vision—and he hoped that ZeniMax wasn't about to make that same mistake with virtual reality.

IN IRVINE, LUCKEY AND DYCUS FLASHED HIGH-INTENSITY SHOCKS THROUGH the backlights of an exposed LCD screen. Some shined, some pulsed, and a few just plain fried. This was but one of many tests that the two of them had been conducting as they searched for ways to reduce motion blur in the DK2.

Motion blur, just like it sounds, is that blurry, shooting-star-like appearance of an object in motion. Sometimes, and with great success, motion blur is intentionally used as an aesthetic choice to help convey speed, velocity, or acceleration. This technique works especially well in cartoons (meep meep!) and can be a nice stylistic touch in live-action films and video games. But that type of motion blur was different from the one that Oculus was trying to solve; in fact, it was almost the complete opposite. Because the problematic blur that plagued those who wore the Rift was not specific to an object in motion (like the Road Runner or Wile E. Coyote) but rather was the result of turning your head and shifting your view of the virtual world. So, in a way, it was like *you* were the object. You were the thing that was moving. And just because you were moving your head to the right or left, it didn't make sense that fixed objects—a dog, a door, whatever—should blur in any way. This didn't happen in reality and, therefore, shouldn't happen in virtual reality.

The reason this *was* happening in VR, and what made a fix so difficult, was that ultimately this was a hardware issue. Specifically, it was a screen issue, because LCD screens—the standard for just about every smartphone or tablet at the time—were particularly susceptible to motion blur. This blur wasn't a big deal when texting friends or crushing candy, but when worn against your face, serving as a surrogate for your eyes, the consequences were significant. Or, as Joe Chen once remarked, "When it's really bad, your entire world defocuses and, for a fraction of a second, it feels like the world around you is disintegrating."

There was some hope on the horizon. Over the next few years, some of the more prominent mobile manufacturers were planning to roll out next-gen, LCD-replacing screens. This next generation of display technology was called OLED (short for organic light-emitting diode) and, among other advantages, was more capable of displaying images without ghosting, smearing, or motion blur. That said, OLED was no simple fix. These screens were "more capable," but the extent of that capability was still an open question. For proof of this, look no further than the PS Vita—Sony's recently released OLED-based handheld gaming device—which performed so far below expectations that they ended up going back to an LCD screen. Nevertheless, OLED had the potential to be much better. But all that potential was almost moot because, for the foreseeable future, it would be nearly impossible for Oculus to get their hands on these cutting-edge screens. At the moment, only three companies had the means to produce them: Sony, Samsung, and LG. And with Sony appearing to harbor their own VR ambitions, that left only Samsung and LG—both of whom Dillon Seo, Oculus's man in South Korea, had met with recently; and both of whom had about zero interest in working with a company as small as Oculus.

Regardless of which display Oculus ended up using—for DK2, CV1, and beyond—this motion blur issue would still need to be addressed. Luckey scoured the internet in search of an expert—someone who cared as much about motion blur as he did about virtual reality. And though it seemed unlikely that such a person would exist, Luckey was pleased to discover that such a person did. He lived in Ontario and his name was Mark Rejhon.

Mark Rejhon was the founder and chief operator of "Blur Busters," a website devoted to eliminating motion blur in high-refresh-rate displays (specifically 120 Hz displays, which were those that displayed 120 images per second). It's difficult to pinpoint what drew Rejhon to such a niche interest, but fans of Daredevil—the blind superhero whose loss of sight helped hone his other senses—might be inclined to think that being born deaf played a factor. Another factor, one that echoed Luckey's own origin story, was encouragement from John Carmack, who, over Twitter, had complimented Rejhon on a project that attempted to reduce motion blur in LCDs by replacing the standard backlight (which was "on" all the time) with a self-made scanning design (which only flashed on for a fraction of the time). This technique—called "low persistence"—may sound counterintuitive. Why would illuminating something *less* frequently actually help produce *more* effective imagery? But as Rejhon's low-persistence experiments revealed, this technique can work; and the reason it does has to do with how our brains work.

When we go to the movie theater and watch a film, we feel like we're watching things move on that magical movie screen. But we're not—we're not really watching characters walk, cars speed away, or paper bags blow in the wind. On some level, we all know this. Because we know that above us, in a tiny room, film is being fed into a projector and that, in actuality, the "movement" we perceive is really just a sophisticated trick, the cumulative effect of watching thousands upon thousands of slightly different images being projected onto a screen. The reason this trick works is the same reason that when we thumb through a flip-book, we perceive a story told in continuous motion. Even with tangible proof in our hands that these are just static images, we interpret this transition from A to B to C as movement. Our brains are wired this way, to fill in the blanks. To the extent that even if you ripped out a few pages from that flip-book, or did away with some of the frames in that film, you likely wouldn't miss a beat. Your brain wouldn't let you; it's too good at building mental bridges between what you are (and what you are not) seeing. And this persistence of vision was the key to why Rejhon had been able to reduce motion blur in LCDs, and why Luckey was hopeful the same would be possible in Oculus's HMD.

Reading through everything that Rejhon had posted on Blur Busters, Luckey couldn't help but think: we need to get this guy to help us immediately. So he emailed Mark Rejhon, hired him to contract for Oculus and, on January 22, Luckey excitedly emailed the team to say, "Our contractor made a registry tweak that lets you enable Nvidia's Lightboost 3D mode on our BenQ display . . . to drive the pulsed backlight properly. I have it working on one of our machines, the results are pretty incredible. I am going to drag everyone over to my desk tomorrow to take a look. Motion blur is completely and wholly eliminated."

This was amazing! Granted, it *was* only on one of the BenQ low-latency game monitors, but the larger point was that the method here could work—it did work!—and surely a similar sort of low-persistence method would work when they moved on to building DK2.

But within hours of this excitement, the mood at Oculus was suddenly shaken when Luckey and Iribe received the email they'd been dreading:

FROM: John Carmack
SUBJECT: Id and Oculus

I have been officially told to not do any more work with VR at all, unless a deal is reached between Zenimax and Oculus. This includes finishing the Doom 3 BFG support, which makes everyone involved look bad.
　Sorry.
　I have enjoyed working with you, and personally wish you all the best—it is an important direction that I feel will make a major impact.

"THIS IS NOT IDEAL," PATEL STATED BETWEEN BITES OF NASI GORENG AT THE Pullman Hotel's breakfast buffet. Mitchell sighed in agreement—impressed by Patel's ability to somehow *always* hold a measured tone—while McCauley groused, noting that he'd always known Carmack was a "loose cannon."

The three of them had been in China for nearly two weeks now, overseeing the pilot run at Berway. This ZeniMax stuff was the last thing they needed given the lingering production issues. There were so many issues that it was hard to pinpoint the most distressing one. The lens cups were wobbly, the screens were improperly positioned, and the light leakage—the amount of natural light that could sneak into a headset—was out of control. These, at least, were the types of issues Mitchell and Patel had expected to face. But what they hadn't expected were things like the screen divider being too reflective (and distracting from the image), the nose wall bopping noses (on 75 percent of the testers, at least) and the foam pads coming unglued (not at first, but subtly, sneakily about two to three days after initial use).

Nevertheless, in coordination with the engineers at Berway and the team back in Irvine, the guys were able to get DK1 into decent shape. Though right now, over breakfast at the Pullman Hotel, it was a little dispiriting to think that the final version of DK1 would likely never get the chance to play *Doom 3: BFG Edition.*

BACK IN CALIFORNIA, IRIBE SPOKE WITH ANTONOV ABOUT HOW THEY SHOULD handle Carmack's Email of Doom. Then he laid out what he thought the team ought to do, sparking a thread of ideas from the rest of Oculus's exec team:

FROM: Brendan Iribe
SUBJECT: RE: Id and Oculus

If we can't strike a reasonable deal with Zenimax (which is looking likely), we should do the right thing and integrate Oculus SDK with Doom 3 BFG ourselves.
  Fans and Carmack will appreciate it.
  Mike's idea, which I totally agree with.
  It will cost us less than buying 8k copies of Doom 3 BFG;-) and earn the love . . .

>>>FROM: Nate Mitchell
SUBJECT: RE: Id and Oculus

Nirav, Jack and I came up with the same idea over lunch. Great minds think alike. We joked that the community will probably do it themselves if we don't.
  This is the beauty of open-source software (*hint, hint*).
  Anyway, let's make it happen. I think Carmack's VR-related code is in the open-source release (someone needs to double check this). Either way, it's more great testing for the SDK.
  Do we want to hire a contractor or handle it internally?

>>>**FROM:** Brendan Iribe
SUBJECT: RE: Id and Oculus

Contractor. We need a super Doom 3 coding enthusiast.

As a lifelong fan of the *Doom* franchise, Luckey could understand where Iribe and Mitchell were coming from. Like them, he hated the idea of a *Doom*-free devkit, and he especially hated that it had been him in the Kickstarter video who explicitly says that Oculus will provide "a copy of *Doom 3 BFG Edition*, the first Oculus-ready game!"

There were few things Luckey hated more than looking like a liar; but when it came to making business decisions, he felt it was critical to put aside his ego and do what was best for the VR community. Giving them a taste of *Doom* would be cool, but it wasn't worth feeding the community something undercooked.

FROM: Palmer Luckey
SUBJECT: RE: Id and Oculus

I think leaving this one to the community might be best. Carmack got that one level working mostly fine in VR, but doing the entire game is a lot of work. We would have to fix all the cutscenes (Something Carmack did not even do), make all the weapon models work well (Only some of the weapons were adjusted for VR), make sure all the effects work, rework some parts of the UI, etc. Maybe some of these things are easier than I think, I just want to be careful of taking on a project we may not be able to do well.

Another thing to keep in mind is that people are having much, much better experiences in our Unity and Unreal demos than they are in Doom. Some of that will be helped by our new tracking code, but at least part of the problem was the dark, claustrophobic environment with a gun glued to your head. On top of that, it is just not that great looking of a game compared to a modern Unreal title.

Paying for 7k copies of Doom 3, paying a contractor to make it work on a very short timeline, and requiring users to patch a copy of a game that may not even be a good experience does not make sense at first glance to me. My thoughts right now are that we should take this as a blessing in disguise and cut Doom 3 from our launch. From there, we can either release something better on our own, or give them a Steam credit like Nate suggests. We have the Unreal sample working pretty well, is there any chance that Epic would work with us to replace Doom 3 with Unreal Tournament 3?

>>>FROM: Nate Mitchell
SUBJECT: RE: Id and Oculus

Unreal Tournament 3 is very old and dated. It costs <$5 during Steam sales. It might come off like we're trying to save money. Maybe we could work out something with Epic to make it more appealing (new VR-features, new VR-level . . .)? One neat thing about Unreal Tournament 3 is that there's multiplayer.

Jumping back to the Steam Credit idea, we could offer a coupon/credit for the Oculus Store ($20 off your next Oculus purchase!). Again, I think we should still offer DOOM 3 BFG to people who demand it, but offering a coupon for future Oculus products would bring people back in and provide a way for backers to "donate" back their DOOM 3 to Oculus in an exchange that feels fair.

Brendan and I actually had discussed providing a coupon like this to all our Kickstarter backers just out of the kindness of our hearts, so this might be a great opportunity to build good will.

>>>FROM: Palmer Luckey
SUBJECT: RE: Id and Oculus

Do you mean we should offer Doom 3 with VR support to people who want it, or just stock Doom 3?

I really love the idea of an Oculus store credit. It would be a good way to jumpstart the store, and I think our backers would love it too.

Antonov agreed—about the value of jump-starting Oculus's inevitable virtual store—but he had trouble moving past what felt like a "bait and switch." He hated the idea of people paying hard-earned money and

230 THE HISTORY OF THE FUTURE

then not getting *exactly* what they had earned. For Oculus to change the world, they needed to be a company of integrity.

FROM: Michael Antonov
SUBJECT: RE: Id and Oculus

I think this discussion misses an important point: There exists an important number of people who bought the device to *SOLELY* play Doom 3 BFG. I don't know how many people that is, but i'd guess its [*sic*] at least 10%, byt [*sic*] may be 25. These people wouldn't have bought the kit at all if the game wasn't available. Offering them future product credit does nothing to address this. The only way to make these people happy is to get Doom 3 working and get them download credit. The second alternative would be provide a different game, but even that won't go completely smooth.

>>>FROM: Brendan Iribe
SUBJECT: RE: Id and Oculus

Well, that wasn't the point of our "dev kit" and it's not really our fault that Zenimax are pricks . . .

>>>FROM: Palmer Luckey
SUBJECT: RE: Id and Oculus

We want to keep those Doom fans happy, but the cost of getting Doom 3 working well could be high, and we do not even know if it would be possible to finish it before we launch. On top of that, we don't know if Zenimax would let us get away with it; they have a pretty open approach to most mods, but they may take issue with us releasing something that they have been promoting as their own . . .

Zenimax announced the Doom 3 support, we took them at their word. Speaking ill of them in public does not make sense, but it needs to be made clear to these Doom fans that the lack of support is entirely on Zenimax, and was a surprise for us as well. Otherwise, it looks like we had a choice in not keeping our end of the bargain. I think the message should be that Doom 3 support was announced by Zenimax (without any strings attached), and that they chose to not go through with it. We don't have to give any reasons for their decision, but people need to understand that this is not a case of us going back on some kind of equity-for-content deal we had with Zenimax from the start.

We may be unable to satisfy every customer, but hopefully we can disappoint them without it looking like it was our fault.

Without the umbrella of id, of *Doom*, of the ability to name-drop John Carmack—a thin cloud of fallibility now hovered over the office. But as luck would have it, before any storms could come, some unexpected news had the team suddenly feeling sunny again; sunnier, in fact, than they'd ever felt before.

"TAKE THESE," JOSH TOPOLSKY SAID, PASSING A PROTOTYPE OF THE OCULUS Rift to the beloved host of NBC's *Late Night with Jimmy Fallon* on January 31.

Fallon's eyes lit up. "This is the . . . *everyone's* talking about this thing!"

Watching from the audience, Iribe smiled.

"Imagine you're putting on a pair of ski goggles," Topolsky suggested to Fallon.

"I don't have to imagine," Fallon replied. "It feels like ski goggles."

Although that had not been a joke, Fallon's comment invariably received some lighthearted laughter from the audience. Under normal circumstances, Iribe likely would have been tempted to look around and make a split-second focus group out of those who had reacted. Did they think the similarity to ski goggles made the Rift seem silly? Or was it charming that such a futuristic product would contain elements that were so familiar? This was the kind of recon Iribe typically loved to do, but as Fallon prepared to put on the Rift, there was something else on Brendan's mind. Please *work*!

There's always a danger that some glitch will cause a tech demo to go awry. But Iribe's anxiety was more than just ordinary doubt. Because even though, by this point, Oculus had demoed the Rift for hundreds of folks at conferences around the world—PAX, GamesCom, CES, and so on—this was the first time they were demoing the Rift outside of a tech-oriented setting. This was mainstream TV.

The biggest obstacle to this portrayal was the PC—specifically the need to find a way that didn't make viewers feel like they had to buy some high-tech supercomputer to power the Rift. This was a challenge because, well, the headset *did* kind of require a computer that was more sophisticated (and expensive) than whatever most nongamers already owned. How much more expensive? How much space would it take up? Was there any way to just upgrade what they had instead of buying a brand-new PC? These were all valid questions, but they were the last thing Oculus wanted the mainstream world to be worrying about right now. For the same reason that when Apple launched the iPod, they didn't go out of their way to advertise that you also needed a computer if you actually wanted to add any music to the device. Granted, a big difference here was that most people already owned, or had access to, a computer sophisticated enough to run iTunes, but even that distinction was somewhat moot.

"So this is the Oculus Rift," Topolsky told Fallon, as the late-night host slipped the headset onto his face. And as the moment of truth approached—so much to gain, so much to lose—Iribe instinctively inched forward and . . .

WITHOUT LOOKING AWAY FROM THE COMPUTER MONITOR AT HIS APARTMENT IN Finland, Steve LaValle reached beside him and felt for the hand of his wife, Anna Yershova, to share this moment with him, to share the good and the bad as they always have. The roboticists awaited Jimmy Fallon's reaction to the Oculus Rift.

JIMMY FALLON
(blown away)
Whaaaaaaat is going on?!

The roboticists glanced at each other, beaming. There was much they wanted to say to each other, but that would have to wait until after the segment. For now their eyes eagerly returned to the monitor where, simultaneously on-screen, real-time footage of the snowy Citadel appeared so that viewers could see what Fallon was seeing.

JIMMY FALLON
Holy mackerel!
JOSH TOPOLSKY
So look; look to your left, look to your right. Alright, now look behind you. Move your body, but don't move your head . . . alright, alright, look forward.
JIMMY FALLON
Wow!
JOSH TOPOLSKY
So this started as a small project that a few guys were working on. They wanted to make $250,000—
JIMMY FALLON
(noticing a trio of virtual knights)
—is that a person?
JOSH TOPOLSKY
Yes.
JIMMY FALLON
I'm seeing people.
JOSH TOPOLSKY
I want you to take this controller. Use the left stick. It's an Xbox controller . . . Walk up to that knight, the guy in the middle. Walk right up to him.
  (as Fallon approaches the virtual knight)
Get right in his face. Now look down.

Fallon obeys, now looking downward: a silvery crotch in his POV.

JOSH TOPOLSKY
Look at that codpiece!
JIMMY FALLON
Why'd you make me do that?!

(cracking up, along with the audience)
Stop that! Why'd you make me do that?! Unbelievable.

Still cracking up, Fallon removes the goggles and exits his first VR experience.

JOSH TOPOLSKY
So this started as a little project. They made $2.4 million on Kickstarter. And now they're making developer kits. They're going out, in March, for $300 to developers. This is going to be coming to consumers and it is . . . mind-blowing.
JIMMY FALLON
It's a game changer.

When the segment ended, LaValle and Yershova had only one thing to say to each other: we should accept Iribe's job offers to join the company before it is too late! It was time, they decided, to move from Finland to Irvine, California

CHAPTER 20

WALL-TO-WALL DRAMA

February/March 2013

NOBODY COULD REMEMBER EXACTLY WHO CAME UP WITH "THE GAME." BUT IT was almost better that way, like some sort of Oculus office tradition that had been passed down from generations ago. The employees loved it, especially late at night—because it was energizing, competitive, and, compared to the work they did all day, stupidly, stupidly simple.

"How do you play?" Julian Hammerstein asked Luckey during his visit to Oculus.

Julian Hammerstein, like Chris Dycus, was one of Luckey's oldest and closest friends from ModRetro. But unlike Dycus, he had turned down Luckey's offer to join Oculus in the days prior to QuakeCon. After shrugging his way through high school with near-superlative apathy, Hammerstein had come to appreciate formal education during his time at the Rochester Institute of Technology (thanks in part to the informal education of living with other DIY doers in the school's prestigious Computer Science House). Thus, turning down Luckey meant another year of RIT; another year of CSH; another of supplementing his raw passion for mechanical engineering with the expertise he hadn't realized how much he'd lacked during his ModRetro days.

For all those reasons and also a few more—his family lived on the East Coast, he had a serious college girlfriend—Hammerstein was glad to have stayed in school. But with the summer fast approaching, he was looking to find an internship. And where better to intern than with the company his old pal PalmerTech had started?

For Luckey, Hammerstein's visit couldn't have come at a better time. There was no shortage of people saying nice things about Palmer Luckey at this time, and having someone like Hammerstein around—someone who knew him way-back-when and who, like Dycus, would never hesitate to bust his balls—was an invaluable check on his ego. That said, friendship was a terrible reason to hire someone, which is why Luckey had absolved himself from the interview process. If Oculus were going to bring on Hammerstein, it would be based purely on merit.

Oculus was particularly desperate for a mechanical engineer who specialized in the design, development, and construction of physical materials. Even more crucially, Luckey needed this person to assist him with the two biggest projects he would be shepherding throughout the course of 2013: Oculus's second development kit (DK2), and motion controllers for DK2.

"How do you play?"

Luckey repeated Hammerstein's question, as he and Hammerstein watched a handful of employees gathered by a large wall in the office. "Oh, it's supereasy. You start with your hand on the back wall. Then, as soon as the person with the stopwatch shouts 'GO' you run as fast as you can to the other wall."

Shortly thereafter, GO was shouted and an employee sprinted from one wall to another.

"I like it," Hammerstein said. "I like anything that makes engineers huff and puff. But I have to say: this does *not* seem like your typical Palmeresque shenanigans."

Hammerstein's implication was that Luckey had gone soft. That the kid who used to scheme about smuggling ice cream onto planes was losing his edge. "Just wait," Luckey said, as Dycus lined up for his turn to run. "Because what you learn really quickly, and this is why the fun comes in"—Luckey paused, someone shouted GO and Dycus dashed across the room—"is that the best way to get a good time is *not* to decelerate."

Then, as if on cue, there was a tremendous *SPLAT.* Dycus had crashed into the opposing wall like a bug on a windshield. For a moment, he flailed on the ground; and then, with a smile, he popped up to celebrate his time.

"Ah yes," Hammerstein noted with glee. "I would like to revise my previous statement. This does indeed seem *very* much like standard Palmer shenanigans."

Suddenly, GO was shouted and Oculus's founder was on his way.

Cheers erupted, and Hammerstein found himself momentarily mesmerized. Palmer Luckey was by no means fast, nor did he move with any particular grace; but when the clock was ticking, and all eyes were upon him, he was propelled by an unexpected and unstoppable fury.

ON FEBRUARY 5, MCCAULEY EMAILED THE TEAM WITH SOME BAD NEWS: THE tooling modifications needed to relieve the nose on the design was going to take longer than expected—pushing the production schedule by another two weeks.

WHILE MCCAULEY DEALT WITH THE LATEST DELAY IN CHINA, LUCKEY LISTENED to Ouya CEO Julie Uhrman deliver the keynote at the 2013 D.I.C.E. (Design, Innovate, Communicate, Entertain) Summit in Las Vegas.

"In December," Uhrman told a captivated audience, "we launched our development consoles and we opened up our developer portal. We've had 15,000 downloads of our ODK. We've had 900,000 views of our unboxing video. 115,000 people hit our website yesterday. But more than anything we have this developer forum that we can't even keep up with any more . . ."

Meanwhile, Luckey noted, our SDK is nowhere near ready and we're still months away from shipping a product worthy of unboxing (although McCauley claimed that later this week, the Irvine office would be receiving an updated batch of headsets that were "almost there"). In the face of Ouya's staggering numbers, Luckey tried to take solace in the devs Oculus had on board. Carmack and id were out, of course, but Valve, Epic, and Meteor Entertainment (the makers of *Hawken*) were proving to be good partners. Furthermore, they were close to signing a deal with Paul Bettner's new venture.

Or so Luckey thought, because then Uhrman said, "And Paul Bettner, from *Words with Friends* fame, has just started a new studio called Verse; he's bringing his first game to us."

THIS DID NOT MEAN THAT BETTNER HAD CHOSEN OUYA OVER OCULUS (HE COULD still develop for both platforms), but it meant that, at the least, he was keeping his options open. For Luckey, it means that Oculus hadn't yet done enough to convince him that VR truly was the future.

Luckey and Mitchell would be seeing Bettner in early March at the South by Southwest Gaming Expo in Austin, Texas. The Oculus guys would be hosting a panel on developing games for virtual reality, and Bettner was one of the three devs they had asked to join them.

"Do you think he's going to bow out?" Mitchell asked.

"Nah," Luckey replied. "Paul doesn't strike me as being that type of guy. And he *was* pretty up front with us about his interest in Ouya."

Since his visit to Irvine, however, Bettner's interest in Ouya had only grown. He was still psyched about Oculus and still planned to do "something" in the VR space, but over the past couple of months a few things had happened that nudged virtual reality to the back burner.

One of those things was a game—an idea for a game, really—that he had his team were excited about. Tentatively titled *Thereafter*, it was a sandbox-style game that valued gameplay over graphics and seemed like a perfect fit for Ouya's low-cost, living room console. The second thing that had happened was that Ouya was no longer the only nontraditional company hoping to disrupt gaming in the living room. Following a flurry of announcements at CES, Ouya was no longer even the only company hoping to achieve this disruption with an Android-based console. They'd be competing against PlayJam's "GameStick" and Nvidia's "Shield" (not to mention the many iterations of Valve's Steam Box, although those would be much costlier). This market space was about to get crowded. And while this competition would make things a bit harder for Ouya, it made things a little easier for developers; because if Ouya didn't work out for whatever reason, there would at least be other comparable options.

As much as Bettner's heart was in VR, he had just founded a new studio—with employees to worry about—and so, to do right by them, he needed to put aside his heart and focus on that which he thought would make or break VR: hands.

HANDS—NEARLY A DOZEN OF THEM—REACHED FOR THE BOXES THAT FEDEX had just delivered.

"Woohoo!" Laird Malamed cheered, lifting one of the first forty units to arrive from China.

"Niiiiiice," Dycus said, grabbing one of these "pilot units" for himself; and another for Luckey.

"Hmmm," Reisse chirped, showing an uncharacteristic amount of enthusiasm.

From a purely aesthetic standpoint, these development kits were perfect. The hardware was polished, the packaging peerless, and, as Luckey unboxed it in his office, he was pleased by how light and slick it felt in his hands. Ultimately, all that mattered was how they performed.

To find out, Luckey hooked up the headset and strapped it on his face. As did the other dozen or so employees, each donning devkits in their offices. And after comparing notes, there appeared to be three significant issues: faint vertical color bars, a preponderance of dead pixels (especially in the center of the screen) and inconsistent dithering patterns (which resulted in the wrong colors being projected at the wrong times)

What did these three issues have in common? They were all related to the display. This made Iribe furious because, as he now wrote in an email to McCauley, "Our concern the whole time was that Berway has never worked with LCD screen technology and might have problems. Looks like this is turning out to be the case."

In the next week or so, after Antonov had an alpha version of the SDK ready, these pilot units would be going out to partners like Valve, Epic, and Meteor Entertainment. Since Oculus was in regular contact with most of these partners, they'd likely be pretty forgiving about these ongoing issues. The bigger concern was the nine-thousand-plus folks who had ordered devkits. Oculus only had one chance to make a first impression with them and, in its current state, their hardware was unacceptable. The following day, the scoreboard hanging in the office that posted how many devkits had been sold finally clicked past ten thousand units. This was a cause for celebration, the latest indication that Oculus was on its way. But beneath those smiles lurked a thought shared by all, even the most optimistic souls: What, exactly, will we be shipping to all those people?

IN THE MIDST OF THIS CRUNCH TO GET DK1 INTO SHIPPABLE CONDITION, IRIBE started making plans for Oculus to try and raise a Series A round of investment. As part of the due diligence process, all employees needed to make sure that they'd handed in any agreements they had made that might impact the company in any way. Which is how—nearly half a year after signing it—Luckey informed Iribe about an agreement he had signed with Mark Bolas.

Apparently, as per this agreement Luckey had signed with Mark Bolas—the professor who had hired him to work in the lab at USC— Bolas was entitled to sizable, *non*-dilutable stake in *any* company that Luckey ever started that had anything to do with VR.

"Palmer," Iribe said, "This is a *crazy* agreement. And I'm upset that Bolas even created an agreement like this and got you to sign it."

In addition to being surprised that this agreement existed, Iribe was also disappointed that Luckey hadn't shared it with him earlier. But he understood—he could see that Luckey was nervous about it. "Palmer," Iribe said, "I'm gonna have to unwind this; fix this. This thing can't live on. We can't do our Series A and do funding with this kind of an agreement out there. We have to fix this now."

"THIS SUCKS!" IRIBE SHOUTED, APPALLED BY THE STATE OF OCULUS'S SDK. HE took another look into the headset, shaking his head in disgust. "Come on. The distortion here just sucks."

"It doesn't suck!" Antonov barked back, trying his best to match Iribe's tone. "It's fine, Brendan. It's fine."

"Mike," Iribe said, now staring directly at Antonov. "The distortion is not right. It's just not. And it's embarrassing."

This was not the first time that Iribe and Antonov had clashed about distortion, although it was the loudest and most public to date. Usually, they argued privately—by phone, over email, or during a shared meal—but with GDC now only seven weeks away, the time for niceties had run out. "The distortion cannot be our priority!" Antonov countered. "We need to get this demo, this documentation, these SDKs ready. There's still a ton of engineering work left to do. And the distortion—okay, maybe not perfect—but it is already working okay."

"Okay is not good enough," Iribe replied.

"THE SPACE IS TWENTY BY TWENTY," CHEN SAID, REFERRING TO THE BOOTH that Oculus would have at GDC. The conference fast approaching, it was time to figure out how that space should be constructed. Chen and Mitchell sat down to formulate a game plan for the show.

"What are the key message points?" Mitchell mused aloud. "What's the key language? What do we want the key takeaways to be?"

The key to all these keys was first defining who Oculus was now—or, rather, who they would be at GDC when they formally launched DK1. This was tricky, since the company had thus far defined themselves by their quest to fulfill a promise: the DK1. But it had only fulfilled a *step toward* that promise.

Mitchell and Chen focused on expectation setting. They wanted to make clear that GDC wasn't a finish line, but rather the start of something new. Even for those who would be blown away by the devkit, it was critical that their excitement race beyond just the hardware in their hands; they needed to also imagine on their own a glimpse of where VR would be headed.

They coined the phrase "Day Zero" as a way to think about GDC. It reset expectations, conjured thoughts of days ahead, and, quite frankly, sounded pretty cool.

There were, however, two big drawbacks to thinking about GDC—and, really, this early phase of VR—as Day Zero. One was that, by inserting a timeline into people's mind, it naturally begged the question: When will the consumer version launch? And at this very moment, Oculus didn't have a great answer.

EVERY TRIP TO VALVE WAS A CAUSE FOR CELEBRATION. BUT ON FEBRUARY 11, Luckey, Iribe and Patel were in particularly good spirits after learning that Valve was going to have their VR version of *Team Fortress 2* ready in time for GDC. That was huge! Probably even big enough to make up for losing *Doom 3: BFG Edition.* Not just from a gaming standpoint—ensuring that devkit owners would have something fun to play out of the box—but from a PR standpoint, too. It would show the community that just because Carmack/id/ZeniMax had abandoned VR, there was no reason to worry. Valve was still on the bandwagon! And sure, there were still only a handful of Valve employees on that bandwagon—most were still skeptical of the company's VR exploration—but that internal divide wasn't known to the world at large. Especially when Valve's biggest champion for the technology was also one of their highest-profile employees: Michael Abrash.

In light of this, the Oculus guys were also thrilled that both Abrash and Joe Ludwig (who, along with Tom Forsyth had been spearheading the VR version of *TF2*) would be giving talks at GDC. Ludwig would be talking about what he and Forsyth had learned about porting games to VR, and Abrash planned to cover the larger, fundamental challenges of virtual reality.

"The overarching message," Abrash said of his talk, "is going to be that VR is hard."

A wave of laughter passed through the room.

"It's true!" Abrash said. "VR is hard. Really hard. And who knows that better than the handful of us in this room? So I want to make clear that 'VR isn't just around the corner.' Because it seems so simple, doesn't it? Isn't VR just a matter of putting a display in a visor and showing images on it? Well, that actually turns out to be really hard."

The three really hard problems that Abrash planned to discuss were tracking, latency, and "stimulating the human perceptual system to produce results indistinguishable from the real world." That said, Abrash also planned to make clear that while we're still a ways away from perfect VR, "good VR" *is* finally here now—in the form of the Oculus Rift development kit—and that the Rift had a good shot of breaking through for a handful of reasons: it's lightweight, it's ergonomic, it's affordable, and so on.

"Most important," Abrash continued, "gaming on the Rift is highly immersive. So there's potentially lots of content in the form of ported 3-D games—not to mention all the new stuff that developers will surely come up with. But again, I don't want this excitement to mask the fact that all this stuff is really, really hard. And since this is a conference for developers, I want them to understand why that's the case, so they can make rational plans for VR game development now and in the future."

"That sounds perfect," Iribe said to Abrash. "Yours, too, Joe. And they'll pair real nicely with the presentations that we're doing."

Like Valve, Oculus would be giving two talks about virtual reality. One by Antonov and Mitchell, about the Oculus SDK and integrating games, and another by Luckey about virtual reality being the "Holy Grail of Gaming."

"Just to be clear," Luckey clarified, "the 'Holy Grail' is not the Rift. Like my talk isn't gonna be me getting up onstage and saying: Behold, the Holy Grail has arrived!"

Before leaving, the Oculus guys stopped by the office next door to say hi to Jeri Ellsworth and her hardware team. Although Ellsworth was always friendly whenever the guys stopped by, it was pretty clear that she wasn't a big fan of Oculus. At first, Luckey and Iribe had assumed that she resented them, distrusted them, or perhaps didn't respect their work. But as they would later learn, Ellsworth's opinion had less to do with them than what Oculus was trying to achieve. In her opinion, VR just wasn't capable of ever becoming a mass market device. AR, on the other hand—well that was a whole other story.

THE FOLLOWING DAY, JERI ELLSWORTH WAS CALLED IN TO GABE NEWELL'S OF-fice. That evening, she tweeted an ominous message: "Wow. I suddenly have a lot of spare time."

"THEY FIRED JERI," LUCKEY SAID, POPPING HIS HEAD INTO IRIBE'S OFFICE.

"No way," Iribe replied, skeptical that Valve would fire anyone—let alone such a high-profile employee.

Although Jeri Ellsworth, thirty-eight, hadn't been in the games industry as long as most at Valve, her ambitious, retro-fueled DIY projects had been turning heads in the tech community for nearly a decade, beginning in 2004, when Ellsworth took the circuitry of a Commodore

64—that beloved home computer, first introduced in 1982—and managed to condense it down to a single computer chip. This chip was then tucked into a joystick, fortified with classic games, then productized by a company called Mammoth Toys, which sold hundreds of thousands of these play-on-your-TV devices on the QVC. "Ms. Ellsworth is demonstrating that the spirit that once led from Silicon Valley garages to companies like Hewlett-Packard and Apple Computer can still thrive," cheered a 2004 profile in the *New York Times*. That can-do spirit led Ellsworth to short stints at tech companies (like Xilient, NewTek, and Ubicom), but it wasn't until 2011 that she ended up at Valve, a place where she could see herself for years to come.

Recruited by Gabe Newell, Ellsworth was brought on to build a hardware R&D lab at the industry's most dominant software maker. With a simple and alluring mandate—"to build a team that will make games more fun"—she was given a wide berth to explore all sorts of things: from crafting consoles and controllers to conducting skin resistance experiments to measure (and tweak) a player's emotional state. This autonomy led Ellsworth to augmented reality, around the time when Abrash joined Valve.

Since Abrash, initially, was focused on AR, he and his group got along well with Ellsworth's hardware team. But as Abrash's interest moved on to VR, this created a minor internal schism between AR and VR. It never seemed to rise to the level of rivalry, but it seemed no coincidence that mere months after the divide, Ellsworth no longer had a job at Valve. And it wasn't just her. Per a *Forbes* article published later that day: "Reports are coming in that between 25 and 30 staff have left Valve in targeted layoffs . . . The story began with a tweet yesterday evening from Jeri Ellsworth, hardware designer extraordinaire . . . Her departure may mean that Valve's focus will be on coordinating third parties producing for the Xi3 Piston announced at CES—not *the* Steam Box, but clearly *a* Steam Box . . ."

"We should hire Jeri!" Luckey exclaimed, upon hearing the news.

Iribe agreed, but, for the moment, was still fixated on what the news of these layoffs might signify. Did this mean that Valve was moving away from doing hardware in-house? If so, would Abrash and his gang be next?

ON FEBRUARY 18, PAUL BETTNER—LIKE A DOZEN OR SO OTHER TRUSTED devs—received one of Oculus's pilot units. Within minutes of unboxing the headset and hooking it up, he found himself on the brink of tears. Because, as he would post on Instagram later that day: "Shit just got real."

Bettner was blown away by the pilot unit, but couldn't help but notice that whenever he moved through the "Tiny Room" demo that Antonov had designed, he kept instinctively looking down, hoping to see his hands. It felt weird, being so immersed in this virtual space, but not having the ability to reach out and touch stuff. As humans, Bettner thought, our primary input devices are our hands. And if he needed any additional proof of this, all he had to do was poke his head into the next room where his six-year-old daughter and four-year-old son were constructing a tall, twisty, seemingly-Seuss-inspired marble run: grabbing plastic pieces, connecting them in various arrangements, then clapping with all-out enthusiasm whenever marbles were dropped into their course.

Given that this was developer kit, and merely the first step of Oculus's long-term mission, it was understandable that virtual hands weren't a priority at this time. But thinking back on his meeting in Irvine, and everything he'd read about Oculus's vision thus far, Bettner began to worry that incorporating hand controllers just might not be that important to Oculus. Or to the industry, for that matter. For the past twenty-plus years—ever since the death of the joystick—gaming had been defined by two types of inputs: gamepads (for console gaming) and keyboards and mice (for PC gaming). So it wasn't surprising that, at least for the moment, Oculus seemed content with either of those two options. But at some point, Bettner believed, they'd need to come up with an input that directly brought your hands into the virtual world.

How expensive could hand controllers be anyway? A quick internet search showed that the best one cost $17,995 . . . per hand. Granted this was a top-of-the-line, twenty-two-sensor, finger-tracking controller, but the astronomical cost was sobering.

"SO . . . I'M ACTUALLY KIND OF SCARED," BRIAN SINGERMAN TOLD IRIBE ON February 20.

Singerman was a partner at Founders Fund, and although it's typically a bad sign when you've managed to scare a potential investor, Iribe couldn't stop smiling. Because what had made Singerman momentarily squirm was a character in the VR demo Iribe was running

"Wow," Singerman said, standing in his office with a DK1 prototype over his eyes. "For a split second, I actually thought that guy was real!"

"WE NEED TO GET THIS STARTED," MITCHELL WROTE TO LUCKEY AND PATEL ON February 28. He was talking about the HD-prototype that they were planning to get ready for E3—in particular the fact that McCauley (apparently) was preparing a competing version of this DKHD headset. "Let's huddle for 10 minutes tomorrow and come up with a plan to knock these out . . ."

"Codename Beanstalk," Patel replied. "Because there are two plans. Jack and the Beanstalk. And Jack keeps trying to chop the Beanstalk down."

THE FOLLOWING MORNING, PROJECT BEANSTALK OFFICIALLY GOT GOING—WITH Patel scribbling out roadmaps for Oculus' various projects on a whiteboard: DK1, DKHD (Beanstalk) and—most important—DK2. Which, as Luckey and Patel would write out in a roadmap over the next few days, would introduce a "significant new feature set to the Oculus Rift platform." Including:

- 1920 x 1080 pixel resolution display for higher fidelity image quality
- Stereo cameras for positional tracking and real-world video pass-through
- Field-programmable gate array (FPGA) for system integration and additional functionality
- Mechanical focus and interpupillary adjustments
- Programmable buttons on the headset
- Motion controllers

Integrating the player's body into the virtual world would be key to DK2. "DK1 immersed the player in the virtual world visually, but did nothing to integrate their physical body," they wrote in the roadmap. "With DK2, we want our players to feel like they're actually Batman when they're playing in Oculus VR."

OVER THE PAST TWELVE YEARS, BRENDAN IRIBE HAD PITCHED DOZENS OF potential investors. Some meetings had gone well, others not so much. But regardless of outcome, they all had one thing in common: nobody had ever fallen asleep while he was talking. That changed in early March, however, when Iribe went to Boston and pitched North Bridge Venture Partners.

The meeting had been set up by Ric Fulop, a general partner at North Bridge, who saw potential in VR and reached out to Iribe to hear the pitch. By this point, Oculus had yet to find any investors for their Series A—with one exception: Founders Fund. But that had only raised $1 million of the $15 million Oculus was looking for, so in the weeks leading up to GDC, Iribe hit the road hoping to raise the necessary funds. Coming to Boston and boring an investor to sleep was definitely not how Iribe had hoped to start off his trip.

This is crazy, Iribe thought. If VR doesn't keep you awake, then what will! But discouraging as it was, Iribe did his best to power through it.

Fortunately, Narcoleptic Investor's partners were much more impressed than he. And before Iribe left, Ric Fulop presented him with a term sheet. Or, uh, maybe not.

"Ric, what is this?" Iribe asked, looking over the document. "This isn't even a term sheet. This is a convertible note. You don't even have a valuation in this!"

"Oh," Fulop replied. "I got you the wrong meeting."

All in all, things couldn't have gone worse. "It was a crash-and-burn partner pitch," Iribe told Founders Fund's Brian Singerman, whom he called right after leaving North Bridge. "This just happened. It's pretty discouraging."

The whole ordeal, however, turned out not to be a total loss. Because Fulop—who felt bad and still sincerely believed in Oculus—connected Iribe with another lead: Santo Politi, who was a partner at Spark Capital.

"Reach out to Santo," Fulop told Iribe. "He's a great person. And also in Boston."

Politi was interested, but currently traveling. That's fine, Iribe said. I'll come to you. And three days after his crash and burn at North Bridge, Iribe found himself in San Francisco, giving Santo Politi a demo in the conference room of an airport hotel.

Back in early the '90s—during the heyday of janky VR—Politi had been an executive at Panasonic, so he had a vague sense of what he was about to see. He knew that this would be significantly better, of course, but he had no idea *how much* better it would turn out to be.

"I'm floored," Politi said. "I'm absolutely amazed."

Iribe smiled. He had been hoping to hear something like that.

"Let's have dinner," Politi suggested. "I think there's a shitty restaurant downstairs."

Over dinner, Iribe talked about Oculus's origin story as well as his own, and Politi did the same, providing a glimpse into his investment philosophy. "We're early-stage investors," Politi explained. "We're really early-stage investors. For us, it's all about product. If you have a great product, and we fall in love with it, you're very likely walking away with an offer and check. That's how it usually goes. We tend to do things slightly crazier than others."

Almost immediately, the two of them hit it off. And by the time Politi had to leave for his red-eye flight, they were already coordinating a time when Iribe could head back to Boston and give a demo to Politi's partners.

Feeling good, Iribe stayed overnight in the Bay Area and the following day he pitched Joe Lonsdale, a prominent entrepreneur who had recently cofounded a hot new growth-stage venture firm called Formation 8. So hot that *Fortune* would call Lonsdale and his partner, Brian Koo, "the hottest VCs since Andreessen Horowitz!"

The pitch to Lonsdale went great and would end up leading to investment. Though as with Founders Fund, it was a relatively small amount and Iribe was still searching for someone willing to go big and lead the Series A.

The firm he was most hoping would do that was Andreessen

Horowitz. He had been dying to work with them for years and thought now might be the right time. To find out if that would be the case, Iribe pitched their newest partner, Chris Dixon, on March 8.

WHILE IRIBE WAS PITCHING TO ANDREESSEN HOROWITZ, LUCKEY WAS IN TEXAS to speak on a SXSW panel with CliffyB, Chris Roberts and Paul Bettner.

"Paul!" Luckey said, excited to get a few minutes with Bettner before the panel. "I've been meaning to talk to you more about motion controllers, but thought in person would be better than over email."

Part of the reason why Luckey felt this was because he could sense that some of the other execs weren't as adamant as he (and Bettner) about adding motion controllers ASAP. "But don't worry," Luckey told him, "I know that getting hands into VR is a make-or-break thing; and we'll be able to make it happen."

Less than an hour later, during the panel, Bettner said with a smirk, "The key input device is hands. And it's up to you guys to make that work!"

Bettner didn't even need to turn his head, he could feel Luckey nodding in agreement.

THREE DAYS LATER—AFTER A GREAT MEETING WITH CHRIS DIXON—IRIBE RE- turned to Andreessen Horowitz three days later, this time to pitch Andreessen himself.

The meeting went well, save for one question. "How long do you think it's going to take you to solve motion sickness?"

The truth was that Iribe didn't know. Nobody knew. And so instead of trying to spin it, that's what he told Andreessen. "I really don't know," Iribe said. "It could be five months, or it could be five years. I really have no idea."

Andreessen and Dixon appreciated Iribe's honesty, but ended up passing.

Under normal circumstances—given how much Iribe had wanted to work with them—this probably would have felt like something of a crushing blow. But time was running out and so were the funds in

Oculus's bank account, so he forged forward and focused his energy on trying to get an offer from Spark and test the waters with a VC from Matrix Partners.

On March 11, Iribe met with that VC—Josh Hannah—and then hopped on a red-eye for Boston to meet Santo Politi's partners at Spark; and, since things had gone quite well with Hannah earlier that day, the senior partners at Matrix (which happened to be based in Boston) wanted to meet with Iribe while he was in town.

Some serious offers appeared to be on the horizon and Iribe was psyched, even more so after things went superwell with Politi's partners at Spark.

"We're all pretty excited," Politi told Iribe. "What would you like from us?"

"I'd like you to lead our Series A," Iribe said.

"You got it. We'll do it. You'll have the term sheet tonight."

Iribe remained perfectly calm until he exited the building and took a celebratory exhale. Onward to Matrix he went, hopping onto the T (Boston's subway) for the quickest way, though unfortunately that turned out not to be the best idea when Iribe was unable to avoid some afternoon snow.

This guy's got some balls, Matrix partner Antonio Rodriguez thought as Iribe walked into his company's office. This guy's taking the T, braving the snow, and coming to see me on such short notice. He must have a lot of conviction about this demo!

Despite that positive first impression, Rodriguez didn't expect much to come from this meeting. He *was* interested in AR opportunities, but VR? Really? Nice try, Virtual Boy.

"This is only going to be like ten to fifteen minutes," Rodriguez told his assistant, and then met Iribe in a conference room to try the demo.

"So what are you going to show me?" he asked, trying to sound excited.

"I'm gonna show you our Tuscany demo," Iribe told Rodriguez. "So you'll be able to escape this cold weather and go relax for a little in Italy!"

"This is amazing," Rodriguez finally said. "I'm *soooo* blown away right now."

Hoping to share this feeling, Rodriguez poked his head out of the conference room and attempted to get the attention of anyone else on the investment team who was free.

Shortly after this visit, Matrix put together a term sheet for Oculus. Iribe mentioned the interest from Santo Politi at Spark (who, it turned out, was a good friend of Rodriguez) and with about ten days to go until GDC, it was decided that Spark and Matrix would colead the round.

BACK IN IRVINE, THE CREW RACED TO GET EVERYTHING READY FOR GDC. THIS was the time for final tweaks, last-minute bug fixes and prayers that all went according to plan. And for Luckey, Mitchell and Antonov—each of whom would be giving talks at GDC—it was also time to figure out what the hell they were going to say.

For inspiration, Luckey watched some old John Carmack keynotes on YouTube. Then, in a moment of utmost awe, he messaged Patel. "Looking at Carmack talks," Luckey wrote, "you realize how brilliant he is and how hilarious it is that thousands of people are going to pile into a room to listen to a speech given by a 20-year-old who got lucky and was blessed by Carmack but has nothing to say that has not already been better said."

CHAPTER 21

GDC

March 2013

IRIBE HAD HOPED TO HAVE THE SERIES A WRAPPED UP BEFORE THE START OF GDC, but as he joined hundreds in line at to the expo entrance—waiting for the show doors to swing open—it was now clear that that wasn't going to happen.

There were worst-case scenarios to consider. Things like a bad show, poor launch, or failure to live up to the hype—any one of which meant impacting the not-yet-signed term sheets with Matrix, Spark Formation 8, and Founders Fund, who were potential backers. Needless to say, Iribe was feeling a little on edge.

He'd been coming to GDC since 1998, running tiny little booths with Sonic Fusion and Scaleform. Snapshot memories from over the years quickly came to mind, as did the visceral pain of how hard it had been to lure developers. And now? What was he hearing . . . Strangers in line talking about his latest company? It seemed almost too good to be true. So Iribe reminded himself that the buzz might just be relegated to this small sample size around him.

Suddenly, with a whoosh, the expo doors opened, and most in line started running. To where? To what? Iribe didn't know, so he started running too. Following the crowd to wherever it was headed. Past Unity and Nintendo, past Wargaming and Turtle, and then lo and behold the stampede reached its destination: Booth 218, Oculus.

"I'D LIKE TO PROPOSE A TOAST," IRIBE SAID, RAISING A GLASS THAT NIGHT, shortly after he, Luckey, Mitchell, Antonov, and Patel sat down for dinner at a ritzy spot in the Bay Area called Amber India.

"I'll keep it short," Iribe said. "But it's been an incredible ride . . ."

As everyone around the table briefly conjured up memories from the "incredible ride," Antonov couldn't stop thinking about the beginning. The *beginning* beginning. Back to a time before he'd ever met Patel or Luckey or Mitchell—back to the days when he and Iribe were working out of that small basement office in Laurel, Maryland, and success seemed so damn far away. After Iribe's toast, Antonov recounted the first time he and Iribe attended GDC.

"So when Brendan and I started working together," Antonov recounted to the group after Iribe's toast, "a very big thing was always going to the Game Developers Conference. Every year we would go religiously."

"I'd been going since *before* Scaleform even," Iribe said. "That was where I met Paul Pedrina for the first time—and came back to campus all excited about doing a windowing system!"

"So, yes, Brendan had gone to the show prior," Antonov continued, "but the first year he took me there—this was just back in the very beginning—and Brendan was like: Mike, I'm heading over to the Game Developers Conference, you should come. Okay, yes! So we fly over to California and we are going to this conference and we land at San Francisco airport. First of all: I've never been to California before. Second of all: I'm from a fairly poor family. For me, a hundred dollars is a lot of money."

"Hey," Iribe clarified, "that was a lot for me too back then!"

Antonov stared at Iribe, as if trying to decide between laughing at Iribe's recollection or launching into a rant about how things *really* use to be. "So there we are, we land and literally we take a taxi and the taxi rolls over a hundred. And I'm like: OH MY GOD! I'm going to have a heart attack!"

Everyone at the table cracked up, including Iribe, who—between laughter—added, "But we made up for it with the motel! Remember that horrible cheapo place?"

"Yes!" Antonov replied. "We, of course, got, like, a little motel *very* far away from the convention center. Across the street there was actually a drive-in movie theater with a parking lot field and screens. . . I will just say this: it was rough. But at the same time, going to the show, it was a very special feeling. You walk in and when you're young you think the tech is so impressive—like, all these colors, all these big monitors, and then there's these people, who love it, so many playing games. I'd never seen anything like this."

The nostalgic, celebratory mood that circled around the table was soon interrupted by the reality that it wasn't yet time to pop the champagne. The hardware was shipping tomorrow—that was locked—but there was still last-minute work to be done on the SDK. The execs needed to put together a license agreement for their SDK, and since this would literally spell out what users could and could not do in their ecosystem, it meant that a decision would need to be made about how "open" Oculus was going to be. Which, whenever it had been brought up in passing, had always been a big point of contention.

"Because it's very hard," Antonov declared at the table. "On one hand we want to be open source; we want to drive the industry forward. But on the second part, we need to protect Oculus. Because we all know it's going to be very easy for someone in China to replicate the headset."

That, in a nutshell, summed up the push and pull of this debate: the more open the system was, the more freedom users would have; but the more freedom that was granted, the easier it would be for competitors to replicate all the work that Oculus had done.

Having been in the middleware business for over a decade, Iribe preferred that Oculus be as closed as possible. Mitchell did, too, believing that's what was necessary to protect themselves. On the other side of the spectrum, Patel and Luckey essentially wanted everything to be open-sourced. That left Antonov in the middle, the deciding vote in a sense. As the guys debated the pros and cons of various strategies, Iribe excused himself to take a call.

"WebKit, Mach, POSIX," Patel said, listing off examples of great open-source software.

Okay, but Mitchell offered situation-specific reasons why open source worked with those examples, but would be much different for Oculus.

"What about this," Antonov said. "We give you a full source, but the only thing you can do is use it with a valid device?"

This seemed like a good compromise, but before anyone could hammer out all the specifics, Iribe returned to the table.

"We just closed," Iribe said, his cheeks busting.

"Really?" Mitchell asked, bursting with excitement.

"Yup. Sixteen million. Led by Matrix and Spark. Closed!"

Handshakes. Hugs. High fives. With Oculus shipping their first product in less than twenty-four hours, it finally felt like time for a little celebration.

PART 3

THE GOOD
OLD DAYS

CHAPTER 22

MOVE SLOW AND BUILD THINGS (AKA FACEBOOK 2.0)

April 4, 2013

SINCE THE DAWN OF THE TWENTY-FIRST CENTURY—SINCE THE INTERNET WENT viral, tech turned sexy, and Silicon Valley became Mecca for a new breed of American dreamers—nobody had been more successful than Facebook founder Mark Zuckerberg. In less than eight years, powered by the mantra "move fast and break things," he had managed to turn a dorm-room project into a global social network that, as of last count, boasted about 1.2 billion users. That said, even a digital tycoon like Zuckerberg had regrets. And stepping onstage to address a mix of reporters and employees, he hoped to address one of his biggest regrets with an exciting announcement.

"Today," Zuckerberg said, "we're finally going to talk about that Facebook phone."

Rumors of a "Facebook phone" had been circling for years, first popping into the public conscious in 2010 with a *TechCrunch* piece entitled "Facebook Is Secretly Building a Phone." At the time, this seemed like somewhat of an odd thing for Facebook to do. Why would a software company, with zero hardware experience, attempt to complete against the likes of companies like Nokia, RIM, and Apple? After all,

Google had just tried this with its Android—putting a ton of resources into its Android operating system—and, by the end of 2009, had only 3.9 percent of the market.

Well, by the end of 2010 Google's market share jumped to 22 percent. By 2011, it hit 39 percent. And by the end of 2012, Android sales—toping 357 million phones—accounted for 69 percent of the mobile market.

MARK ZUCKERBERG
We're going to talk about how you can turn your Android phone into a great, simple device. You're going to be able to transform your Android phone into a great, social phone.

For those in the audience, this clarification made the announcement a tad underwhelming. Facebook, it became clear, would be releasing nothing like the "Facebook phone" that had been rumored for years. This wasn't a platform play so much as a can't-beat-'em-join-'em decision to try and mold Google's mobile OS in Facebook's friend-first image.

MARK ZUCKERBERG
Today: our phones are designed around apps, and not people. So we want to flip that around. Now we want to bring this experience—of having a home, of always knowing what's going on around you—right to your phone.

Even though this wasn't the big-bet scenario that those covering the event had hoped it might be, this *was* still interesting: a phone designed around people instead of apps. That sounded good. It seemed to make sense. But what, exactly, did it mean?

MARK ZUCKERBERG
The home screen is really the soul of your phone. You look at it about 100 times per day. It sets the tone for your whole experience. And we think it should be deeply personal. So today we're going to talk about this new experience for

your phone. It's a family of apps. And you can install it and
it becomes the home of your phone. So we're calling this:
Home.

As Zuckerberg's introduction segued to a video demo of "Home"
in action, several of the reporters on hand began nodding with enthusiasm. So, too, albeit with supreme stoicism, did a product designer seated
near the back of the room. This designer—a thirtysomething woman
with a high and tight haircut; her bangs stylishly quaffed upward—was
Caitlin Kalinowski. She nodded not because she was dazzled by Home,
but because after fourteen years in Silicon Valley, she finally felt like
she'd found a company that felt like home: Facebook.

Kalinowski had only been with Facebook for a couple of months
now, but her journey there began more than a year earlier. In Portland,
in November 2011, at the annual Grace Hopper Celebration of Women
in Computing where Kalinowski crossed paths with a recruiter who
expressed interest in bringing her to Facebook. At the time, Kalinowski
was at Apple, where she was the technical lead on the company's funky-
looking, coffeemaker-inspired cylindrical Mac Pro. Though Kalinowski
greatly enjoyed working on that project, she had been at Apple for several years and was open to the idea of making a change, especially if
that meant going to a company with a reputation like Facebook; and
even more especially upon learning that they were looking to hire her
for a "supersecret project."

Unfortunately, however, the supersecret project turned out to be
a phone. So when she was extended an offer to join Facebook, Kalinowski said, sweetly but bluntly, "I want to join, but I can't. Because
you're building a phone and I don't think that's a good idea!"

At Apple, she had seen exactly what went into building a phone,
and Facebook's ambitions seemed naive, if for no other reason than
the fact that they had never shipped a product before. They were unfamiliar with phrases like "change off" or "ID lock"—they'd never gone
through the process of building stuff for people. Those wooing Kalinowski conceded these points but explained that this was exactly why
they wanted to hire her: to help them understand hardware and how to
build real-world products.

"I'm flattered," Kalinowski replied. "But if you hire me, I'm just gonna convince you not to do that! And I think that would be a little silly—that you'd hire me and I'd just tell you not to do what you hired me to do."

A couple weeks later, Kalinowski received a call from an unfamiliar number. Skeptically, she answered: it was Mark Zuckerberg. "Caitlin," he said, "we're really serious about hardware." It wasn't just a line to lure over a hardware lover like herself, but something much greater than that: a critical piece of Zuckerberg's vision for the future.

Zuckerberg had come to believe that a big technological shift was coming—a disruption that would be similar in scope to mobile, but involve some combination of VR, AR, and AI. Having missed out on that last shift, he didn't want to this time around. His competitors were ready; or, at least, they all had vastly more experience when it came to building consumer electronics. Apple had been doing it since the late '70s. Google was already dominating the mobile market. Even Amazon, whom nobody thought of as a "hardware company," now had years of experience after successfully producing Kindle e-readers since 2007.

In order to compete, Facebook would need to move beyond their "move fast and break things" model. This was the beginning of a new Facebook, a new post-IPO Facebook; and to best position themselves for the challenges ahead, they would need to move slow and build things. "So we need you," Zuckerberg told Kalinowski. "We need you to come here and help teach us."

After highlighting the features of Facebook Home to the excitable attendees, he announced that the Taiwanese electronics manufacturer HTC would soon begin producing a line of phones called the HTC First that would come preloaded with Facebook Home.

For the reporters in attendance, this collaboration with HTC represented what appeared to be merely the beginning of Facebook's hardware ambition. But little did they know that the HTC first would be Facebook's last attempt to build hardware in the mobile space.

CHAPTER 23

NINE STORIES

April 2013

"CHECK THIS OUT," LUCKEY SAID, SHOWING DYCUS AN EBAY WEB PAGE LIT-tered with results.

Skimming the items, Dycus couldn't help but grin: Oculus's $300 devkits were in such high demand that some were going for over a thousand bucks.

By this point, the first week of April, it was now clear that the launch of DK1 had been an unmitigated success. Secondhand units were selling for three times their price; tech journalists were publishing glowing reviews; and Luckey's in-box was flooded with affection and admiration—comments like this one, from a Korean fan, proclaiming that Luckey was "going to be a historic human in 21c." But as cool as all that was, none of it compared to the fact that developers all over the world were starting to receive their devkits in the mail; and over the next few months, these devs would get to work and begin building incredible things . . .

1. JUSTIN MORAVETZ
Santa Monica, California

In seventh grade, Justin Moravetz and his classmates were asked to give a presentation about "The Future." So Moravetz decided to talk about the technology that excited him most—virtual reality!—and even cobbled together a makeshift headset using a VGA monitor and two Game Boy Screen Magnifiers. For the next two decades, he waited for VR to finally arrive in all that glory he had imagined as a boy. But year after year yielded disappointment after disappointment.

From Forte's VFX1 to eMagin's latest Z800 3-DVisor, Justin Moravetz had seen it all. From his experience as a 3-D animator at Sony Computer Entertainment, Moravetz had insight into what a powerhouse was doing with virtual reality. And, well, it wasn't much. He came to believe that the only way virtual reality could ever really take off would be for a small, scrappy outside force to come in—resurrect the technology from the ashes of its failures—and force big companies (like the one he worked for) to jump into the fray.

Would this ever happen? Moravetz did not know. But he felt hope—glorious, long-lost hope—as he tore open the package containing his devkit and read the welcome letter inside.

Unlike almost every other developer, Moravetz had *already* come up with an idea. In fact, using Unity, he had already spent the past six months creating various levels for a retro-inspired, Arkanoid-style brick-breaking game he called *Proton Pulse*.

It was challenging to make an Oculus-ready game without an actual Oculus headset, but Moravetz's lifelong passion for VR coupled with his animation expertise gave him the ability to develop something that he thought would be viable. Now, after six months' worth of nights and weekends spent developing this game, he was about to find out how viable it would really be.

2. DENNY UNGER
Qualicum Beach, Canada

Within moments of putting on DK1 for the first time, Denny Unger started crying. He couldn't help it. Ever since trying a Virtuality VR system at Edmonton's Klondike Days expo in 1992, he had been infatuated with virtual reality. It wasn't that the Virtuality demo had been especially good (he left the demo sick to his stomach) but because, he assumed, that if VR hardware progressed on the same trajectory as, say, video game consoles, then it would only be a few years until virtual reality started to take off.

That, of course, didn't happen. But eventually, Unger found solace in a community of fellow VR diehards called MTBS3D. It was there he met Palmer Luckey who—as Unger would fondly recall years later—"had kind of cracked the code of how we could do this affordably using current technology." So when Luckey announced that he was planning to sell an open-source HMD kit on Kickstarter, Unger was so excited that he volunteered to design the logo for this little venture called "Oculus."

At the time, Unger had no expectation that VR would become interesting to anyone outside the hundred or so hobbyists who frequented the forums on MTBS3D. But his perspective began to change after John Carmack wound up demoing Luckey's prototype later that year at E3. Oh shit, Unger thought. There's a big opportunity here that no one understands yet! In late 2012, Unger quit his job (designing tabletop games) and recruited a few tech-savvy friends (Christopher Roe, Matthew Lyon, Aubrey Erickson) to start a studio (Cloudhead Games) focused on creating "deeply immersive" gaming experiences.

"We've got this golden opportunity to really do something special," Unger told the team during one of their early meetings in his garage-turned-game-studio. What he had in mind was a game that combined the adventurous spirit of films like *The Goonies* and *Indiana Jones* with the hidden challenges and mystery-island feel of his all-time favorite puzzle game: *Myst*. To help finance this built-for-VR puzzle/adventure game—*The Gallery: Six Elements*, they called it—the Cloudhead Games team launched a Kickstarter campaign in late March 2013. And that campaign had all but hit its goal by the time Denny Unger received his devkit in the mail.

There was a lot riding on the Tuscany demo he was about to try. Which is why, within moments of putting on DK1 for the first time, Denny Unger started crying. Tears of joy—of pure, unadulterated, dream-come-true joy. Because this moment felt like every Christmas ever; because even though he *knew* he was still in Canada, he *felt* like he was all the way in Italy; because if this was just first-generation hardware, could you imagine what the future would bring? It was all so wonderful and overwhelming. But before allowing himself to fully imagine those possibilities—for himself, for Cloudhead Games, for the future of human interaction—Unger had the urge to email an old friend. "Thanks for bringing us the future," he wrote.

3. OWLCHEMY LABS
Boston, MA

"I think that this is the beginning of a forty-year-industry," Devin Reimer, the CTO of Owlchemy Labs, told Owlchemy founder and CEO Alex Schwartz. This was months earlier, right after Oculus's Kickstarter had launched. Reimer—being a big fan of John Carmack—had been following the lead-up to the launch, and somewhere between E3 and QuakeCon, he had become convinced that VR was about to upend the gaming industry. "I think that this is the beginning of a forty-year-industry. And I think we should do something here."

At the time, Schwartz didn't know much about VR—Owlchemy's focus, after all, was mobile gaming—but he trusted Reimer's instinct and agreed they should order a devkit. Then the two of them got back to work and didn't discuss VR again until a devkit showed up at their office.

Intrigued by the headset and impressed with Oculus's SDK, Schwartz and Reimer decided to take off a month from their current obligations and create a VR version of a base-jumping game they had published in 2011 called *AaaaaAAaaaAAAaaAAAAaAAAAA!!! For The Awesome.*

Since *Aaaaa!!!* had originally been built using Unity, it only took two days to get their port working. But then it took the rest of the month to get it working in a way that was fun and functional and could be played end to end without taking off the headset.

During this time, the Owlchemy team learned many things. Things like: the camera angle is sacred, controllers (not keyboards) are essential, and motion sickness—believe it or not—is addressable. Though perhaps the biggest lesson they learned was that if you tell anyone—friends, family members, fellow developers; anyone—that you're thinking about "betting the future of the company on VR," every single one of them will inevitably say the same thing: YOU'RE CRAZY.

4. CCP
Reykjavík, Iceland

"I know it's a bit crazy . . ." wrote Sigurður "Siggi" Gunnarsson, a senior web developer at CCP Games, in a late 2012 company-wide email that he hoped might lead to the creation of a VR team. It didn't need to be a big team—even just a few would signify a great start. "I know it's a bit crazy to spend your free time doing something you spend your working hours on, but the goal here is not to create a commercial success, but to have fun, learn new things, and try out new ideas."

Shortly after this call to arms, a small crew of collaborators came together. In addition to Gunnarsson, there was graphic artist Andrew Robinson, software engineer David Gundry, and a pair of QA (quality assurance) analysts: Ian Shiels and Louisa Clarke. On paper and in person, this small skunkworks team had little in common. Except for two things: they all had relatively low-level roles working on CCP's hit space adventure MMORPG *EVE Online*; and they all had been fascinated by VR for many, many years.

"For me, it started in college," Gunnarsson recounted to the team over beers, as they met at a local bar to discuss potential game ideas. "Games like *D&D* and *Shadowrun*. I was always just imagining how would it be if you could make a computer game in the future where you and your friends could go into VR and just go somewhere else and work together as a group."

The crew really liked that idea—of group play; of something social; of creating a VR game that thrust multiple people into the same shared experience. But what should that experience be? A roller-coaster ride? A rocket race through space? Maybe a wormhole battle? *Wait, what the hell was a wormhole battle?*

Andrew Robinson took notes on a bar napkin and drew little sketches of these ideas. Just scribbling, just screwing around, until Louisa Clarke threw out a concept that sounded like a game that they all suddenly very much wanted to play.

5. CHRIS GALLIZZI
Los Angeles, California

Like many gamers in 2013, Chris Gallizzi was obsessed with *The Elder Scrolls V: Skyrim*. After all, what could possibly be better than playing as Dragonborn, training with the Greybeards, and battling Alduin in an epic, open-world civil war? Well, actually, Gallizzi thought, there *was* one thing that could make *Skyrim* even better: actually *becoming* Dragonborn.

As the head of R&D for Hyperkin—a hardware manufacturer best known for cloning retro consoles—Gallizzi wasn't afraid to get his hands dirty. So he used an open-source 3-D driver called Vireio Perception and began modding a version of *Skyrim* that would work on his DK1.

When the mod was in a semiplayable state, Gallizzi called up the head of Hyperkin to tell him how incredible it felt to be immersed inside his favorite game. "I want you to see," Gallizzi said, and then took in his PC and devkit for everyone in the office to see.

Unfortunately, Gallizzi's initial demo didn't go so well—leading several of his colleagues to feel nauseated—but that only inspired him to make his mod better. To perfect the warping and stabilize the experience, which he did during his off-hours over the next two weeks before demoing it all again. This time, the reaction was totally different; this time he actually made believers out of a few people. And now that his colleagues were starting to see VR through rosier-colored eyes, Gallizzi made his move.

"I think we should get into VR," Gallizzi told the head of Hyperkin: CEO Steven Mar.

"That's a nice idea," Mar replied. "But we're *solely* focused on retro-gaming."

This was true, of course. The company's claim to fame was their RetroN 2, which was a two-in-one console that could play cartridges for both the NES and SNES systems. Getting into VR wasn't exactly a lateral move. "But," Gallizzi explained, "when Hyperkin first started, retro was a small niche market. And now it's kind of exploded. VR is niche now, but I think it's about to blow up in a similar way. So let's get in at the ground level!"

Mar wasn't yet ready to devote many resources to this new niche, but he did give Gallizzi permission to work on VR on the side. Perfect! That was all he wanted. And then almost as if a reward for his gumption, Gallizzi's *Skyrim* mod started blowing up online. An article on Kotaku ("Here's *Skyrim* Running on the Oculus Rift VR Headset") led to parroted pieces on IGN, Polygon, GameSpot, and dozens more. But for a true believer like Gallizzi, all that press—while humbling—was nothing compared to a vote of confidence from Cymatic Bruce.

6. CYMATIC BRUCE
San Jose, California

By day, Bert Wooden was a program director for Galileo Learning's summer camp initiative; by night—under the alias "Cymatic Bruce"—he was a beloved VR evangelist.

With warm greetings to "Rifters," "VR Heads," and "Fellow Purveyors of Virtual Worlds," Cymatic Bruce's YouTube videos chronicled his daily VR adventures—quickly becoming the go-to streaming source for those wanting to learn about the latest Oculus mods, demos, and games. And on April 13, Bruce recorded what would become one of his most-viewed videos to date.

"All right!" he said, starting off the stream. "This looks wild . . ."

The "this" was Chris Gallizzi's *Skyrim* mod and for the next minutes, Cymatic Bruce took viewers on a journey through a VR version of the game. Then after climbing a tower and avoiding the fiery breath of an angry dragon, Bruce proclaimed, "*Skyrim* is fantastic, people. It's fantastic . . . you have to use mouse and keyboard, unfortunately (unless someone has a hack to use both controller and the mouse at the same time), but yeah: this is great, this is really good. I feel great afterwards as well. And all of the stuff when the dragon was popping out is just . . . really shocking."

Cymatic Bruce wasn't the only Rifter streaming VR videos, so what was it that elevated him above the crowd? His earnestness? His curiosity?

His ability to toggle between the technical and the cool? It was all these things, yes, but there was another thing, too: consistency. Just about every day—during this exciting, unprecedented period where thousands of devs were playing with VR for the first time—he put up a new video on YouTube. New mods, new demos, new games.

So it was no surprise when, the following day, Cymatic Bruce was back to work: "All right! Hey, Rifters. Welcome to another video . . . *Half-Life 2.*"

7. PAUL BETTNER
McKinney, TX

"I played *Half-Life 2* on the Rift," Paul Bettner told the small-but-growing team at Verse. They were now up to sixteen devs. "It is a life-changing experience. No exaggeration."

The artist inside Bettner wanted to follow this up by proclaiming that his company would now be shifting all its resources from developing for Ouya to Oculus, but the businessman in him knew that just wasn't practical right now. Because even if Ouya flamed out, the game they were building for Ouya's console—a first-person adventure game called *Thereafter*—could easily be ported to PC (and that PC version could probably even be enabled to work in VR). Whereas, unfortunately, the reverse just didn't seem true: anything designed with VR in mind seemed unlikely to port easily for consoles or PCs. And while it was very impressive that Oculus had sold 20,000+ devkits, that was literally less .001 percent of the install base for console and PC gaming.

"It's a chicken-and-egg thing," Bettner told his friend Nabeel Hyatt over the phone.

Bettner had known Hyatt for years and they shared a special bond: both had sold their gaming companies to Zynga in 2010; and then both served as VPs at Zynga for two years until they could no longer resist the siren song of start-ups. However, they went about doing so in rather different ways: Bettner, of course, left Zynga to go and start Verse; and Hyatt left to go and help start several companies as a VC at Spark Capital. Which, in this instance, was actually what had prompted the conversation: Hyatt was doing due diligence for Spark's investment in

Oculus. And at the end of the call—after Bettner expressed how impressed he has been by Luckey, Iribe, and the rest of their team—the conversation turned personal and Hyatt asked if he was developing a game for the Rift.

"I've reassembled the team that made *Words with Friends*," Bettner replied. "Sixteen of the most talented developers, artists, and designers I've ever worked with. We're creating a brand-new first-person adventure game. A game that also happens to work in VR although it's not designed specifically for that. So, yes, we *are* developing a game for the Rift, but it isn't the game I *really* want to make. The market is just too new, so we needed to mitigate that risk and design something that'll work on multiple platforms."

Hyatt understood. He was no stranger to battles between passion and logic.

"But . . ." Bettner said, as their call came to a close. "I kinda wish this wasn't the case. I kinda wish we could somehow take the craziest risk. I wish we could make a game that could *only* exist in VR."

Even more specifically, what Bettner *really* wanted to make was the "Mario of VR." He wanted to do for VR what *Super Mario Bros.* had done for the 8-bit NES; or what *Nintendogs* had done for the Nintendo DS; or what *Wii Sports* had done for the Wii. The artist in him wanted to plant a flag in this new medium, but the businessman whispered that his growing game studio was in no position to take such a risk.

If only there was a way to satisfy both sides—to be a pioneer, but a responsible pioneer. And then shortly after his conversation with Hyatt, Bettner had a crazy idea; so crazy, in fact, that he thought it might actually work.

8. JOOHYUNG AN
Seoul, South Korea

Joohyung An enjoyed playing games. But to him, the magic of VR was about so much more than that. Maybe it harkened back to his college days, when he was majoring in architecture and first learned how VRML—the virtual reality modeling language—could be used to visualize buildings, communities, and all sorts of virtual environments.

That was all now years ago, but it had stuck with An enough that between the time when Oculus's Dillon Seo gave him a demo of *Hawken* and his DK1 arrived in the mail, his mind had moved away from games to movies. What he was thinking specifically was how great it would be create a virtual movie theater. And the beauty of such a place would be that—with just the six-inch display in the headset—you could actually feel like you're looking up at a sixty-foot-wide screen.

9. GOROMAN
Kawasaki, Japan

"This is so dangerous," tweeted Yoshihito Kondoh: aka GOROman. "I feel like I'm in the movie theater."

GOROman loved Joohyung An's VR Cinema3D. And he loved seeing what other devs were coming up with. Every day, it seemed, there was something new to try! And every day, it seemed, there was some new reaction video sweeping the internet. His favorite was the one where a ninety-year-old grandmother freaks out: "It's so real . . . oh Lordy!"

That video had already been viewed over a million times. But it wasn't just the buzz and cool content that had GOROman feeling bullish about VR. It was also the sense of community—this feeling that he was a part of something special. And nowhere was this feeling summed up better than on the Kickstarter page for Cloudhead Games' upcoming puzzle/adventure game:

CALLING ALL OCULUS RIFT DEVELOPERS!
We see all of you as brothers-in-arms, brave pioneers of the new frontier! We want you involved, traveling with us on this journey. We would like to invite you into our process, to help us identify issues and to find creative solutions. We want you to learn from our trials and tribulations so that you can turn around and create amazing content for the Oculus Rift.

The camaraderie and creativity of the dev community inspired GOROman. It made him even more eager to share and revise the pet project he'd been working on: *MikuMikuDance* in VR.

"Miku" was Hatsune Miku, the turquoise-haired, teenage pop sensation who was taking the Japanese music world by storm. Also: she didn't actually exist. At least not in the conventional sense. For she was not human, but rather a piece of Vocaloid software developed by Crypton Future Media, in conjunction with the Yamaha Corporation. Marketed as "an android diva in the near-future world where songs are lost" and first released in 2007, Hatsune Miku quickly resonated with a generation of musicians and music aficionados who fell for her sweet, not-too-robotic synthetic voice. And since her debut, Miku has voiced dozens of hit songs, starred in several video games and—as a performing hologram—sold out concerts around the world.

But there was one thing Hatsune Miku had not yet done: appear in virtual reality. So GOROman set out to change that—to create a way for fans to finally "meet" the pop diva. To look at her; to be looked at by her; to spend time with a fictional character that somewhere along the way had become real.

CHAPTER 24

THE FUTURE OF GAMING

April 2013

"I THINK I NEED TO HOP OVER TO ICELAND," MITCHELL TOLD IRIBE.

"Iceland? What's in Iceland?"

"Well . . . apparently . . . CCP!" Mitchell said, explaining how, without quite realizing where CCP was based, he had been corresponding with the company's senior director of business development—a guy named Yohei Ishii—and agreed to come out (and bring along some devkits) for some kind of a VR demo that CCP was planning to unveil at that year's Eve Fanfest. With Iribe's blessing, Mitchell soon found himself on a red-eye flight to Reykjavík.

Traveling alone (though soon to be joined by Chen in Iceland), Mitchell was surprised to run into a familiar face on the plane: veteran game journalist Ben Kuchera.

"What are you doing here?" Kuchera asked, already knowing the answer. But what he didn't know was if Mitchell was headed to Fanfest as a fan, or as part of an Oculus-related endeavor. "You play *Eve?*"

"Uh, I've *played* Eve," Mitchell began, now realizing that if Kuchera reported running into him, it would likely ruin CCP's VR demo surprise. So he quickly pivoted. "And I *love* Eve! It's the best!"

Kuchera immediately understood Mitchell's ploy. But instead of "outing" him, Kuchera posted a vague-but-upbeat tweet: "There is a super-secret EVE Online tech demo for press later. I figured out what it is, and I'm VERY excited." It was a gesture that Mitchell appreciated (almost as much as he appreciated this reply from someone in the Netherlands: "If it's Oculus Rift, I'll pee myself").

After arriving in Iceland, Mitchell met up with Chen at their hotel. Both were jet-lagged, undercaffeinated, and generally unsure about CCP's VR plans, but they were just stoked that CCP had anything planned at all. Unlike most massively multiplayer online (MMO) games, which attract players on release but then lose them over time, *Eve: Online*—that year celebrating its tenth anniversary—was on an upward trajectory. Just two months earlier, the game surpassed five hundred thousand paying subscribers. Therefore, any association with *Eve Online* was bound to be a win for Oculus, which is why, when a pair of devs came by the hotel to pick them up, Chen and Mitchell were eager to find out how big of a win this might be.

"After sending out that email," Gunnarsson recounted, "I pulled together a small team. We started meeting over beers and discussing what sorts of game ideas would be fun to make in VR. I was personally pushing the idea of something like *D&D*. But then—well, you should say . . ."

"So I had this idea," Louisa Clarke said. "Why don't we just fly the fighters in *Eve*?"

In essence, Clarke's idea was to take the drone "fighters" from *Eve Online*, put players inside those ships, and then create around that a multiplayer dogfighting game. Conceptually, this idea quickly caught traction with CCP's VR skunkworks team. But just as quickly, they began to encounter all sorts of challenges unique to designing a game for VR.

For example: simulator sickness was a big issue. As was acclimating players to their virtual environment (while also "deprogramming" them from always looking straight ahead, like in a traditional game). But working around the clock and living by the motto "Don't Stop 'til Sunset" enabled the team to put together a viable demo a few weeks before Fanfest was set to begin.

"We showed what we were working on to Yohei and Jon [Lander, the executive producer of *Eve Online*]," Clarke explained, "who then showed it to Hilmar [Pétursson, the CEO of CCP]. They were all quite pleased and gave us the support to keep going."

"If you like the prototype," Gunnarsson said, "we would like to show it at Fanfest."

By the time Mitchell and Chen were brought back to CCP's office, they were fired up for the demo. A playable demo . . . set in the *Eve* universe . . . that Yohei and Hilmar had put their support behind: holy shit!

"This is mind-blowing!" Mitchell raved, trying out the demo at Gunnarsson's desk. It was, without hyperbole, the greatest VR experience he had had to date; capturing the visceral thrill of space battles that he'd only ever seen in films like *Star Wars* or *The Last Starfighter*. And in that respect, being in that virtual cockpit was like the realization of all his childhood dreams.

Shortly after this realization—that Mitchell had helped create something that enabled a team in Iceland to create something that delivered the magic of a lifelong dream—Jon Lander came over to say hello to their guests.

"Oh my God!" he said. "It's great to meet you! What do you think of the demo?"

"This is totally blowing me away," Mitchell replied. "This is really what VR is all about." Lander was hoping Mitchell might say something like that. Because, with Oculus's permission, CCP wanted to set up demo stations for the final day of Fanfest; and also wanted to add a tweak to the end of Hilmar Pétursson's keynote presentation. Would that be okay?

Of course. Of course! Holy shit: yes, of course!

Less than twenty-four hours later, CCP's CEO addressed a crowd of adoring fans. "Now that we start here," he said, "on the ten-year birthday of *Eve Online* and we're talking about the second decade and all those things, it's relevant to come and show something that is basically from the future. So an amazing sort of grassroots initiative happened about sort of bringing something really new into the EVE universe—using a technology that is about to become available in the real world . . . and I'm now going to play a trailer, which they put together."

Suddenly, cockpit images from an epic space battle, and set to thumping music, filled the screen behind Pétursson. *Eve-VR*, this game was to be called.

ON MAY 7—IN AN EMAIL ENTITLED "THE FUTURE OF GAMING NEEDS YOU!"— Luckey reached out to the engineer who had once upon a time inspired him to begin modding: Ben Heck.

"You have a very particular set of skills," Luckey wrote. "Skills you have acquired over a very long career. Do you have any interest in working at Oculus VR?"

Luckey wanted Heck to work for something called the "Z-Team," which—as conceived by Luckey and Patel—was to be a group that rapidly brainstormed, built, and iterated on virtual reality technologies. Led by Luckey, this prototyping team would ensure that Oculus continued to innovate outside the box as the rest of the company focused on what would be *inside* the box of DK2.

On paper, this sounded great: a dream team of pie-in-the-sky prototypers. But from the get-go, Patel was worried about how such a team might impact the culture at Oculus. "I don't like the idea of having a team that only prototypes and a rest of the company that primarily ships products," he told Luckey. "We also have a room full of talented and creative people who have long lists of ideas they'd like to prototype but are spending all of their time on what Oculus needs immediately: shipping the SDK, DK1, DK HD, and DK2."

Luckey loved this about Patel: the guy was always thinking about unintended consequences. In this case, he was probably right.

"Another option may be to do a prototype jam," Mitchell suggested, meaning some kind of an internal company-wide hackathon. Or maybe—echoing the now-famous idea set forth in Google's 2004 IPO letter: "We encourage our employees, in addition to their regular projects, to spend 20% of their time working on what they think will most benefit Google"—Oculus ought to institutionalize some sort of passion project proviso. "Something along the lines of Google's 20 percent time," Mitchell said. "Although 20 percent might be a lot for us at this stage. Ten percent?"

"As far as I am aware," Patel told Luckey and Mitchell, "it rarely occurs at Google, and when it does, it is more like 120 percent time.

The context switch requires too much overhead, and since you are still at your desk, the 20 percent project takes a backseat to actual work."

So instead of emulating Google's 20 percent program, or putting together a formal Z-Team, Patel proposed a hybrid idea: Anyone in the company can propose a project, either solo or with other employees. All proposals would go to Palmer for feedback, approval, and supervision (as he would be the only permanent member of the Z-Team). And, if approved, the employees would then move to a designated area and spend up to two weeks rapidly iterating their idea.

Mitchell liked this idea—provided that proposals also be vetted by him, Patel, and Iribe—and Luckey liked it, too—provided that Oculus brought in some additional hardware firepower. So he put together a list of potential recruits that included special-skill-set Ben Heck, ModRetro alum Julian Hammerstein, and no-longer-at-Valve-maker-extraordinaire Jeri Ellsworth.

Ellsworth, as it turned out, would not be an option. As she told Luckey and Patel a few days later, she was starting a company of her own, Technical Illusions, which would be a competitor to Oculus and which planned to sell a headset that Ellsworth would eventually bill as "the most versatile AR & VR system."

A bummer though it was to miss out on Ellsworth (and, apparently, to prepare for some competition), Oculus wound up wooing someone else from Valve—someone who would end up playing a critical role in coding DK2.

CHAPTER 25

ANDREW REISSE

May 2013

"MIKE AND BRENDAN HAVE KNOWN EACH OTHER *FOR*EVER," PETER GIOKARIS explained to Steve LaValle over lunch at Chipotle, giving his new colleague the lay of the land. "I heard that back in the day, when they were starting their first business, they even lived together at Brendan's mom's house for a while. So you can imagine how close they are. They're like brothers, basically."

"Ah!" LaValle replied, quickly replaying in his mind the short history of his relationship with Iribe and Antonov, which now suddenly made a bit more sense. "And what about Nate? He's significantly younger than them, isn't he? Maybe by ten years or so? So he couldn't have been there then, but he seems to have a very close connection with the two of them."

"Yeah, exactly," Giokaris said. "Nate was definitely later to the party—I don't think he hooked up with them till 2008 or 2009—but he's *definitely* a member of that 'Scaleform Mafia.' He's superclose with Mike and Brendan. Especially Brendan; those guys are almost inseparable. And Andrew [Reisse] is part of that group too. From the very beginning. He was actually Mike's old roommate in college. Before they all dropped out and started Scaleform."

"What's Andrew like?" LaValle asked. "I've only interacted with him on a handful of occasions, but he seems difficult to get to know."

"Tell me about it!" Giokaris replied, now chuckling. "I'm doing the

Unity integration, and he's doing the Unreal integration, so you'd think we'd hit it off. But nope, not at first. At first I thought the guy just plain didn't like me! But I talked to Mike a bit and realized that's just kind of Andrew's way. He's very quiet and very selective of what he says. And when he speaks, he picks his words out carefully, like he pauses for a few seconds before speaking."

LaValle had heard fragments of these details before, but never had it all coalesced together so nicely. For this, LaValle was grateful to Giokaris.

"I'M NOT EXACTLY SURE HOW TO PHRASE IT," LAVALLE TOLD GIOKARIS OVER yet another Chipotle lunch, "but, in a sense, my underlying curiosity is essentially this: What's the deal with Valve, man?"

Peter Giokaris started cracking up.

"What?" LaValle asked. "Is that weird for me to ask?"

Giokaris assured him that it wasn't weird. It was just funny, the way he asked it—like everyone was in on some secret and LaValle wanted to understand the joke. Giokaris also realized it must be kind of hard to understand the sacrosanct devotion that Valve typically received. Valve didn't just have fans, they had fanboys and fangirls; save for Joe Chen (who didn't really play games), everyone at Oculus fit this bill.

Prior to Oculus, Steve LaValle had never heard of Valve. But given how often his new colleagues talked about this beloved game company, he set out to learn as much as he could. This proved quite difficult, at least initially, since Valve was a notoriously tight-lipped company. "One of the most secretive (and lucrative) studios in the business," *PC Gamer* had even written earlier that month. But eventually LaValle found resources that provided a background into what made Valve so unique.

The most important was Valve's famous employee handbook, which opened up the keyhole into what made them so revered. Desks on wheels? No bosses? Nobody had ever been fired for making a mistake? Even if this wasn't actually how things worked, the existence of that handbook was proof by itself that, at the very least, Valve was different.

"So I *kind of* get it," LaValle said to Giokaris. "Lots of interesting personalities. Lots of freedom to create. And there's an inspiring backstory there—a mythology, it almost feels like—so I get why someone like you, or even like myself, would want to work there. It sounds like a very interesting environment. But it's the fan stuff I don't fully understand. Because, yes, they've made some very successful games, but at the end of the day—unless I'm reading this wrong—they are basically a digital distribution company. They seem to make most of their money from selling *other people*'s games. And while that seems like an excellent business model, kudos to them for sure, I can't think of any other company that's so beloved for taking a cut of work they did not create. In fact, it's usually the opposite! Isn't it?"

Once again, LaValle had asked a strangely up-front but reasonable question, one that made Giokaris ruminate on.

"Have you ever heard of the PC Master Race?" Giokaris asked. "Or sometimes it's referred to as the 'Glorious PC Gaming Master Race.'"

The PC Master Race was a tongue-in-cheek term used by PC gamers—meaning, sorry, Steve, those who played games on personal computers—to express, what this crowd fervently believed, the obvious superiority of gaming on PCs over consoles. The phrase was first coined in 2008 by Ben Croshaw, then a reporter for *The Escapist* who, when reviewing *The Witcher*—an action RPG that was only available on PC—snarkily quipped that "those dirty console-playing peasants" would therefore not ruin the fun for "the glorious PC gaming master race."

This epithet was originally meant as an insult, an attempt to mock what Croshaw saw as "an elitist attitude among a certain kind of PC gamer." Yet despite its derisive origins, the phrase managed to strike a chord with many in the PC gaming community. Maybe they saw some tribalistic truth in the phrase, or maybe they just liked how it sounded like bad dialogue from a shitty action movie. Whatever the case, the phrase was reappropriated and, through the magic of memes, became a rallying cry, a badge of honor, and, frankly, an easy way to make what most PC gamers thought was an accurate point: that anyone who gamed on consoles over PCs was kind of . . .

"Let's just say *missing out*," Giokaris said. "I'm more of a console guy, but I do play PC games from time to time. I don't want to say there's, uh, a 'herd mentality,' but there's a camaraderie that comes from being a PC gamer. Especially because it wasn't until somewhat recently that PCs could go toe to toe with consoles. So there's, you know, an underdog sort of thing going on."

"But it's not really an underdog situation any more, right?" LaValle asked. He had done research into the PC gaming landscape, since this would be Oculus's consumer base. Just one year prior, the PC gaming industry was estimated to have generated sales of over $20 billion.

"You are mostly correct," Giokaris said. "And a lot of that's because of Valve. They really pushed PC gaming at a time when others . . . I don't want to say that others didn't care, because there's *always* companies that care, right? But they went full force when others would not."

LaValle really liked that, since, in a way, that's what Oculus was doing, too. Throwing their hat in the ring while others waited on the sideline. And if things broke right then maybe, quite possibly, Oculus would end up becoming the "Valve of VR."

GIVEN THAT GOAL, LAVALLE FIGURED IT WOULD BE WISE TO KEEP LEARNING about Valve. And as luck would have it, he was given the perfect opportunity when Tom Forsyth—who had just been wooed from Valve to Oculus—came down to Irvine to spend some time with his new team. Forsyth's Valve experience, though, while generally positive, did not quite match up with the haven that LaValle had envisioned.

"So Valve is a very strange company," Forsyth explained. "I'm sure you've read the *Employee Handbook* and stuff? Well, the sort of terrifying thing is that all that is 100 percent real. And it sounds really interesting. But there's a couple of weird things, weird secondary effects."

"Like what?" LaValle asked.

"One is that since no one is your boss and you can change what you're working on at any time, that also means that you can't necessarily plan much. Like you can't say: okay, Alice, Bob and Carroll will be working on X for the next six months. That just can't happen. And so planning is not just hard, but essentially impossible. It's just a waste of your time. The other thing is that if you want someone to work on your project you have to, you know, *romance* them. You have to go to them at lunch and say: Hey man, there's this really cool project that I think you'd enjoy working on! And meanwhile, the project they're already on, those guys are maybe saying: No, man, don't go over there. Those guys are losers!"

LaValle laughed.

"It's all a bit funny, right?" Forsyth said, now laughing himself. "Anyway, a bit strange."

Strange indeed. And frustrating, too, it must have been. But given the freedom that Valve supposedly gave their employees, LaValle was still a little unsure why Forsyth had ultimately decided to leave.

"Because when it's all said and done," Forsyth explained, "I just . . . I like shipping stuff. Partly because of the discipline it requires—that discipline to answer questions like: What is important? What can be discarded? What can be put off until the next version? You learn a huge amount by doing that process. But the real big payoff is seeing people use your thing. And being a graphics engine person, my job is to bring a fake world into realization. I enjoy watching people use my stuff and knowing that, hey, without me, this wouldn't work—but with me, when done right, they'll think that no effort went into this at all. That's magic."

LaValle understood, though he never knew that game development—nay, shipping—could be so beautiful.

"I'M GONNA GO PICK UP CHIPOTLE," CHEN TOLD LUCKEY.

"Oh! Can you get me something?" Luckey asked.

"No. I mean, yes. I mean I'm going to go pick up Chipotle for *everyone.*"

It was about nine p.m. on May 29—just a typical work-all-night Wednesday evening—and with E3 only thirteen days away, everyone in the office was in crunch mode, too busy to think about food. But busy or not, the gurgle of grumbling stomachs could occasionally be overheard.

When Chen returned with the food, everyone moved into the conference room for what the team had started referring to as "family dinners." As was often the case, this tiny respite of time was spent bonding over fun and nerdy things. Like Batman vs. Superman, the viability of cryptocurrency, and whether or not upcoming games like *Wolfenstein: The New Order* would "suck" or "rule." These often heated, laughter-peppered conversations were, at once, both stupid and wonderful—adding up, over time, into a blurred banter that recalled the early days of Oculus. But by virtue of that blur, it became hard to remember who said what exactly, and impossible to tie those comments to specific days. But on this evening, there was one thing that Chen would always remember: Andrew Reisse thanking him for getting dinner—not once, but twice.

"Dude, relax," Chen said. "I just went on a food run, it's no big deal. We all do it."

"Yes, but thank you," Reisse said again.

So earnest and sweet, such a silly exchange this should have been. This was the last conversation they ever would have. Because that week, for whatever reason, Reisse took off Thursday (instead of Wednesday) for his weekly respite from the office to rest his hands. On that Thursday afternoon, May 30, 2013, Andrew Reisse was struck by a vehicle and killed.

The vehicle, a Dodge Charger, was driven by a Santa Ana gang member, a gang member who was attempting to elude police at nearly 100 mph on a residential street. At some point, the Charger ran through several red lights, collided with a truck, ricocheted into another car, and then, spinning out of control, hit a pedestrian—a brilliant thirty-three-year-old pedestrian named Andrew Reisse—who had just eaten lunch at Chipotle and was waiting for the light to change at a crosswalk on South Flower Street.

CHAPTER 26

IDEAS SO CRAZY
THEY JUST MIGHT

June/July/August/September 2013

"E3 IS NEXT WEEK," IRIBE SAID, SPEAKING TO A DOZEN OR SO SAMSUNG EX-
ecutives in San Francisco on June fourth. "We'll be showing off our
latest prototype: DK1 HD, which—as the name suggests—is basically an HD version of the DK1. So we're psyched about that. And
like you saw in the investor deck, we were nominated for Best Hardware at E3 last year, and earlier this year—at CES—we took home a
ton of 'Best Gaming' and 'Best in Show' awards. So we've got incredible momentum . . ."

By this point, Iribe had given a pitch like this a hundred times or
so. And of all those, this one was probably the hardest. Not just because
of what was at stake—the chance to build a relationship that could help
Oculus get the OLED screens they wanted for DK2—but because it
had only been five days since Reisse had passed away.

Of everyone at Oculus, Iribe and Antonov took Reisse's death the
hardest. After all, they had known him the longest and, somewhere
along the way, stopped thinking of him as a "friend" or "colleague"
but much closer to something like a brother. So from the moment they
first found out they'd been a mess. They'd start bawling when they saw
each other sometimes, and just generally still couldn't accept that he
was gone. So Iribe and Antonov did what they always did when in the

throes of personal conflict: threw themselves into their work, this time harder than ever before.

"This is my third company with Michael," Iribe continued. "Michael Antonov, our chief software architect. The last two exited over $400 million. We're confident this will be the best yet. We'll be creating innovative and disruptive technology that brings an entire new generation of video gaming and interactive entertainment to the mass consumer. Ultimately, it's all about how to hook players to the platform via the Oculus App Store. Basically: our version of Steam."

The Samsung executives appeared impressed.

Thank God, Iribe thought. He'd been dying to make traction with Samsung for months now—trying to find a way to purchase OLED displays—but the people there didn't have the time of day for production runs as "small" as 250,000 units, let alone the 25,000 or so that Oculus would be looking to order for DK2. Iribe had all but given up hope before Formation 8 cofounder Brian Koo (who was heir to the LG family) was able to arrange this meeting. None of these execs even worked in Samsung's Display division, but Iribe hoped he could maybe offer some advice.

"Now that you've heard our presentation," Iribe said, speaking afterward to one of the execs—a guy named Chan Woo Park, "what do you think it would take to get OLED panels from you? We need those OLED panels!"

A mischievous grin emerged on Park's face as he said, "I'd like to show you something privately. It's top secret."

Woo walked Iribe down the hall and into an office where he picked up a small cloth bag and told Iribe, "We have an idea."

Out of this little bag, Park pulled a 3-D-printed plastic phone holder that had pieces of foam glued onto one side. Park then picked up a Galaxy S4, dropped it into the janky holder and voilà: a cheap and easy way to view immersive content on your phone. "We're thinking that we could partner with you on this," he said. "You can help us make this. We can make a product together."

Iribe tried this makeshift immersive experience and, well, it was not great. High-latency and low frame rate, it didn't even have distortion correction. But then again, that was exactly why Park had suggested they work together.

"We can certainly make this better," Iribe said.

"I will help you get an introduction and a close relationship with the display team," Park offered, "if you will help us with this Galaxy S4–based device."

MEANWHILE, IN THE MIDST OF ALL THIS SAMSUNG STUFF, IRIBE RECEIVED AN email that made him very happy.

FROM: John Carmack
DATE: June 21, 2013
SUBJECT: RE: Oculus technical role

Ok, I just got home from my last day under contract at Id, and I haven't signed anything with anyone, so everything is on the table at the moment.

I hope we can formalize and announce some kind of technical advisory arrangement soon; so you need to meet my wife Katherine Kang to discuss specifics . . . Start from scratch with her, laying out the options as you see them and what you think the prospects are. She values my opinion, but she hasn't drunk the VR Kool Aid, so expect some level of skepticism . . .

Within days of Carmack's email, Iribe found himself across a table from Kang. And though she was a fierce negotiator, and though her husband clearly had many good options (i.e., Valve, SpaceX), not only was Carmack itching to get back to VR, but he also happened to believe that the future of VR was mobile, and he badly wanted to spearhead the mobile project with Samsung.

ON JUNE 22, PAUL BETTNER CONTACTED LUCKEY, IRIBE, AND MITCHELL TO share the "crazy idea" that he had come up with shortly after his meeting with Nabeel Hyatt.

"The iPhone afforded us [an] opportunity to risk it all on an unproven platform," Bettner wrote. "Because they solved the chicken and the egg . . . They had killer hardware and they shipped the killer apps. Safari, email, google maps, the camera—they kickstarted the platform with killer 1st and 2nd party software that created a vibrant platform that others (like us) could then take a risk on. This is what Nintendo has done so well too (historically at least). So, the crazy idea: what if—instead of seeking a typical publishing deal—we, Verse, did something radically different. What if we worked directly with Oculus (as publisher) to develop true 2nd party software for the first consumer Rift. What if we helped build those killer apps, those titles that could only exist in VR?"

"Let's talk more ASAP," Iribe replied, almost immediately. "Believe it or not, we had a similar hope and wanted to approach you guys."

Five days later, Luckey, Iribe, Mitchell, and Malamed flew out to visit Bettner's studio in McKinney, Texas. There, Bettner walked Oculus through his vision to build a game studio capable of becoming "the Nintendo of Virtual Reality." And with that inspiration in mind, Oculus hired Verse to develop a collection of short, mini-games that the two sides excitedly referred to as "the Wii Sports of Virtual Reality."

IN LATE JULY, IRIBE AND PATEL TRAVELED TO KOREA TO MEET WITH THE SAMsung Display team.

Most of the trip consisted of meetings to discuss the design of the first Gear VR prototypes and how Samsung could (and should) integrate Oculus' DK1 tracker tech. These meetings generally involved us going to Samsung Digital City, which is effectively a Samsung company town in Suwon, entering a big conference room, and sitting across from ten to twenty Samsung engineers from different departments trying to extract as much as information as they could from us while revealing as little as possible.

"Is it just me," Patel said to Iribe, "Or do they always seemed a little short of competent in meetings, but then they would turn around and build something impossibly quickly."

Iribe started cracking up.

"I'm serious!" Patel claimed. "And this is another point that I don't understand. Samsung owns basically everything from the machines that dig the iron ore for components out of the earth through retail stores to sell their products, and they somehow use that to cheat time and build products on schedules that shouldn't work."

That said, the highlight of the trip turned out to be Iribe and Patel showing off something that "shouldn't work": a 95hz Samsung phone that Valve's Monty Goodson had hacked to get working in low-persistence mode (and shared with Iribe to share as proof of concept to Samsung).

"So Nirav started by explaining the theory of low persistence and what the prototype was doing to achieve it," Iribe told Mitchell when he got back from Korea.

"At first," Iribe explained, "the Samsung Display engineers refused to accept and just shook their heads and repeated (in Korean) 'that's not possible.' We then showed everyone the prototype and reiterated how it worked. Again, their engineers shook their heads. This went on for at least a half hour to hour. At one point, I called Monty (around one a.m. Seattle time) to verify to Samsung that it was indeed operating at 95hz the way Nirav described. It also felt like the right thing to do. The experience was so surreal and frankly exciting, we wanted Monty to (at least partially) share in the fun. Luckily, he was still in the office!"

"That's amazing!" Mitchell said.

"It really was," Iribe replied, nodding his head. "It was a magical moment. One of those rare times that you know you're experiencing something so special and important that you'll remember it forever."

BY A WEIRD COINCIDENCE, THE 3-D-PRINTED HOLDER THAT SAMSUNG HAD BEEN using for their mobile VR demos had been produced by their research team in Dallas. Of all the 300,000 people at Samsung, John Carmack thought, it's remarkable that their team here in Dallas is the one doing their initial exploration. And as a result of this good fortune, he was able to check in on their mobile project before flying out to Irvine in late June.

"So what do you think?" Iribe asked his new technical advisor during that visit.

"Looking at it," Carmack said, "it's like all the other terrible phone-based VR things; like the Hasbro View Master and that one that Palmer did: FOV-2-GO. So, okay, this is no good right now. But I've done work on mobile devices, I've done iOS and Android—I understand the GPU's on there very well; and the sensors—and I think I can make this work."

Iribe grinned, as wide as his face could bear. Though that grin would soon be challenged by the one that came after what Carmack said next: "I'm probably, literally, the best person in the world to make this project come together. I can do this. I can make it really good."

For Carmack, his passion to pull this off this novel technical challenge was largely driven by his belief that Mobile VR could be very, very big. "Because," he described to Iribe, "it's not a long stretch to think of a world where these holders are the equivalent of, you know, phone cases. Where a billion people buy these. So this very likely could be our best way to get a billion people into VR."

Iribe responded enthusiastically, though in the weeks and months ahead Carmack would come to realize that this enthusiasm did not equal agreement. "For a long while," Carmack would later say about this period, "the mobile project was just this thing Oculus was doing to make Samsung happy so we can get the screens." Sometimes Carmack would find this lack of faith frustrating; but for the most part, he relished the freedom that came from this apathy. So much so that Carmack soon sought to expand his role at Oculus—to formally leave id Software, turn down enticing job offers from Valve and SpaceX and go all-in (albeit remotely) on Oculus and the future of virtual reality.

Luckey, Iribe, Mitchell and the rest of Oculus' exec team could hardly believe that they appeared to be on the brink of obtaining maximum VR effort from the John Carmack. Not just because he an icon and inspiration to each them for as far back as they could remember, but because it seemed unfathomable that Carmack would actually leave the company that he had been with for twenty-two years. And yet, here they now were. But there was one potential snag: to hire Carmack, his wife/manager made clear, Oculus would need to make him their CTO.

To Luckey and Patel, this seemed no snag at all. But Iribe, Mitchell and Antonov each had concerns—highlighted by Iribe emailing Luckey on July 9 to say that he was "not convinced Carmack is the best 'CTO' for Oculus given he'll be remote and not really into business decisions." Luckey pushed back. And after re-quoting something he had said to Iribe the first time they ever met— "When Jesus asks to borrow your clothes, your boots and your motorcycle, you say yes!"—he and Patel were able to persuade the others and get them to say yes to John Carmack as CTO.

ON THE SAME DAY THAT CARMACK WAS OFFICIALLY NAMED CTO, BETTNER emailed the guys at Oculus to let them know that his game company would be changing their name. "We are now Playful Corp! I liked Verse a lot, but I yearned for a name that was more fun, like what Newtoy was . . . Playful perfectly suits the nature of our studio and the products we're creating."

Five days later—with a new name and new logo now adorning the office—Bettner welcomed fellow Texan John Carmack to check out a few of the VR prototypes that Playful had been cooking up. Prototypes like:

- **TELEPORT TOWERS** (Inspiration: *Portal, Unreal*'s Teleport Gun)
  Lob a tiny device to ledges high above you, teleporting to it whenever you like. Try to reach the key to escape the room in as few throws as possible. Skill and planning take the front seat in this cerebral physics game.

- **MISSILE COMMAND** (Inspiration: *Space Invaders*)
  Aliens have massed and are bombing the city! Take control of the last turret in the city to blast their bombs out of the sky and to return fire on the encroaching hoard of spaceships. Just try not to level the city while you're protecting it, ok?

- **SPACE TWISTER** (Inspiration: *Tempest, Gyrus*)
  Strafe your agile fighter to dodge incoming meteors, enemy ships and bullets while returning fire with your trusty blaster. Enemy ships zoom right past your head in this immersive new perspective on a classic arcade shoot-em-up.

"Each one of these has enough of a hook that I want to keep playing even after getting the basic gist of them," Carmack said. "I think you guys are off to a great start!"

THROUGHOUT THE SUMMER AND INTO THE FALL, LUCKEY HAD A FEW CRAZY ideas of his own in an effort to answer this question: what does the ultimate universal VR controller look? Or perhaps more accurately: what is the best set of compromises we can make to create hand controllers that will allow developers to create any experience they can imagine?

"First and foremost," Luckey said to Patel, Dycus, Hammerstein and Heck (who had agreed to do some contract work for Oculus), "you *have* to have your hands in VR. You have to feel like your actual hand is in VR and *still* preserve the traditional inputs that are required to preserve to play what people think of as a game. Like you *have to* be able to do simple things—like wave at somebody, or point at something—but at the same time you *have to* be able to use this controller to shoot guns and perform typical in-game tasks."

As discussions between the Controllers Team continued, the word that kept coming up was "natural." And over time, in pursuit of that objective, Luckey's list of have-to's expanded to include fine-grained interactions. "Like if you're throwing a grenade," Luckey said, "you *have to* be able to do something as specific as pull the pin out of that grenade."

Pin-from-a-grenade precision presented numerous design challenges, not the least of which was that this would essentially require finger-tracking. How else would it be possible to naturally create the sensation of pinching thumb and finger against the top of a grenade? Well, this got Luckey thinking—about a prototype controller that Virtuality had made in the '90s.

"They made a prototype controller that had a hard plastic shell. And for the controller grip, it had molded rubber for your fingers. And then *underneath* the rubber cover for all that, there were several soft rubber hoses in the grooves. And then those ran to a pressurized air bladder and a barometer inside, which created this really clever capacitive touch system; so basically if you squeezed the controller on any of your fingers, it could detect which finger was pressing it."

Guided by that capacitive touch inspiration, but combined with modern features like a thumbstick, analog trigger and scrolling wheel, the Controllers Team started designing something that Luckey called the "Oculus Virtuflexitron 3000."

OVER IN TEXAS, BETTNER'S STUDIO BEGAN RAPIDLY PROTOTYPING MINI-GAMES.
Games like:

- **MONOLITH** (Inspiration: *Tron, Myst*)
  Solve runic puzzles to unlock the gates surrounding a magical
  tower. Once freed, you can soar across the fantastic landscape,
  exploring at your own pace.

- **HAUNTED MANSION** (Inspiration: *Luigi's Mansion,
  Ghostbusters*)
  Track down supernatural disturbances in an old abandoned
  house. Your trusty flashlight can help you see hidden ghosts
  that you can knock from their hiding places. Once discovered,
  you'll need to keep your proton beam centered on the ghost to
  wear it down before it escapes.

- **SUPER CAPSULE BROTHERS** (Inspiration: Mario games, *FEZ*)
  Super Capsule Brothers reinforces a beloved style of gameplay
  with a compelling new perspective. Jumping and dashing has
  never been easier, and the towering heights and yawning pits
  lend a sense of urgency and immediacy not present in 2-D
  games.

WHEN IT CAME TO SOFTWARE, IRIBE'S PERSPECTIVE WAS GENERALLY RE-
spected internally at Oculus. After all, he had been a programmer once
upon a time and had over a decade's worth of experience managing
software teams. But when it came to hardware, well, Iribe had developed a reputation for what some had started referring to as "Seagull
Management." Meaning that Iribe would "swoop in out of nowhere,
seeming to possess minimal knowledge of a project, and then just start
shitting all over it." Oftentimes, these seagull sightings would end with
a "suggestion from on high" that didn't really seem to make a lot of
sense—like one of the most memorable examples, which occurred on
August 8.

By this point, the Controllers Team was making good progress on the Virtuflexitron 3000. So it surprised the team when Iribe dropped in to suggest that "my current feeling is that the two controllers really should be symmetric and work in the left or right hand the same."

After receiving swift rebuttals from Patel ("Definitely not") and Luckey ("The comfort will be subpar"), Iribe explained his reasoning with a line that neither of them would ever forget, "Jobs would ship ONE magic wand."

"Jobs was smart," Luckey replied, "but it took a shouting match with Jony Ives to convince him into allowing native apps on the iPhone. He sometimes valued form over function, something we cannot afford to do. We need a great VR controller, not a cool-looking one."

Over the next few weeks, Luckey and Patel were able to get Iribe on board with their way of thinking with regards to the Virtuflexitron 3000—culminating in a September 3 email where Iribe told Luckey "this should be your sole mission in life."

With a green light from Iribe, Luckey hit the ground running. And by the end of that day, he had already found the industrial designer who he thought they ought to work with in bringing their project to life.

"Carbon Design," Luckey wrote to Patel. "We should work with these guys . . . they do a lot of handheld things. Controllers, racing wheels, mice, etc."

CHAPTER 27

THE ROOM

September 2013

"HEY, BRENDAN," MICHAEL ABRASH SAID, PHONING IRIBE IN MID-SEPTEMBER, "we have something in the office at Valve that you're gonna want to see . . ."

When Michael Abrash says you need to see something, Iribe thought, you go and see that thing ASAP! Within days, Iribe took a flight up to Washington to see what had Abrash so excited.

"Nobody's gotten sick yet," Abrash explained, walking Iribe down a hallway toward a room to demo Valve's latest prototype. Something they tentatively referred to as "AtmanVR."

"Not even you?" Iribe asked.

"No, I haven't gotten sick either," Abrash answered.

Iribe and Abrash were, by far, the most sensitive with regard to simulator sickness at their respective companies. Each could only spend about a minute in DK1 without feeling queasy. So as much as he admired Abrash, and knew Abrash to be above technological exaggeration, Iribe was skeptical that Valve had put together a demo that wouldn't turn his stomach. That skepticism heightened when Abrash brought him into a small, low-ceilinged room that was empty save for the following:

- Dozens of black-and-white QR-code-like fiducial markers plastered up and down the wall

- A bulky headset with exposed circuit boards and loose
  cables
- A man seated in the corner of the room, sitting behind a
  computer: Atman Binstock

"I'm going to run you through the demo," Binstock explained, then typed a few commands into his computer and suddenly—*whoosh!*—Iribe was transported to a virtual space where hundreds of small cubes hung suspended in the air.

Iribe swiveled his head and then looked up and down. It was a beautiful environment—this room of shimmering cubes—but thus far nothing extraordinary.

"You can move around," Binstock suggested.

Iribe obliged—crouching forward, savoring the freedom to move without a controller—as he inspected the underbelly of a glowing cluster. "Whoa," he murmured. Or maybe he hadn't even said it aloud. It was becoming hard for Iribe to differentiate between what *was* real and what *seemed* real.

What made this experience possible was a small camera on Iribe's headset—a camera that scanned the room, read fiducial markers, and then used that spatial information to track Iribe's position and properly orient what he saw in the virtual world.

Whoa.

Binstock had an entire sequence of demos. Next, he transported Iribe to a room that had what appeared to be web pages on each wall. Yes, web pages, as if Binstock were really trying to baffle his subject.

"Pick a word," Binstock said, meaning this literally. He wanted Iribe to reach forward and simulate grabbing a word from one of the web pages. Once again, Iribe obliged, and again, he was stunned. No matter how he oriented himself—no matter how quickly he turned his head or twisted his body—the word didn't smear. Nor—still turning and twisting—did the environment around him jitter or blur. Throughout the entire sequence of rooms that Binstock guided him through that afternoon, nothing did.

"Again," Iribe said, as soon as the demo was over.

"Sure," Binstock replied. *Whoosh!*

After going through the whole sequence another time, Iribe took off the headset. "Man, I feel completely comfortable. How long has it been?"

Abrash and Binstock checked the time and told him it had been about thirty minutes.

Whoa.

"I felt *totally* comfortable," Iribe proclaimed. "But there was this other weird sensation I was wrestling with. Like I truly felt like I was *there*, does that make sense?"

"Yeah," Binstock replied. "That's *presence*." That really was the perfect word to describe what he had felt.

Iribe was tight on time (he had a flight to catch), but Abrash and Binstock quickly told him about what made that sense of "presence" possible. Low-persistence and distortion correction were important, but more than anything the key was great positional tracking. Without that, your brain starts to reject what you are seeing. "Especially for sensitive folks like us," Abrash added with special acknowledgment.

"Thank you guys for showing me this," Iribe said. "I've gotta think about what this means. But right now, I've got to run to the airport."

"Want me to give you a ride?" Abrash asked.

When Michael Abrash offers you a ride to the airport, Iribe thought, you say yes ASAP! Not only would this give him more time to talk about the room demo, but it'd also give him some one-on-one time to take Abrash's pulse about maybe leaving Valve and coming to work at Oculus. Over the past year, Iribe had half-heartedly floated the idea on several occasions.

Abrash was flattered by Iribe's compliments during their ride, but quick to credit Binstock. "Atman probably knows more about what it takes to make VR work than anybody on the face of the earth. He was key to going and looking through literature, and he is the person who most understands how VR systems work."

"I agree," Iribe replied. "I need both of you guys at Oculus championing this thing with me. So what will it take to make that happen?"

Abrash chuckled, but, as he would say several times during their ride to the airport, he wasn't really interested in making a move at this time.

Naturally, Iribe seized on the phrase "at this time," as his mind swelled with permutations of when might be better, and what he might be able to dangle in the future. Equity? Autonomy? The chance to do something special with Carmack once again?

"You know, Brendan, I think the company that's going to make VR really successful is going to be a big company. Because the capital that's going to be required to do the custom displays, the custom hardware, the custom sensor systems, and all this work—to build out the full headset—it's going to be very expensive to really do this right."

"Don't worry," Iribe told him. "We've already raised $16 million through our Series A and we'll go raise a lot more."

"Sixteen is great," Abrash said. "But you're gonna need so much more."

As an example, Abrash pointed to Microsoft and their in-development AR headset (HoloLens). He had heard that Microsoft was spending many hundreds of millions of dollars, perhaps even billions, on their AR project. And the HoloLens wouldn't even be consumer-ready for years!

"Just trust me," Iribe pleaded. "If that's what it's going to take to do this thing right, then I'll go out and raise that money."

This wasn't enough to settle Abrash's skepticism. But the good news was that unless there was a big company out there who was ready to try and do VR for real, then Iribe was more than happy to continue his research at Valve; and more than happy with everything Oculus was doing to push that boulder up the hill. Even if Oculus never managed to be the "company that's going to make VR successful," what they were doing was still critical to the larger revolution.

By the time Iribe was dropped at the airport, he was already putting together a mental list of potential investors for a Series B. Then, after imagining some of the things he might say during those Series B pitch meetings, he reflected on that incredible sense of "presence" he had just experienced at Valve and emailed the execs at Oculus:

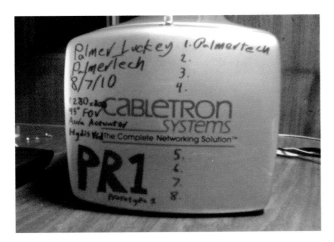

In 2009, Luckey builds his first prototype headset: the PR1.
(Photo courtesy of Palmer Luckey)

In 2012, Luckey founds Oculus and poses with Chris Dycus (who will soon become the company's first employee).
(Photo courtesy of Chris Dycus)

VR developer Denny Unger creates a logo for Luckey's VR startup; but little do they know that this logo (and everything else about Oculus) is about to change . . .
(Photo courtesy of Denny Unger)

In mid-2012, after selling a pair of high-profile startups, programmer-turned-entrepreneur Brendan Iribe looks for his next challenge.
(Photo courtesy of Jack McCauley)

In June 2012, Luckey dines with Iribe and other members of the "Scaleform Mafia."
(Photo courtesy of Palmer Luckey)

In July 2012, Iribe is blown away by Luckey's virtual reality headset.
(Photo courtesy of Michael Antonov)

Software architect Mike Antonov is also blown away (but only after he remembers to remove his glasses).
(Photo courtesy of Michael Antonov)

Oculus obtains the ultimate vote of confidence from Valve founder Gabe Newell, and legendary game programmer Michael Abrash.
(Photo courtesy of Palmer Luckey)

After launching one of the most successful Kickstarter campaigns of all time, Luckey heads to QuakeCon for a panel with his childhood hero John Carmack.
(Photo courtesy of Palmer Luckey)

In December 2012, *Words with Friends* co-creator Paul Bettner decides to take a chance on virtual reality.
(Photo courtesy of Joe Chen)

In China, factory workers build Oculus' iconic DK1 headset.
(Photo courtesy of Nirav Patel)

To manufacture their headset, Oculus relies on Guitar Hero Jack McCauley.
(Photo courtesy of Jack McCauley)

Joe Chen, whom Luckey calls the "ultimate foot soldier," serenades his new colleagues at the 2013 Consumer Electronics Show.
(Photo courtesy of Joe Chen)

Oculus founders Palmer Luckey, Nate Mitchell, Mike Antonov, and Brendan Iribe autograph their first product.
(Photo courtesy of Joe Chen)

Before, during and after, Oculus steals the show at GDC 2013.
(Photos courtesy of Joe Chen)

The Oculus gang unveils their Crystal Cove headset at CES in January 2014.
(Photo courtesy of Joe Chen)

Palmer Luckey gets photographed for a feature in *WIRED*.
(Photo courtesy of Joe Chen)

Palmer Luckey and hardware guru Nirav Patel contemplate leaving the tech industry to pursue careers in male modeling.
(Photo courtesy of Palmer Luckey)

Luckey loves playing Valve's *Team Fortress 2* in VR.
(Photo courtesy of Joe Chen)

On March 25, 2014, Facebook finalizes a deal to acquire Oculus for $2.7 billion.
(Photo courtesy of Heidi Westrum)

Mighty Morphin Power Tinkerers.
(Photo courtesy of Nicole Edelmann)

Luckey and Edelmann cosplay as Quiet from *Metal Gear Solid V: The Phantom Pain*.
(Photo courtesy of Yui Araki)

Luckey and Dycus spend a weekend working on "side projects."
(Photo courtesy of Palmer Luckey)

Among friends and fans at Machi Asobi in Japan, Palmer Luckey is ready for what's next.
(Photo courtesy of Yui Araki)

FROM: Brendan Iribe
SUBJECT: Valve's Holy Grail

It's hard to fully put into words, but Valve has done the impossible—
they've created a VR experience with zero simulator sickness.

For me, the demo was the second most memorable and significant
experience since we began, second only to see the Rift running
Carmack's tech demo for the first time.

As everyone knows, I'm super sensitive to simulator sickness and
every Oculus demo so far has made me feel ill after some amount of
time, usually not long.

I'd like a group of us to visit Valve and experience the demo first
hand.

The magic? Getting everything, namely positional/orientation
tracking and distortion/fov really dialed in. No exceptions . . .

When Iribe got back to Irvine, he couldn't stop raving about his
transformative experience. "This changes *everything*," he told the team.
"Suddenly, it really clicked: VR is going to be for way more than gaming . . . and leaving there, I thought: wow, consumer VR is actually
going to work."

The folks at Oculus were psyched that their trusty leader was
psyched. They'd never seen him like this before and were glad he was
now a true believer. But on the other hand: What the fuck? Their
own CEO hadn't been a true believer this whole time? And the thing
that "finally" made him one had been built by a company other than
Oculus?

For several employees, Iribe's obsession with Valve had finally gone
too far. There was still so much further he was willing to go.

CHAPTER 28

JOCKEYING FOR POSITION

October 2013

THE MORE IRIBE REFLECTED ON VALVE, THE MORE HE YEARNED FOR OCULUS to have a demo like that as well. On October 2, after getting permission from Samsung to share some specs with Valve, Iribe emailed Abrash and Binstock to offer a trade: "We can send over single 120 Hz and dual 90 Hz samples, dev boards with screens, and our schematics for the Note 3 driver" in exchange for "one of Atman's VR room demos."

"Interesting thought," Abrash replied, saying he'd get back to Iribe.

"THIS IS A *TERRIBLE* IDEA!" LUCKEY SHOUTED ONE DAY LATER. THIS WAS THE first time he had ever raised his voice at Oculus, but he couldn't help himself. Days earlier, the team had received a proposal from the firm Luckey had contacted about doing the ID for Oculus' motion controllers (Carbon Design) outlining a process that would cost about $750,000. Luckey thought that was a good price—a *great* price, even— but instead of moving forward, Iribe and Mitchell had decided to pull the plug.

"Controllers are just *not* critical to Consumer v1," Iribe told Luckey and then reiterated over email on October 4. "Controllers are simply not the top priority. They're important and we want to have them, but there's a high chance they won't come until v2 . . ."

Not only did Luckey find himself on the other side of Iribe on what he believed was a critical decision, but he also found himself there alone.

"I am 100 percent behind holding off for consumer 1," Malamed said. "That seems wise to me."

"Good," Carmack added. "I have never felt like this was a good idea for the first consumer device, and the effort has distracted from things I think are more important and easier, like ergonomic iteration and built-in audio."

Meanwhile, in the midst of killing the controller project, Iribe heard back from Abrash about the trade he had offered with regard to the Note 3. Abrash thought this could work, but also wanted similar assets for Samsung's S4. Iribe was amenable to that. Over the next week, Iribe and Abrash negotiated a trade. Valve—after signing an NDA with Samsung—would get Oculus's driver specs and datasheets for the Note 3 and S4; and in exchange, someone from Valve would head down to Oculus and install their "VR room."

With a deal now in place, Iribe enthusiastically emailed the team: though this enthusiasm was sincere, few at Oculus shared it. And worse: any attempt to curb Iribe's enthusiasm would fall largely on deaf ears. Like this exchange between Iribe and Tom Forsyth:

FROM: Tom Forsyth
DATE: October 9, 2013

Just a quick reality check. I know the fiddy system very well—I wrote a fair chunk of that code (though Atman may have re-re-written it since). The precision and accuracy are significantly lower quality than are coming out of [Oculus' nearly finished tracking system] . . . So while it's nice to have a working system, it is a long way from any sort of "gold standard". On the other hand, it's nice to know that even with all its considerable faults, the wall-fiddy system gave you a good experience.

>>>**FROM:** Brendan Iribe
DATE: October 9, 2013

That's all fine and dandy . . . but nothing I've seen works nearly as well
as what I saw at Valve and we simply don't seem to have the bandwidth
to get it there . . .

Once again, Forsyth tried to kindly push back on Iribe. "There is
some serious 'grass is always greener' stuff going on here. I'm trying to
tell you . . . It's not that Valve have magic tech, it's that they're careful
about their calibration. We need to strike a balance somewhere between
our mad rush to ship something and Valve's mad rush to never ship
anything ever."

But again, Iribe was unmoved. Even though the demo that Valve
put together cost tens of thousands of dollars and was absolutely un-
productizeable, Iribe continued to praise its glory, as he constantly
demeaned what was being done internally. Especially the proprietary
tracking system that Oculus was putting together.

"For months now," Antonov vented to a colleague, "all he says is 'You
don't know what you're doing!' No, Brendan, ours is gonna be better.
'Not better than Valve! They are so good! They are so smart!' Brendan is
seeing Abrash and Valve guys as these VR gods. Come on, what are you
even talking about? We have brilliant professor Steve LaValle here and
I'm working and Dov is working: we all have this thing going. Trust me.
But he does not. Because I am not them. It makes me kind of sick."

"I'VE GOTTEN SICK EVERY TIME I'VE TRIED [THE RIFT]," IRIBE CONFESSED TO AN
audience of industry leaders at the first annual Gaming Insiders Summit in San Francisco.

Iribe had never actively hid this fact, but he hadn't ever publicly
announced it either. Today, however, he felt comfortable doing so. Because, after his trip to Valve, this was no longer the case.

"Every time until recently," Iribe continued. "In the last couple
weeks, I've tried a prototype internally where I did not get sick for the
first time, and I stayed in there for forty-five minutes."

This breakthrough, he explained, came from the team's recent improvements to latency. The result of Oculus's persistent quest to bring latency under five milliseconds. And as a result of the progress that had already been made, as well as that which seemed within reach, Iribe felt confident enough to proclaim, "We are right at the edge where we can bring you no-motion-sickness content."

Naturally, these words stoked the interest of game execs, developers, and reporters on hand. Over the next twenty-four hours, stories like "Oculus: Motion-Sickness Is Solved" (*IGN*) and "CEO Promises Oculus Rift Won't Make You Sick" (*Forbes*) would spread throughout the tech community. This was great, exactly what Iribe had hoped would happen, but even better than that—even better than what would happen when these stories went live—was what happened as soon as Iribe stepped off the stage and turned on his phone: he received a message from Brian Cho, one of the partners at Andreessen Horowitz.

Cho was at the Summit and had been intrigued by Iribe's talk. If motion sickness really had been solved, then perhaps Andreessen Horowitz could play a role in Oculus's series B. This was music to Iribe's ears; he had wanted to be in business with Marc Andreessen since the start of Oculus. Before that even. Since the '90s, really, when Andreessen—fresh off IPO-ing Netscape—appeared on the cover of *Time* magazine for a piece about the so-called Golden Geeks.

Shortly after receiving Cho's email, Iribe wrote back to say, "We have a new prototype in the office which you guys really need to see. It ties everything together for a comfortable experience that proves VR is very close to mass market ready." From there, Iribe reconnected with Chris Dixon and then directly with Andreessen: "We're ready to engage," he wrote.

GIVEN THE WHIRLWIND OF DRAMA THAT THE PAST COUPLE YEARS HAD BEEN, IT was rare that any single development—positive as it may be—could propel Luckey past his default cautious-but-contagious enthusiasm toward a state of over-the-moon giddiness. Nevertheless, this did happen from time to time, and whenever it did Luckey savored sharing these developments with close friends. Which is why, on October 28, Luckey had a big smile on his face as he emailed Patel the following news: "It is back."

"What is back?" Patel replied.

"The McRib," Luckey answered, referring to McDonald's famously elusive barbecue-flavored pork sandwich.

"No fucking way," Patel wrote back, skeptical of this big delicious news.

But instantly, when Luckey replied that he was "eating one right now," Patel knew that the news must be true. Because some things were just too sacred to joke about. For Luckey and Patel, the McRib was one of those things.

As magical as the McRib tastes, something even bigger happened three days later: Marc Andreessen, Chris Dixon, Brian Cho, and a fourth partner (Gil Shafir) visited Irvine and left telling Iribe, "We are fully converted believers." So much so that on November 5, Andreessen even emailed an old friend—Facebook CEO Mark Zuckerberg—to ask, "Have you seen Oculus?" Assuming not, Andreessen then proceeded to tell Zuckerberg about what he'd just experienced. "Blew my brain wide open," he said. "The key seems that they have added John Carmack, co-founder of id, co-creator of *Doom* and *Quake* and essentially the inventor of 3D computer games and one of the all-time great hackers as their CTO. He is completely obsessed with every detail of it. I wanted to just give all my money to him on the spot."

It's rare that a firm would pass on Series A opportunity, and then later participate in future rounds of funding. But Dixon, who managed the relationship, told Iribe he wanted to "correct that mistake." He and his partners wanted to do so in a big way: leading a $75 million Series B.

In fact, it was so important to them that this work out that Marc Andreessen put Iribe in touch with Mark Zuckerberg to put in a good word.

"What do you see as the 'killer app'?" Zuckerberg asked Iribe during a short reference call on November 19.

"Gaming," Iribe replied. "But it's gonna be huge for Communication too."

They proceeded to talk about the past—about Andreessen, about whom Zuckerberg couldn't have had nicer things to say.

Not that Iribe needed much (or any) convincing. From the get-go, Andreessen Horowitz had been the VC firm he most wanted to work with. If anything, Iribe wanted more insight into what changed *their* mind about Oculus. And part of the answer had to do with a phrase Dixon coined called "the Bat Signal Effect."

"The way I think about it is like this," Dixon began. "Let's just take Elon Musk. There were all these people around the world who wanted to build electric cars. Like [for example] there was some smart MIT grad who was working in the Innovation Group before whose dream was to work on electric cars. And the only place to do it prior to Tesla was to work at a regular car company. In some back room. They don't really take the project seriously. And then you maybe see this person— Elon Musk—is starting this company and you're intrigued, but you're not sure where it's going. But then maybe you see that they sold some cars, or raise some money, or recruited some talented person [laughs], and you realize that this sort of Bat Signal goes out. And they realize: all right, this is going to happen; my dream is going to happen. And then all these people come out of the woodwork and you become this company; and the company becomes kind of the talent magnet for all these people. This sort of latent body of talent out there. It's not as if SpaceX and Tesla created all these great space engineers and electrical engineers; they were just sidelined. They just weren't the protagonists, right? It took these entrepreneurs to come and unlock that, right? I think that's a lot of what happened with Oculus."

After leading Oculus's Series B—giving them enough funding to cover DK2 and finance stuff like Luckey's controller project—so began a mutually beneficial relationship between Dixon's firm and Oculus.

CHAPTER 29

ZUCKERBRO INTRIGUED

January/February 2014

ON JANUARY 14 ALEX SCHWARTZ AND DEVIN REIMER FROM OWLCHEMY LABS took the stage at Steam Dev Days—Valve's every-few-years-or-so developer conference—to give a talk called "The Wild West of VR." The "Wild West" they were referring to was the process of making games for a medium that doesn't yet have many rules or standards. But there was also another "Wild West of VR": the business side of this brand-new medium. And minutes after Schwartz and Reimer finished speaking, a pivotal shot was fired in that business-side land grab when Valve's Joe Ludwig announced "Steamworks VR API."

"This is a new addition to the Steam platform," Ludwig said, "that we've written to expand on the services that are provided by the Oculus SDK and by other hardware-specific APIs."

This move was far from a "checkmate," but it did put Oculus in a terrible spot. Valve was essentially saying to developers: Why write your game for Oculus when you can just use our platform instead? Your game will still run on Oculus's headset and also potentially on other headsets down the line. To the hundred-plus Steam users who already knew and trusted Valve, it sounded like a compelling argument.

The guys at Oculus were pissed. It was as if they had spent the past eighteen months creating a toolkit, and now Valve was essentially offering a toolkit of toolkits.

WHERE ARE THE NEXT BILLION USERS COMING FROM?

This was the question that drove Facebook heading into 2014. And it was the question-behind-the-question when, after returning from Irvine, Cory Ondrejka grabbed Mark Zuckerberg and asked him, "How do you think about VR?"

Like most geeks who grew up in the '80s and '90s—those whose childhoods included cameos from things like Virtual Boy and *The Lawnmower Man*—Zuckerberg had a longtime (but limited) fascination with virtual reality. It was that cool thing that was always supposed to happen, but for some reason had not. Why was that?

The answer, like many things in tech, was complicated. But mostly it came down to the fact that: the tech was not yet ready. But, in Ondrejka's estimation, it now was about to be.

He knew that it would still take several years (and several billion dollars) to create a solid consumer-ready product; and that it would take several more years (and billions) before such a product could maybe go mainstream. But regardless of cost and time frame, the bottom line was this: VR was about to become a viable play, so what did that mean to Facebook?

Potentially: a lot. "Because my assumption," Ondrejka explained, "is that whatever the next big consumer revolution's going to be, it's going to be closer to the eye."

"This sort of matches my thinking," Zuckerberg replied. One of the main lessons that he and Ondrejka had learned from the smartphone revolution was that convenience is king; this, single-handedly, was how smartphones dethroned desktop computing.

And what was even more convenient than a smartphone? Well, perhaps, a smartphone that you didn't even need to hold—a device that interfaced directly with your eyes. Maybe that meant glasses? Or contact lenses? Zuckerberg and Ondrejka both believed that these options had the potential to be the next major way that humans interact with computers, a revolution along the VR/AR continuum. It was probably time Zuckerberg check out some of the stuff that Oculus was doing.

"Their current tech is the best I've seen by quite a bit," Ondrejka said. "But you have to actually go down and see the Room demo they have in Irvine."

"Can they come up here and do a demo?" Zuckerberg asked.

"Yeah. But it'll be comparatively shitty. I mean, it'll probably still blow you away, but it's just not as good as the Room."

Zuckerberg was intrigued . . . but still not enough to rearrange his schedule.

"I'M GONNA FLY UP TO MENLO PARK AND GIVE A DEMO TO MARK ZUCKERBERG," Iribe said.

"Nice!" Luckey replied. "Any particular objective in mind? Or just because, you know, it's Mark Zuckerberg?

It was mostly because it was Mark Zuckerberg, but Iribe and Malamed had also started throwing around a crazy idea: What if a company like Facebook led our Series C? It wasn't the type of thing Facebook had done before, but, hey, there's a first time for everything, right?

"TURN TO YOUR RIGHT," IRIBE SUGGESTED, WALKING ZUCKERBERG THROUGH A demo in Facebook COO Sheryl Sandberg's conference room.

Donning one of Oculus' Crystal Cove headsets and trying out the Tuscany demo, Zuckerberg tentatively swiveled his head.

"Now try leaning forward," Iribe said, highlight Crystal Cove's positional tracking.

Beside Zuckerberg was Ondrejka, gleefully watching his boss amble through virtual environments. And beside him was Facebook's CTO Mike "Schrep" Schroepfer, and Facebook's VP of Product, Chris Cox—both eager for their turns.

For a split second, as he returned back to reality, Zuckerberg was delighted, despite some residual dizziness that had kicked in. Nevertheless, he was impressed. So were Schrep and Cox, whose reactions each included multiple utterances of the phrase of "Holy crap!"

"And if you liked that," Ondrejka told them, "you gotta get down to Irvine and go check out 'the Room' demo." It was clear from their reaction that this wasn't the first time that Zuckerberg, Schrep, and Cox had heard Ondrejka rave about it.

Zuckerberg agreed it was important that he make time to visit soon. Maybe the following week? And with tentative plans on the horizon, the Facebook guys spoke with Iribe about his vision for Oculus, the critical role that all four of them believed that virtual social spaces would play.

LEAVING MENLO PARK, IRIBE FELT GREAT ABOUT HOW THINGS HAD GONE AT Facebook. He didn't like getting ahead of himself, but there was a sense that this could be the beginning of a big, down-the-road collaboration. And that feeling grew even stronger the following day when Iribe received a promising follow-up email from Zuckerberg. So promising that he immediately passed it along to the executive team.

FROM: Brendan Iribe
DATE: January 24, 2014
SUBJECT: Fwd: Oculus

FYI—Thought you'd enjoy Zuckerbro's follow up email after yesterday's demo . . .
——Forwarded message——
It was great hearing your vision and seeing the demos yesterday. I was a little dizzy right after taking off the headset, but it's clear where it's all heading and it's amazing.

As I said yesterday, my immediate reaction was that a huge amount of the value will be in the social space created, and the most valuable social space will be the one that establishes a massive active network first.

I think we could help out on both of these fronts and it would be great to figure out a way to work together closely on this.

As a standalone developer, it probably won't make sense for us to invest in building for this until it reached a much larger scale—maybe for four or five years.

But if we could help build the core social world and help you distribute these a lot faster, then that could be exciting. Given our experience designing all kinds of social experiences that more than one billion people use almost every single day, I think we could do some great work together here.

I'm going to find an afternoon when I can come visit you in Irvine soon. If you have ideas and want to talk sooner, just give me a call.

MARK

TRUE TO HIS WORD, ZUCKERBERG ARRANGED TO VISIT OCULUS FIVE DAYS later.

In the meantime, Amin Zoufonoun—Facebook's VP of corporate development—reached out to Iribe to let him know that if Zuckerberg liked what he saw, then there was a chance he might want to acquire Oculus. "Are you open to being bought?" Zoufonoun asked.

"Probably not," Iribe said. "We'd always planned on staying independent, at least until we get our first product to market. But, of course, if the number's right, we'd be stupid not to talk."

As scheduled, Mark Zuckerberg came down to Irvine on January 29. By this point in the life of Oculus, employees were used to hosting "VIP" guests, but Zuckerberg visiting was particularly, well, cool. Mostly because they assumed that if he liked what he saw, it could lead to some sort of Social VR collaboration with Facebook (but also partly because he showed up with a to-go bag from McDonald's, and that was a nice man-of-the-people move).

With limited time, Zuckerberg demoed the Room, checked out some prototypes of DK2 and said hello to someone he had wanted to meet for some time now.

"Hey, Palmer," Zuckerberg said, popping his head into Luckey's office. "I'm Mark."

"I know who you are!" Luckey replied, standing up to shake Zuckerberg's hand.

"You have some really amazing stuff here."

"Totally agree. Virtual reality is going to change everything."

Zuckerberg smiled. It was a smile that Luckey recognized by now: that of a true believer.

"Look," Luckey said, "I'm a huge fan and I love what you've done with Facebook, but honestly I need to get back to work because I've got something important. But we should chat some other time."

"Oh yeah," Zuckerberg replied. "I totally understand. Back to work."

That evening, an Oculus fan posted the following scoop to Reddit: "My friend works in the same building as Oculus, and he ran into Mark Zuckerberg taking the elevator to Oculus' floor. Do you think he was just checking it out? Or is there [something] more devious going on?"[1] The post was immediately met with skepticism—"If your friend really worked in the same building, he could at least provide proof"— and then downvoted to oblivion.

The next night, on January 30, Iribe and Zuckerberg met for dinner—as Zoufonoun had hinted at—Zuckerberg elevated the conversation from collaboration to acquisition.

"I think you'd be in a meaningfully stronger position doing this with us rather than doing it independently," Zuckerberg told Iribe. Then throughout their dinner—and even more explicitly over email two days later—Zuckerberg outlined "some of the most important ways" he believed that Oculus would be positioned if they were to be acquired by Facebook:

- "You'd have a well-designed version of the Metaverse with a social Facebook experience for v1. This will likely be one of—if not the single biggest—killer app of the platform. Keep in mind that 1.25 billion people use Facebook and it is >20% of all time that people spend on their phones. It is by almost 10x the app people spend the most time with on phones, tablets and computers, so I expect it to be in VR as well."
- "If you work with us, we'll develop this experience exclusively and not build any Facebook experience for any other VR platform at all. This will establish Oculus as the standard and make it very difficult for other VR platforms to catch up to yours. On the flip side, if we don't work together, then it won't make sense for us to build for your platform for at least ~5 years until it reaches 50-100 million units."

- "Similarly, we will help bring other game developers and studios to the platform as well. Lots of folks are interested now, but moving them from experimenting to really committing is hard. We will help through signaling that we're investing heavily, by offering incentives to developers and by growing the platform faster. This will further coalesce development around your ecosystem. Without us, the rest of your content platform will also be weaker for at least the next few years."

- "We can reduce the cost of your product and therefore help distribute it faster. We could cut the device cost by almost half by not requiring a profit upfront and by doing a lot of the marketing ourselves through our partners. We could cut the all-in-one device plus computer cost meaningfully by getting our ODM partners to build your spec and sell it to us at barely above cost. If you do this independently, you will have to charge higher prices and the platform will grow more slowly."

- "We can supply capital so you can build more units and distribute them much faster. If you do this independently, you'll ever need to raise a lot of money—which will be very dilutive—or you'll need to wait until you have enough profit to reinvest in building more units—which will be slow."

- "We can significantly de-risk your product development and help you ship v1 faster by enabling you to use our payments engine, platform software and other systems like the voice recognition, internationalization engine. These are all proven systems that we use in production today and you could just plug in. If you do this independently, you are relying on a team of contractors in China."

- "We can significantly grow your engineering team with very high quality people. You'll of course be able to hire some high quality people without us, but this will be much easier with us. We also have good managers that can join your team to help you scale. This is the #1 thing that Kevin

Systrom says has helped Instagram since joining us. You can also use our customer service, legal and finance teams without having spending much of your time on this, if you want."

Later that night, and throughout the next day, the Oculus exec team held discussions to discussed whether this was something that they were seriously interested in exploring. As Iribe had anticipated, the execs didn't have much interest in selling the company. Though, like him, they thought Facebook made a surprising suitor—given that Facebook had no footprint in the game industry—one that could provide Oculus with resources they'd need to bring affordable VR to gamers; and then extend their reach to a broader, mainstream crowd.

"The word Mark kept using was *turbocharge*," Iribe relayed to the guys. By which he meant that Facebook would provide Oculus with the resources to expedite their mission. That was a compelling argument to many on the team, but even more compelling to some—such as Luckey and Patel—was the idea that if Facebook bet big on virtual reality, it would have a ripple effect that would all but ensure VR didn't fail this time around.

Over the next couple of days, as the execs weighed the pros and cons of an acquisition, the perspective evolved from "disinterested in selling" to something akin to "open to the idea." The price had to be . . . right.

FOUR BILLION DOLLARS.

"Please keep the offer confidential within the board," Iribe wrote to Oculus's board members on February 2. "We told them we'd rather build than sell unless it was 4. Pretty sure they'll pass for now."

"I am sure this was a very hard decision," Spark Capital partner Santo Politi replied. "I wanted to let you know that I really appreciate you and your teams commitment to seeing through to our potential. This offer is a very strong validation to what we are doing. I am very happy that we are not selling early."

"Well said, Santo," added Matrix Partners' Antonio Rodriguez. "I totally agree that this is a platform bet that we should go long on and I commend the size of your balls for seeing it that way."

FROM: Mark Zuckerberg
DATE: February 1, 2014
SUBJECT: Some Thoughts

I'm disappointed the conversation with your investors has increased your price expectations to a point where it may not make sense to discuss this further. I felt like we were still in the stage of getting to know your company and team, and the discussions were very exciting . . .

Now, you may be thinking that from a mission perspective this makes sense, but financially you don't think this is the best opportunity. I actually disagree with that. Even if we assume upfront that you will get to your $3-4 billion valuation when you hit your goals, after you take into account the present value of that future valuation, dilution you'll incur on the way there, additional equity grants you'd get from us and the potential growth of Facebook's stock, it turns out that a much lower valuation today is equal to the price you're looking for at the end of 2015 or beyond. If we can discuss and agree on some of these assumptions, then there is probably a deal structure here that makes sense.

At the end of the day though, the reason to do this should be about what will give Oculus and VR the greatest chance to become the VR standard for the world. You'll have a much greater chance of achieving this with our resources and infrastructure, and when you look back on this decision then, I think that's what's going to matter most.

If you're still open to discuss this, let me know. I'd still love to spend some more time with your whole team and have them visit this week. I'd also love to take you through my valuation math if you want to give me a call tonight or tomorrow morning.

MARK

IRIBE WOUND UP GIVING ZUCKERBERG A CALL, BUT SOON IT WAS CLEAR that—at this point—there was no deal to do done. Nevertheless, Iribe wanted to preserve his relationship with Zuckerberg, and he decided to send him an email in which he stated, "There's no one we'd rather change the world with than you, Chris, Cory, Amin, and the team at Facebook . . . If now's not the right time, the team here is excited to discuss partnering post-v1. Facebook and Oculus will conquer the world then."

Even though the acquisition talks had come to an end, Zuckerberg was still very interested in virtual reality. So on February 15, he and Ondrejka flew up to Stanford and visited Jeremy Bailenson's Virtual Human Interaction Lab. Then later that day, Zuckerberg texted Amin Zoufonoun.

> **MARK ZUCKERBERG:** I just went to this VR lab at Stanford and it was totally awesome.
> **MARK ZUCKERBERG:** It also confirmed for me that Oculus is miles ahead of everyone else . . .
> **AMIN ZOUFONOUN:** any other interesting learnings about vr? i guess we can talk about it next time.
> **MARK ZUCKERBERG:** . . . It's going to be hard to get Oculus done at lower than the amounts we've discussed. The markets are just willing to pay more for it right now independently.

In particular, Zuck was talking about a recent announcement by Jawbone that they'd be raising money at a $3.3 billion valuation.

> **AMIN ZOUFONOUN:** you are probably right—especially with the peak of attention and buzz they are currently experiencing.
> **MARK ZUCKERBERG:** . . . Also, just talked to Reed Hastings about Oculus and he's positive about doing it . . .
> **AMIN ZOUFONOUN:** interesting.
> **MARK ZUCKERBERG:** His view is we should make a couple of big bets like this at our scale, and this seems like a reasonable one.

SHORTLY AFTER CONVERSATIONS WITH FACEBOOK ENDED, IRIBE AND MITCHELL traveled north to Bellevue, Washington, to take Abrash and Binstock out to dinner at a restaurant called Thai Chef—a restaurant that, they would soon learn, had personal significance to Abrash.

"This is the same place where I met John in 1994," he said, meaning John Carmack. "It was actually the second time I met him in person."

The first time had been in 1993, when Carmack had come up from Dallas to visit his mother (who lived in Seattle). While he was in town, he and Abrash met for lunch; and during that lunch, Carmack offered him a job at id.

Had it been a different period in his life, Abrash might have been tempted. But at the time, he was working at Microsoft—serving as the graphics lead on Windows NT—and had no interest in making a change (especially not while his Microsoft options were still vesting).

"Well, John came back a year later," Abrash recounted. "And we went for dinner here"—at Thai Chef—"and he just started talking about his vision. His vision of how he was going to enable people to build these persistent servers and those servers could link to each other and you could add on to them; it would basically accrete into cyberspace. I would estimate it was about two hours that John talked about his vision. And then he said: so, do you want to come work at id?

"And I actually *still* said no."

Iribe, Mitchell, and Binstock all cracked up.

"He asked why not. I said that I had all these Microsoft options. He said that he would go talk to his partners to get some equity in the company. And when he did that, I said: you know what, I had read *Snow Crash* somewhat recently. So everything John was describing just seemed interesting—this opportunity just seemed interesting—and you know what? It *was* interesting! But I probably wouldn't have said yes, and I probably wouldn't be sitting here right now, if I hadn't read *Snow Crash* and fallen in love with the idea of the Metaverse."

This was music to Iribe's and Mitchell's ears. They, too, wanted this dinner to end the same way it had in this same restaurant twenty years before: with Abrash saying yes to a new opportunity. Over the next couple of hours, Iribe and Mitchell pitched their deal hard, offering equity, autonomy, and the chance to *actually* ship an incredible product.

This last bit was a jab at Valve's unwillingness to productize a VR headset of their own. Even if Valve did get serious about shipping a headset, they'd probably just end up outsourcing most of the risk to other hardware partners, just as they had done with the Steam Box over the past year. Did Abrash and Binstock really want to go through that déjà vu? Why settle for that when they had a chance to be part of something truly special?

"We turned down Mark Zuckerberg!" Iribe reminded them. "Abrash, you said you think it was gonna be Microsoft, but I think it can be Oculus, and we're committed to getting that money to go do it. Andreessen's committed to helping us go find more money. But we need you both to join us."

By the end of dinner, Iribe and Mitchell felt they had effectively conveyed their message. As well as things might have gone, though, neither Abrash nor Binstock was willing to pull the trigger. So, not long after, Iribe decided it was time to call in some reinforcements: Chris Dixon and Marc Andreessen.

Andreessen and Dixon were more than happy to assist. They flew up and helped Iribe and Mitchell make Oculus's case over a meeting with Abrash and Binstock at the Seattle Hyatt.

Even with the assistance of these top VCs, Abrash and Binstock were still on the fence. Then Iribe said something to Binstock that seemed to push him through, finally.

"You know for VR to succeed," Iribe said, "the Rift *has* to be as good as it can be. And without you being on board, there's a *chance* we may not make Rift as good as it could be if you were there. How would you feel about that?"

"I'd never forgive myself," Binstock replied. But in that milli-second—on the brink of saying yes—he felt a loyalty to his team at Valve; he would still need to think about this some more.

CHAPTER 30
BLUE'S CLUES

February/March 2014

Mitomete ita okubyou na kako
Wakaranai mama ni—

—WITH A TIRED TAP TO THE ALARM ON HIS PHONE, LUCKEY SHUSHED THE song that he awoke up to each day: the theme song to *Sword Art* On-line. And as Edelmann had once pointed out, it felt fitting that this—the anthem to an anime about virtual reality—was literally what got him out of bed every morning.

Typically, after waking up, Luckey would shower, brush his teeth, maybe grab a bite and then head to the office in his Honda Insight. But on February 21, things went a little different—because Edelmann, who had gotten a job at Disneyland the previous month, needed to get to work early that morning, and therefore taken their car.

"Joe," Luckey said, calling up Chen who lived nearby, "I need a ride to the office."

"Too late! I'm already here."

"So what you're saying is . . . you're awake and in possession of a car. Great! You can come pick me up."

"Nice try."

"Come on! I live so close!"

Luckey *did* live pretty close, only about four miles away. But Chen wanted to make his friend sweat it out. "I have a better idea," he said.

"Since, you know, you're becoming a fatass: why don't you *run* to the office?"

Chen, of course, was being sarcastic. But Luckey processed the suggestion and concluded that since he and his colleagues *were* in the midst of a companywide weight loss effort—the first annual "Oculus Weight Loss Challenge"—and since he *had* done some cross-country running years ago, this was actually a great idea.

"Okay!" Luckey replied. "See you in a half-hour or so."

"Wait! What? Just chill, I'll come pick you up in, like, ten minutes."

"By then, you'll be too late. I'll already be out on the road, running like the wind!"

With "Eye of the Tiger" now playing in his head, Luckey dressed for his sweaty "commute." This, however, was a bit problematic; since his wardrobe consisted of little more than Hawaiian shirts, cargo shorts and flip-flops. He didn't even own a single pair of proper shoes. No, wait, he remembered: Chen had bought him a pair of black sneakers at CES!

So—*Risin' up back on the streeeeet*—Luckey grabbed the sneakers— *Did my time*—threw on a gray T-shirt and—*took my chances*—he stepped into pair of comfy American Flag pajama pants—*just a man and his will to survive.*

Jogging away from his apartment, Luckey felt pretty good. Especially when passing cars honked in support of his effort; with some drivers, perhaps feeling extra-patriotic due to the Winter Olympics going on in Sochi, rolling down their windows to shout, "U-S-A! U-S-A!"

As he hit a rhythm, his mind began to clear and, jogging through the streets, Luckey found himself thinking about Facebook; a little bit about what might have been had Oculus been acquired, but much more about the seemingly out-of-the-blue acquisition agreement that Facebook had announced just two days earlier—to buy WhatsApp for $19 billion.[1]

ON FEBRUARY 24, JOURNALIST DAVID KIRKPATRICK INVITED MARK ZUCKERBERG onstage at the Mobile World Congress in Barcelona to have a keynote conversation. Originally, they had planned to discuss Facebook's Internet.org initiative (and, eventually, they'd get into that), but Kirkpatrick had to at least begin by asking about the news that had shaken up Silicon Valley.

"It's been on everybody's lips for the last week or so," Kirkpatrick said. "You bought WhatsApp for $19 billion. Which, once we got over our shock at that, some of us feel like we understand it. But tell us, here at the Mobile World Congress, which is really the world's major gathering of mobile communications, which is an industry that WhatsApp is a big part of; why did you do it and what does it mean?"

"Well, WhatsApp is a great company and it's a great fit for us," Zuckerberg explained. "Already, almost half a billion people love using WhatsApp for messaging. And it's the most engaging app that we've ever seen exist on mobile, by far. About 70 percent of people who use WhatsApp use it every day . . . So when we had the opportunity to be part of this journey, I was just really excited to take [cofounder and CEO] Jan [Koum] up on that and help him realize his dream of connecting a lot more people."

From his office in Irvine, Iribe watched Zuckerberg's keynote at Mobile World Congress. And though he appreciated the rah-rah rhetoric of Zuckerberg's initial explanation, it was what he said next, and throughout the rest of his keynote that especially piqued Iribe's interest.

MARK ZUCKERBERG
In terms of fit for Facebook, when Jan and I first met and
started talking about this, we really started talking about what
it was going to be like to connect everyone in the world . . .

Connect everyone.

MARK ZUCKERBERG
It wasn't really until we got aligned on that vision, between
Facebook and WhatsApp, that we started talking about
numbers and decided to make a deal. But it's that vision that
I think makes the company such a great fit: the shared goal to
help connect everyone in the world.

Prior to this keynote, Iribe had known that Facebook wanted to "connect everyone." After all, that was explicitly their mission. But over the next forty-five minutes, he began to fully internalize what that mission really meant—how it elevated beyond just words—and how Facebook actually had a plan to try and accomplish that lofty endeavor.

MARK ZUCKERBERG
Today what I really want to focus on is Internet.org and how we can build this model for this industry that can deliver the internet to, you know, 5–6 [billion]—ultimately everyone in the world. And in doing so build what is going to be a profitable model—a more profitable model—with more subscribers for carriers and get everyone on the internet in a much shorter period of time . . .

Internet.org was a partnership between Facebook and mobile carriers to deliver the internet to those in less developed countries. The goal, as Zuckerberg would go on to describe, was to make it so that for an affordable price (ideally: free) everyone in the world could have access to "basic services" online—services like food pricing, weather reports, messaging, and social networking.

DAVID KIRKPATRICK
So it's a kind of gateway drug?
MARK ZUCKERBERG
Yeah. We think about it as an "on-ramp."

The more Iribe thought about Facebook's mission and their biggest acquisitions, the more he harped on this: there's never been a service in the world that reached a billion people every day. None of the television networks, not even in their heyday, ever came close to reaching a billion people. CBS? NBC? They never got to sell commercials to a billion people every day! And here now, in this incredible new world, not only did Facebook have 1.2 billion users but—with WhatsApp (500,000 users) and Instagram (200,000 users)—they might be on their way to having *three* of these billion-reach companies. Maybe even *four*, with the rate that Messenger was growing.

As Iribe fantasized about what it would be like to get a *billion* people into VR, an audience member asked Zuckerberg if Facebook was still interested in trying to acquire Snapchat.

"I mean, look," Zuckerberg replied, "after buying a company for $16 billion, you're probably done for a while."

Iribe hoped that was not the case, as he now planned to reopen discussions with Zuckerberg.

"QUICK UPDATE," OCULUS' HEAD OF DEVELOPER RELATIONS, AARON DAVIES, wrote on March 5, sharing some good Mobile VR news with the executive team. "Max and I made a visit to Blizzard yesterday . . . great meeting and good complement to build top-down momentum to match the bottom-up from the engineers. Most obvious and relevant opportunity, also most likely to get traction, is Hearthstone . . . I already got a follow up email about how they've been talking about VR coming sooner than they thought and what that means for them."

Davies was pleased by how things had gone. With Mobile VR now scheduled for release in 2015, Hearthstone seemed like a potential killer app. Unfortunately, Iribe didn't share Davies's enthusiasm. Leading to a thread that once again brought to the forefront the ongoing internal struggle between PC and Mobile VR:

> **FROM:** Brendan Iribe
> **DATE:** March 5, 2014
> **SUBJECT:** RE: Blizzard Update
>
> I was expecting you'd wait for me and Nate to visit. We need to coordinate better and make sure the message is correct.
>   I don't want to turn too many PC game developers on to motion sickening mobile without being very up front and honest.
>
> **>>>FROM:** Max Cohen
> **SUBJECT:** RE: Blizzard Update
>
> We're not running a misinformation campaign here, Brendan . . .

>>>**FROM:** Brendan Iribe
SUBJECT: RE: Blizzard Update

We've been courting Blizzard for a while. You shouldn't have gone
without us . . .
     You should come back to the office and focus on the mobile SDK.
It's embarrassing right now and isn't in a state we can ship.
     PRO UP!

Pro up? That pissed off Cohen. As had the many *many* times Iribe
had slighted Oculus's mobile project over the past few months. But
before he could respond, Iribe had more to say:

FROM: Brendan Iribe
SUBJECT: RE: Blizzard Update

Samsung's product is neat and a gimmicky toy. It's a taste, but it's not
the magic of VR.
     Oculus is designing, building and delivering the magic. We will be
evangelizing that.
     Please come back to the office and we'll fix this messaging
problem.

>>>**FROM:** Max Cohen
SUBJECT: RE: Blizzard Update

I left my wife at 4:45 a.m. this morning to fly to San Francisco to make
deals happen for this product, and the trip was a huge success. If you
want to talk in person, I'll be in tomorrow morning.
     We don't have a messaging problem. We have an executive support
problem.

Things escalated from there until David De Martini—Oculus's
head of world publishing—stepped in to play a role he'd been playing
more and more as of late: peacemaker.

"What is great about this thread is everyone's passion to do something great," De Martini wrote to Cohen and Iribe. "In all honesty and as we"—a "we" which by the way included Iribe—"discussed on Monday, Mobile is an easy sell because it is pretty good and way better in small doses than people expect, it has a date, a huge partner, and a plan. Everyone that Aaron, Max, and I speak with will be interested and that is a VERY good thing. It is a very good thing, because Mobile will be the 1st place that they put there [*sic*] games and apps and then Max, Aaron and I have a plan to WALK THEM UP THE LADDER TO THE REAL VR GREATNESS—PC . . . There really isn't a story we can tell to get people to commit to PC right now and that is ok . . . This is [a] good thing, and Aaron, Max and I are communicating tightly like brothers and selling the true reality of Mobile VR (flaws and all)."

De Martini's comments were enough to end the sniping. But unfortunately, closure was yet to be found for this growing internal schism.

MEANWHILE, ABRASH AND BINSTOCK WERE INCREASINGLY FEELING LIKE THEY were wasting their time at Valve. They both appreciated the freedom that the company had given them to explore virtual reality, but as each became more invested in this technology, it was harder to ignore the fact that this investment did not seem mutual. That, financially speaking, Valve just wasn't willing to commit much to VR. And after about eighteen months of working with this technology, Abrash and Binstock felt like Valve was perpetually dithering and it was time for them to piss or get off the pot.

In an ideal world, they wanted Valve to produce a VR headset of its own. They knew that this was a big ask—a commitment in the hundreds of millions, if not more—so, for the time being, all they really wanted was an infusion of cash to build some display prototypes. Or if that wasn't doable right now, then they at least needed an indication of when that might be feasible.

Basically, what Abrash and Binstock wanted was some assurance that this work they had been doing was headed somewhere. That this was more than just a cool science project; that it was more than just an oo-lalala to impress visitors. In other words, as Abrash had told Binstock at a coffee shop before any of this had started: if you want a revolution, and you think you are capable of contributing, you should be actively pushing it forward.

So where did Valve stand? Did they actively want to push forward a VR revolution? To find out once and for all, Abrash and Binstock spent the first week of March meeting with Gabe Newell and Valve's board of directors.

The following week, in a surprise to most, Atman Binstock decided to leave Valve and accept Iribe's offer to join Oculus. When asked by a colleague what had led to this decision, Binstock explained with a sigh that, ultimately, "Valve is like this jolly fat man who just keeps getting more money and jollier but isn't willing to take any risks. They aren't pushing for VR to happen; in fact, I'm not even sure if they care at all whether VR succeeds or fails. Whereas Oculus is different. Oculus is this rocket that is either going to deliver VR or explode spectacularly. And I want to do everything in my power to help ensure the former."

ON MARCH 10, HOURS AFTER HIRING ATMAN BINSTOCK TO BE OCULUS'S NEW chief architect, Iribe emailed Valve CEO Gabe Newell to say, "Sorry if we caused pain. We love Valve and want to maintain an awesome relationship with you guys. You're an inspiration to me, Palmer, and the crew. There's no one we'd rather change the world with . . ."

The following day, Newell sent back a short reply: "Yep, we look forward to continuing to work with you all and Atman. I'm moving onto the VR team for a bit."

Iribe felt relief, if at least for a moment. "Our relationship with Valve is still intact," he told his colleagues, forwarding his email exchange. But when his colleagues read it, they interpreted it quite differently.

"This is an implicit threat!" Luckey told Iribe. He was far from the only executive to feel that Iribe was being naive—that he was continuing to let his affection blind him to what Valve really was: a competitor with more resources, more contacts, and an SDK that they were now actively pushing developers to use *instead* of the one made by Oculus.

Still, Iribe opted to see things differently. "I think having Valve promote VR right now is a good thing," he told the team on March 14. "No one will use Steam VR unless it's a LOT better and fully compatible with our full set of hardware, which it simply won't [be]. Let's figure out the right way to message this now, knowing this won't be a problem in the future."

"Having fought this war with Steam once already at EA," David De Martini tried to caution Iribe, "they have the world's most popular platform for PC games and they are trying to become the de facto platform for VR. They are now the 1st mover in that space and if we intend to have an open platform, we should negotiate with them now so that we get a piece of their VR sales in addition to ours. They are not bad people, but they are intense businessmen."

Iribe appreciated De Martini's perspective, but replied that he "wasn't worried at all" because, he said, "we control our own destiny." Iribe then suggested they pick up this conversation after GDC (only two days away), though he did clarify a few things in the meantime:

- "I don't want to engage Valve around Steam VR rev share. That implies we're supporting them owning the customer."
- "We will own the customer. Our product, our platform, our customer."
- "Every step we take should be focused on delivering the best VR experience."

In the midst of this back-and-forth, Iribe received an email invitation from Mark Zuckerberg to visit him at his home at Fair Oaks that Sunday so they could talk in person about an acquisition. Bring along anyone you'd like, Zuckerberg said. He ended with five words that Iribe loved hearing: *I WON'T WASTE YOUR TIME.*

"MARK KNOWS WHERE WE STAND," IRIBE SAID, ADDRESSING THE EXECUTIVE team in a conference room. "And he said he's not gonna waste our time . . ."

Even so, it seemed unlikely that Zuckerberg would suddenly be willing to pay $4 billion. So what did they expect him to offer? And more important: how high did the number have to be for everyone to be on board with selling to Facebook?

"What if it's a billion?" someone asked the group.

No way, too low.

What about $1.5 billion?

"Absolutely not," Iribe said, "We're not getting a number without a two in front."

Luckey and Malamed glanced at each other—both thinking that, uh, there were plenty of great numbers with a "one in front"—but Iribe wanted more, and he believed he could get it. He believed he could maybe even get Zuckerberg up to $3 billion, which would essentially double the ask of everyone else in that room. Everyone except Mike Antonov, who not only thought that Oculus was worth way more, but didn't really want to sell the company at any price. The more he thought about it, the less inclined he was to give up the freedom of remaining independent.

"But we've already given up that freedom," Iribe countered. "We already have investors who own giant stakes in the company."

This was a fair point, Antonov conceded. But still, Iribe knew what he meant. They'd been through this before. It was only three years ago that they'd sold their baby to Autodesk, and Antonov was wary of losing that freedom again.

Iribe understood where Antonov was coming from, and assured him that it wasn't as if a final decision needed to be made that day. This was a process. There would be plenty of time for further decision. All that mattered right now was that Iribe had everyone's blessing before he met with Zuckerberg that weekend to see what the Facebook CEO had to say.

ON SUNDAY, MARCH 16—ONE DAY BEFORE THE START OF GDC AND THREE BE-fore preorders for DK2 went live—Iribe flew up to San Francisco. But before heading to Zuckerberg's home, he sat down for coffee with the man he'd asked to join him for part of the day: John Carmack.

Why Carmack? Why not Luckey or Mitchell or his longtime partner Antonov? Because, Iribe reasoned that if he were Mark he'd want to hear from the "godfather of PC gaming"—the guy who'd been through platform shifts before—and from the purest, most technical deep programmer sense.

"I think it's important for you to share your vision for thirty minutes or so," Iribe told Carmack, as they game planned at a coffee shop. "Mark will probably have a bunch of questions, you and I will answer those, and then you can head back to Dallas and I'll finish the discussion one-on-one with Mark. Sound good?"

It did. Especially because Carmack had thought that Iribe should have accepted Facebook's offer the first time around! Confident about what they were walking into, Iribe and Carmack took a cab to the home of Facebook's CEO.

"Thanks so much for coming!" Zuckerberg said, welcoming them. If Iribe or Carmack had felt any nerves, Zuckerberg's congenial demeanor vanquished them. "I just ordered a pizza," he said with a smile, and then walked his guests out to a patio where—in the shade of a small gazebo—they sat and ate.

"Why don't you tell Mark where you see all this going," Iribe suggested to Carmack.

"I see a world in the future where there are no displays or monitors," Carmack said. "Because we have glasses. Everyone has virtual displays—everywhere, all the time. And soon we'll look back and it'll be almost 'primitive' that we had these old computer boxes, and game consoles, and televisions up on walls."

Carmack was an incredible speaker. The guy was famous for his keynotes and could go for hours at a time without seeming to catch his breath. He had a mental model for how all the technical details would work. As he delivered this overview of what he envisioned, Iribe looked down at his arm and noticed he had goose bumps.

"Wow, yeah, I agree," Zuckerberg said. Carmack, soon after, left Iribe and Zuckerberg alone to talk business.

At this point, it was clear that Zuckerberg wanted to do a deal, but he still wasn't willing to hit the number that Iribe had floated out in January. Four billion outright was too much. He *was* willing to pay nearly $3 billion—and to structure the deal in a way that would especially reward and incentivize Oculus's employees.

"I'm committed to doing this," Zuckerberg said. "I need to get my team on board. In the meantime, I have one ask: that you don't go talk to anybody else, that you don't go shop this deal."

"WE SHOULD *TOTALLY* SHOP THE DEAL!" LUCKEY JOKED, AS THE EXECUTIVE
team began to discuss the deal, one they had internally code-named
"Blue's Clues." But with GDC kicking off and preorders for DK2 going live, there was hardly any time at the moment to talk Blue's Clues.

A dinner at Zuckerberg's was scheduled for later in the week. The
execs were excited to break bread with Zuckerberg and key members of
his team. But before that would happen, their excitement was tempered
by a betrayal at GDC.

"VR has been a dream of many game creators since the computer
game was invented," explained Shuhei Yoshida, the president of Worldwide Studios for Sony Computer Entertainment. "Many of us at Play-
Station have dreamed about VR and what it could mean to the games
we create." Then, minutes into a cryptically described Sony Panel, Yoshida unleashed the big reveal: "I am thrilled to introduce Project Morpheus, a virtual reality system for PlayStation 4."

The crowd went wild. This was big news, of course, but it was by
no means the betrayal. That would come forty-four minutes later into
Sony's presentation, when Anton Mikhailov, from Sony R&D, announced that they were "very excited to be working with CCP to bring
their *Eve: Valkyrie* game to our devkit." Especially because "it's one of
the first games to be built ground up for VR and it shows."

Yes, it was one of the first games to be built exclusively for VR . . .
because it had been built for, and largely funded by, Oculus! In exchange, CCP had given Oculus the exclusive rights to distribute their
game. So what the hell had just happened?

ONE OF THE FINAL HURDLES IN CLOSING THE DEAL WAS GETTING CARMACK TO
sign off. Which, as with hiring him at Oculus, required going through
his wife and business manager Katherine Anna Kang. She was under
the weather that weekend, which made it tough to close the deal—but
even tougher were her requests for Carmack to receive over ten million
dollars more up front; and for him to receive indemnity from Facebook. So that if his former company filed a suit against him, Facebook
would be responsible for bearing those costs.

"Oculus poked a beehive," Kang wrote to Facebook VP Amin
Zoufonoun on March 23. "And we should not be responsible for it. If

there is a judgment by a crazy judge and/or jury, we are not willing to risk our personal assets."

"The Carmack thing concerns me a bit," Zoufonoun texted Zuckerberg over WhatsApp the following day.

"Yes, it concerns me too," Zuckerberg replied. As did the fact that he found Carmack to be "socially awkward in person" (and Kang to be "crazy").

Nevertheless, Facebook agreed to include the indemnity provision that Kang wanted for Carmack. Not only that, but—over a weekend of negotiating at Facebook's campus in Menlo Park—Iribe was also able to obtain a similar indemnity provision for himself as well.

As for the additional upfront compensation, Iribe asked Luckey if he would be willing to reallocate some of his signing day money to Carmack. "Essentially," Iribe explained, "you'd just be moving money from one bucket to another; so the money you'd get today, you'd now just get a few years later—at the end of the vesting period."

"So I'll still get the same amount in the end, right?" Luckey asked.

"Yes. I mean as long as you're still at Oculus . . ."

"I'm going to be at Oculus for the rest of my life," Luckey said.

"I know!" Iribe replied. "That's why I figured this wouldn't make a difference to you."

Shortly after Luckey agreed to the reallocation—on the afternoon of March 24—Mitchell texted Iribe to ask, "Did we get JC?"

"Yes. Signed," Iribe said. "Five minutes ago."

Perfect. Mitchell with just two words: Mission Accomplished.

ON MARCH 25, IRIBE CALLED FOR AN AFTERNOON "TOWN HALL" MEETING IN Oculus's kitchen. By this point, given all the big developments over the past year, last-minute town halls like this weren't an uncommon thing. But what *was* uncommon was the time: 1:30 p.m., which was much later in the day than normal, after market trading hours. After a few employees made that connection, they deduced that Oculus was being acquired; and so, by the time everyone gathered in the kitchen, there was just one question on everybody's mind: Sony or Microsoft?

"We're partnering with Facebook!" Iribe announced.

Facebook? Wait, what? And what exactly did "partnering" mean?

"It means we're selling to Facebook!" Luckey blurted. "And we're all going to be rich!"

As everyone digested the news—and, after hearing the sale price, calculated what this news meant to them—Iribe, Mitchell, and Luckey expressed their excitement and explained what made Facebook the ideal partner to execute Oculus's long-term mission.

"I'll meet with everyone individually over the next few days to discuss what this means in terms of integration," Iribe assured the team. "In the meantime, know that things couldn't be better. No team, ever, has had a better shot at delivering on the dream of virtual reality. This will be the team that solves the hardest problems and delivers the final platform."

"This is truly a special moment," Iribe continued. "The work we've done has captured the world's attention and changed the perception of the medium forever. Congratulations, guys. Now let's get back to changing the world!"

But before everyone could attempt that, Iribe introduced Mark Zuckerberg, who emerged from just outside the kitchen and raved about what Oculus had built.

"People describe it as, like, a religious experience!" Zuckerberg said. "They go into this world"—by putting on a pair of supercharged goggles—"and when they take it off they're, like, sad to be back in reality."

After praising Oculus, discussing Facebook's grand vision, and geeking out over this long-sought-after technology that he believed would dominate the future, Zuckerberg opened the floor to questions from members of his newly acquired team.

"Hey, Mark," Chris Dycus began, asking what would soon become the most famous question in Oculus lore, "I know that you know that people think Facebook is evil . . . so I'm wondering how that will affect the perception of Oculus."

While Luckey checked his phone, greeted by a slew of death threats, fuck yous, and—what seemed to be the outlier—congratulatory words, across the room, Iribe also checked his. As he responded to some family and friends, including one friend who had asked what it felt like to make $2 billion, he replied: "It will leak soon that the deal was actually $3 billion to the founders, and I'll probably burn in hell for this, but it was 4 billion to me.:)"

CHAPTER 31

THE BACKLASH

March 2014

ONLINE, THE BACKLASH TO FACEBOOK'S ACQUISITION OF OCULUS WAS HARSH and immediate, with some of the following Reddit and Twitter posts representing just the tip of the iceberg—most of which was directed squarely at Luckey:

Fuck you, you sellout piece of shit.

FUCK YOU, you fucking SELL OUT.

Traitorous asshole.

YOU WERE THE CHOSEN ONE.

We trusted YOU with our dreams. We paid you to build this. Not Facebook.

Hope the swimming pools of money are worth selling out the VR dreams of every gamer in the world.

So is it true that 2 billion dollars washes the taste of Zuckerberg's asshole out of your mouth?

Over/under on how much longer it takes Facebook to fully become Skynet?

By gamers for gamers, huh?

thanx for the fake hope: Oculus Rift 2012-2014 RIP

#Betrayed

#GiantZuckingSound

#backstrokingbillionares

#FuckYou

"People still angry?" LaValle asked later that day, running into Luckey by the elevators.

"Oh yeah," Luckey said.

LaValle nodded. He felt bad for Luckey. The kid looked exhausted. But at the same time, Luckey appeared to look almost enlivened by the anger.

"I don't mind the hate," Luckey explained, as they stepped into the elevator. "I mean, I don't like it. Obviously. But I get where people are coming from. And the good news is that it'll go away when they realize the acquisition is ultimately a *very* good thing for VR. Still . . . I could *definitely* do without the death threats!"

"Death threats?" LaValle asked, suddenly feeling the most heightened version of a strange sense that had been with him since moving to California.

"Not a ton," Luckey said. "Just like ten or so. But I haven't checked my other email yet, so maybe it's a lot more."

"Should we call the police?" LaValle asked.

Luckey just about laughed at the suggestion. "I mean," Luckey explained, "people on the internet have threatened me with death literally hundreds to thousands of times before I even started Oculus. Like, just playing video games. That's kind of normal. That's just what you do."

"*That* is what you do?"

"Well, *I* don't do. But I'm used to it. When you grow up in multiplayer FPS communities, it's par for the course."

LaValle found this somewhat shocking. Then he compared it to what was said on the field during sporting events when he was a kid, and it didn't seem *that* much of a departure.

"The reality is that none of these people are actually going to come after me," Luckey said. "In the history of the internet, I don't think there's been a single case of someone making a threat online and then following through in real life. So you can't take these things seriously. These people are just venting and showing rage in the tiny way they know how."

"I see," LaValle said. "As long as you're okay, that's all I care about."

"I appreciate it," Luckey said. "It's nothing to waste time worrying about. The funny thing is that for every death threat I received today, I've gotten two—maybe even *three*—messages from devs, thanking me for what this is gonna do for them."

"Really?" LaValle asked.

"Oh yeah. From people already working on VR to people in big studios who want to work on it themselves. They're thrilled!"

"Wow. Maybe you could ask them to say something publicly? Maybe that would stymie the negativity?"

"Nah. It's not worth it to them. Because the knee-jerk reaction from people with absolutely no skin in the game is so strong. Besides, it probably would just get lost in the shuffle; the media only really cares about amplifying negative reactions. Just look at the Notch thing."

The "Notch thing" was a reference to comments made earlier that day by *Minecraft* creator Markus "Notch" Persson. "We were in talks about maybe bringing a version of Minecraft to Oculus," Notch had tweeted. "[But] I just cancelled that deal. Facebook creeps me out." Notch then elaborated on this in a lengthy blog post, stating that "Facebook is not a company of grass-roots tech enthusiasts. Facebook is not a game tech company. Facebook has a history of caring about building user numbers, and nothing but building user numbers . . . I definitely want to be a part of VR, but I will not work with Facebook. Their motives are too unclear and shifting . . . there's nothing about their history that makes me trust them, and that makes them seem creepy to me."

At this point, Luckey reasoned that the only chance Oculus had to stem the tide (and he conceded it was only a chance) would be for him—the face of Oculus—to go online, face the mob, and explain why he had decided to sell the company that he loved. Starting with a short, heartfelt essay that Luckey had written with Mitchell about how Facebook was an ideal partner, how this was a "special moment for the gaming industry" and, ultimately, how "We won't let you down."

Any goodwill Luckey's essay might have generated was immediately nixed by some of the comments that Mark Zuckerberg made during a public investor call that afternoon. Comments like "Gaming is the first big opportunity . . . [but] this is really a new social platform," "We're clearly not a hardware company . . . we view this as a software and services thing" and "Of course we will continue to focus on our extremely important work of building out our advertising platform as well, as part of this."

With outrage still mounting, Luckey decided to spend the evening personally responding to Redditors' concerns. In response to someone worried that Oculus would require users to get a Facebook account, Luckey said "that would be lame" and that one "will not need a Facebook account to use or develop for the Rift." In response to someone who was concerned that this deal would kill Oculus's commitment to openness, Luckey vowed that "We won't change. If anything, our hardware and software will get even more open, and Facebook is onboard with that." And in response to someone who was concerned that *he* would be less open—that he'd stop being "the Palmer, and Oculus we all know and love," he promised that would never happen. Then, as if to prove the point, Luckey replied to someone claiming he had failed as VR's Skywalkeresque "Chosen One"—failed to "destroy the Sith" and "bring balance to the force"—by noting that "Anakin went through some rough times, but he *did* bring balance to the force!"

The following day, a reporter from *Time* collected the "assurances" Luckey had made. Individually, perhaps, none of these assurances would have been enough to assuage doubts about the acquisition. But collectively, even the biggest Facebook-haters had reason to feel hope.

Maybe this really *was* "a special moment for the gaming industry."

Maybe Palmer and the Oculus crew *weren't* going to let down the VR community.

But while these assurances assuaged some, Iribe was now feeling a little outraged. "Palmer," Iribe wrote. "Did you clear these statements with Nate and me before posting? WE (Oculus) have not made concrete decisions around some of these topics and the combined efforts with Facebook have also not been decided. Please stop making promises that we haven't discussed internally. I already see us being set up to fail on one or two of these."

Seriously, Brendan? Are you fucking kidding me? Luckey thought. When was the last time that your in-box was inundated with death threats? When was the last time that you were suddenly loathed by the entire VR community? And where even were you when I was out there—facing the mob—and trying to defend our name?

For several minutes, Luckey's mind whirled, until he realized something he hated to admit: his partner might be right. Stunned into near silence, he replied with just one word: "Okay." Then, as if needing to clarify, he followed up with a short apology and an admission that he "should have handled it better."

He meant it, too. He knew that the messaging right now was critical, and that his partners deserved the opportunity to weigh in. But at the same time, one thing still bothered him about Iribe's email: the fact that, apparently, Iribe didn't think that the assurances Luckey made to the online community were true. And since Luckey felt certain that all these things *had* been discussed and agreed upon internally, it suddenly begged a much larger question: Were there things about the acquisition that his partner had not yet told him?

Luckey rid his mind of this thought and in an almost zen-like way focused on the final lines of an email he received from David De Martini and would think about often over the coming years. "It takes many people to make something like this happen," De Martini wrote. "However, I just want to make sure that I tell you that I am happiest for you. One person had to have a vision as to how this could all come together, and the guts to do whatever it took to make it happen and that person was you. More so than anyone, you made this dream become a reality and you changed the lives of almost 100 people forever. I'll never forget your story and this is how legends are built in life. This isn't the greatest thing you will ever do, but it is ONE great thing that you did. Congrats and make sure you finish the job as part of FB!"

Finish the job, finish the job, finish the job . . .

CHAPTER 32

THE NEW NORMAL

March/April 2014

THE EVENING OF THE ACQUISITION, PAUL BETTNER WROTE AN EMAIL TO HIS team.

>>>**FROM:** Paul Bettner
SUBJECT: FB <3 Oculus

So Facebook is buying Oculus. We had no idea this was happening, we heard about it over the news today just like the rest of the world did.

Facebook and Oculus have both put out press releases attempting to answer the obvious question: "Why??" To me, this kind of news always reads exactly the same - "We did it because this is good for everyone! Good for Company X! Good for Company Y! Most importantly good for our customers!!!!" One simplistic way to look at it: one of the world's largest/fastest growing companies just made a very public $2 billion dollar bet on the future of VR. In that sense, it's highly validating.

But we'll see - only time will really tell. I'm cautiously optimistic. I've met Zuckerberg a few times. I do believe he has the potential to successfully realize a long term vision without compromising. It could be a really good thing for the emergence of VR as a truly mass market platform. But I'm also intimately familiar with how difficult acquisitions can be. There are so many ways for it to go wrong and it could turn into a mess.

What we do know on our end is that Oculus pushed very hard to get under contract with us a week ago and we just received a big milestone payment from them this morning. Obviously they internally knew about this potential Facebook deal even (or especially) as they were pushing us to sign this deal with them a week ago. That would seem to indicate that the relationship they've forged with our studio and the game we're creating for them is important to their future plans together with FB . . .

All that said, I do hope today's news ends up being good news in the long run. These technologies - VR/AR - are truly poised to change the world, sooner rather than later. Just like the iPhone did 5 years ago, these new platforms will let us reach our players in ways we've never even dreamed possible and I am thrilled by that potential and by how well we've already positioned our studio to build on these upcoming opportunities.

Let me know what questions/concerns are on your mind about all this. You guys know I'm an open book:-)

Bettner was a true optimist, and he truly believed in VR. But having gone through a high-priced acquisition himself, he feared what this would mean for his friends at Oculus.

MEANWHILE, IN HAWAII, A PAIR OF ENTREPRENEURS WHOSE PARTNERSHIP HAD recently dissolved had a heated discuss over Skype:

"Been trying to reach you for a few days," typed one of the entrepreneurs: Ron Igra. "I'm going after Oculus . . ."

"My views on taking Palmer to court are the same as before," typed the other entrepreneur: Thomas Seidl. "That it makes bad financial sense for you and me."

In response to this, Igra called Seidl—who answered only to immediately hang up.

> **RON IGRA:** stop being such a jerk
> **THOMAS SEIDL:** stop being such a schmuck
> **RON IGRA:** we can make millions together if you just work
> together

THOMAS SEIDL: stop being such a lying deceiving fuck . . . you
   make one bad decision after the other
RON IGRA: you made a deal with Palmer . . . there is 2 billion
   dollars that have been made from our idea . . .

3,700 MILES AWAY—WRITING AN EMAIL FROM HER OFFICE IN DALLAS—
ZeniMax CFO Cindy Tallent was also thinking about Facebook's acquisition of Oculus.

"Facebook just bought Oculus for 2 billion in cash and stock! Unbelievable!" Tallent wrote to ZeniMax COO James Leder. "And to
think," she finished by saying, "this could have been part of [ZeniMax]
had we played our cards differently."

If only ZeniMax had been willing to invest $6 million, they could
have had 15 percent of Oculus . . .

If only they had been willing to let Carmack serve as a technical
advisor and taken that "free" 2 percent stake in Oculus . . .

A COUPLE DAYS AFTER THE ACQUISITION ANNOUNCEMENT, AS THE BACKLASH
still continued, Carmack checked in on Luckey. "Holding up ok?" he
asked in an email.

The primary answer to Carmack's question was, naturally: "I'm
doing great! I'm about to be rich and, more important, the odds of
VR succeeding this time are now higher than they've ever been." But
although most of Luckey felt that way, a fraction of him was struggling
with the aftermath. Carmack was one of the few people who he felt he
could open up to about that.

>>>**FROM:** Palmer Luckey
TO: John Carmack
SUBJECT: RE: Holding up ok?

I am holding up fine, probably taking this harder than most people
would or should. Thing is, I understand their response, and knowing
what they know (and lacking the behind the scenes big picture), I
would probably feel about the same way. I am guilty of leading this
kind of tribal behavior in the past, as well—about five years ago, I

started the ModRetro forums by getting people riled up about perceived administrative abandonment of the BenHeck forms. We went on to get thousands of members, lots of tech media coverage, and millions of monthly pageviews, but it all started out as the crusade of an angry teenager hiding behind a computer screen. I don't have years or decades of experience to lean on, so blowback from the community feels very personal. I know you have dealt with similar stuff in the past, hopefully I can grow a thick skin like you.

  I agree that people will probably be happy a year from now, when they can judge our actions instead of our words . . .

On second thought, perhaps it wouldn't even take that long for the tide to turn. Because just one day after Carmack's email, Oculus shared some exciting news: Michael Abrash would be joining the company.

"Hope has returned!" decreed one member of the VR community.

"What a shocking twist!" declared another.

So shocking, in fact, that even the folks at Oculus were surprised to learn that Abrash would be joining. Including Iribe, who once called Abrash "the final piece of the puzzle." Now he was finally there, officially Oculus's chief scientist. Which begged the question: What was to become of Oculus's current principal scientist?

"WE'RE NOT JUST PUTTING TOGETHER THE BEST VR TEAM ON THE PLANET," Iribe explained, "but we're cutting off Valve's head and offering it to Zuckerberg."

If that were truly why Iribe had hired Abrash (and Binstock, for that matter), then LaValle knew that he couldn't object to this decision. Because that decision, ultimately, was a business decision and that was outside of LaValle's realm of expertise. But having seen the breathless doe-eyed way that Iribe had spoken about Abrash over the past year, LaValle couldn't help but feel like that infatuation also played a role. And if that was the case, then LaValle realized his being replaced had been inevitable from the start.

Because, LaValle now realized, no matter what he had done for Oculus—no matter how critical he had been in helping this small start-up develop Fortune 500 tracking; no matter how much he had taught them about quaternions, pushed them toward lasers, or had to hold his tongue when Iribe praised techniques that Valve used—there would always be one thing he could never do: go back in time and work on a game that Iribe had loved as a child. And so, of course it all made so much sense, that his days at Oculus were numbered.

As Iribe reiterated the importance of serving Valve's head to Zuckerberg on a silver platter, Steve LaValle again felt displaced—he imagined himself the Cowboy in *The Big Lebowski*—and staring Oculus's Jackie Treehorn in the eye, he thought: Well, a wiser fella than myself once said, sometimes you eat the bar; and sometimes, well, the bar eats you.

THINGS WERE FINALLY, SORT OF, RETURNING TO NORMAL. THE BACKLASH WAS dying down, work was starting to feel like work again and as part of that normalcy, a certain level of silliness returned to the office again. Most notably, Simon Hellam had put together a "special" headset in honor of the acquisition: the Oculus Poke, he called the thing, which was basically just a six-inch, 3-D model of the Facebook "Poke" icon affixed to the front of a DK1.

Luckey and Mitchell loved it, and they loved how it lightened the mood. And with April Fools' Day only a few days away, they suggested that they release a video "announcing" the Oculus Poke! Iribe and Chen were in—sounded like a fun idea.

But Patel strongly disagreed. "It may sound lighthearted and friendly to us," he said. "But it'll make us look ridiculously out of touch . . . like we are trivializing their concerns."

Luckey realized Patel was right. Oculus Poke wasn't worth the risk. But that didn't stop him from simply replying with three words: "No Fun Nirav."

Okay, things really did feel like they were getting back to normal.

On March 31—or what *could* have been Oculus Poke Day Eve—Luckey received an email from Mark Zuckerberg.

FROM: Mark Zuckerberg
TO: Palmer Luckey
DATE: March 31, 2014

Now that things are calming down a bit, I wanted to send you a note to thank you for supporting our partnership and say how excited I am to work together.

I really believe in your vision for VR and my goal here is to make sure you guys have all the resources and support you need to bring this to the whole world.

It's a bummer that the initial reaction from the community has been negative, but it seems like you've been doing a really nice job of explaining the benefits of us working together. It feels like you're starting to win folks over.

When the deal formally closes, I'd love to spend some more time together and understand how you're thinking about the future. I'm also happy to answer all of your questions about ads and our business.:)

It's an honor to work with you on this mission. We're going to build something historic.

Well, Luckey thought, that was really nice of Marky Z to write. But what was up with that part about ads? That seemed a little weird to drop in there.

On April 7, a whole new backlash began. Initially, it had absolutely nothing to do with Oculus but rather that some online people were mad that, six years earlier, Mozilla's new CEO (Brendan Eich) had donated $1,000 to an organization that had been against Prop 8 in California.

That outcry wound up involving Oculus briefly when John Carmack tweeted that he was glad that Eich's response wasn't to crawl on his stomach in search of forgiveness. It wasn't even that Carmack agreed with Eich's alleged views; it was that no one even knew what Eich's actual views were (hence their being alleged!). Maybe he hadn't fully realized who he was donating to? Maybe it was an obligatory "donation" to, say, attend a $1,000 dinner? The point was that a lot of information was still absent and Carmack appreciated Eich not cowering to the online mob machine. However, a bunch of folks didn't appreciate Carmack's appreciation. Folks like Chad Elliot, a developer at Waypoint Software, who emailed Palmer Luckey to say, "Dude, I have great respect for John's accomplishments and ability, but applauding

Brendan Eich for sticking to his guns? Ugh. I don't work for Oculus, but please take my advice as someone who worked under public scrutiny for a decade. Have a serious chat with everyone about preserving your company's reputation before you even get off the ground."

At the time, Luckey thought nothing of his reply to Elliot. But years later—after his life would unexpectedly turn upside down—this would become an email he'd never forget:

>>>**FROM:** Palmer Luckey
SUBJECT: RE: Carmack Tweetastrophe

We don't control what employees say on their own time. We are not going to become one of those megacorps that fires people for having opinions contrary to those of the company or public.

There is no way I would limit my own public communications to technology when I am at Oculus (which should be a very, very long time). "If the king does not lead, how can he expect his subordinates to follow?"

CHAPTER 33
ENTER HTC

May 2014

TYPICALLY, PHIL CHEN TOOK PRIDE IN AVOIDING WISTFUL SENTIMENTS, THE kind where you get lost in memory. As the head of business development for Taiwan's High Tech Computer Corporation—or, as it was better known to its millions of customers: HTC—it was Phil Chen's job to look forward, further. But recently, as the company he loved had fallen on hard times—crashing back to earth with an even greater velocity than that which had fueled HTC's fairy-tale rise—Phil Chen found himself thinking more and more about the past. About the golden years of HTC. How incredible it was that the company went from producing 1.5 million mobile devices in 2007 to 50 million in 2012; how they rocketed from being a niche manufacturer to a $35 billion heavyweight; and how much that corporate success meant to him personally—as someone born and raised in Taiwan—being able to prove that a Taiwanese company could go toe to toe with giants from China, Japan, and Silicon Valley.

At first, reflecting on the innocence and optimism of those glory days felt almost like punishment. Like a daily reminder of what he and his colleagues had squandered. But at some point in 2013, after finally conceding that those days were now long gone, Phil Chen found himself emboldened by a new mission: save HTC.

To accomplish this noble endeavor, he reasoned, HTC would need to replicate what it had done before: identify a technology that few realized was about to change the world and then productize that tech into an affordable consumer product. To try and replicate this formula that had worked so well with mobile, Phil Chen enlisted the help of Doug Glen—a tech-savvy, Hong Kong–based entrepreneur who had previously been the CEO of Imagi Animation Studios—and together they embarked on an around-the-world trip they dubbed their "Technology Pilgrimage."

Israel, London, Portland, and San Francisco—these were but a few places they traveled to meet with influencers at tech companies big and small in their quest to identify that next big thing. From computational agriculture to telematic robots, Phil Chen and Doug Glen encountered many incredible things on their journey. But it wasn't until May 2014 that a meeting with Harmonix CEO Alex Rigopulos led Chen to believe he might be onto something.

"If you're asking what that next trend is," Rigopulos said, "I really think it's virtual reality."

Wait, Chen thought. That stupid thing you put on your face?

"Have you tried the Oculus Rift?"

Chen had heard of Oculus before, but he'd never tried the Rift. And though he had trouble imaging that virtual reality was going to be the thing that saved HTC, he valued Rigopulos's opinion enough to investigate this further. As soon as he returned to Taiwan, he would order a Rift.

"That's a good idea," Rigopulos said. "I would also recommend that you go visit Valve if you have time. Charles [Huang, the cofounder of the *Guitar Hero* franchise] was over there recently and he saw a demo that blew him away. Want me to make an introduction?"

CHAPTER 34

OUT OF THE WOODWORK

May 2014

"PALMER LUCKEY'S FIRED!"

"Did you *hear* me?"

"I just FIRED Palmer Luckey!"

Rapidly and aggressively, these words shot forth from the man in Oculus's lobby. And despite cordial requests from Heidi Westrum—to please stop shouting, to just explain what he wanted—the man continued to shout with an unhinged air of authority.

"I'm your BOSS!"

"I *own* Facebook!"

"I *own* Facebook and THIS IS MY PROPERTY!"

With each absurd utterance, more Oculus employees funneled toward the lobby to see what this commotion was about.

He was Jamur Johnson, a disgruntled Bay Area resident who, that past January, had filed "patently frivolous" complaints against Apple and Facebook. Though he was the first to physically show up at Oculus's office since the Facebook acquisition was announced, he was by no means the first to come out of the woodwork with a seemingly delusional claim of ownership.

The Rift was my idea!

No, it was my idea!

Actually it was I who sent Palmer Luckey on his way!

Given the size of Facebook's offer, Luckey figured it was only a matter of time before blasts from his past started stepping forward with claims like these. And though they were always a pain to deal with—to refute, correct, or ignore—there was also something amusing about them as well. Maybe amusing wasn't the right word . . . Intriguing? Revealing? American? Who, Luckey couldn't help but wonder, would next emerge from the shadows?

Take, for example, the claims of Bill Wallace. Or, as he soon came to be known around Oculus, Briefcase Man. The reason why he earned that nickname (as opposed to his many online handles, like *martinlandau* or *cleverusename*) was because he liked to tell a tale in which he met Luckey at SID Display Week in Boston and gave the young inventor a briefcase. Inside that briefcase, he claimed, were $10,000 cash and blueprints for what would become the Oculus Rift.

When pressed on this claim, Wallace eventually explained that the briefcase was a metaphor; in which the $10,000 represented money he had spent "going to conferences to help him and the community" and the blueprints represented advice he had given Luckey years ago. This much is true: Back in 2009, when Luckey first introduced himself to the MTBS3D community, Wallace suggested that he "check out the leep VR full FOV HMD" to learn about its optical solution. "You get something together similar to that," he later added, "and do a full FOV HMD and I think you can make yourself a lot of money with people here willing to buy."

Both then and now, Luckey was appreciative of that advice. But that was it. That was the singular, five-years-ago piece of advice (which, by the way, was posted frequently on the forum!) that, in Wallace's mind, made him the one who "birthed the original idea to Palmer." Leading him to say things like "Palmer has wronged me, taken a good idea, and wants to write me out of his history" and "There are so many lies to correct, what can you expect from a used car salesman son?"

Of all those with claims against Oculus, ZeniMax was the only one (at this time) to take legal action against him. Which was something that Luckey first learned about on May 1 through a cover story in the *Wall Street Journal*.[1]

It took a lot for Luckey to become visibly angry, but this article did the trick. It was a textbook scummy move—to wage a war in the press—and it felt even scummier because of two things that happened next. The first was that Luckey's address, along with that of his girlfriend, was "mysteriously" leaked. This led to months of unexpected visits from people who either wanted Luckey "dead," because of the Facebook deal, or just wanted to meet "the guy who brought back VR!" The second thing that happened was that, almost immediately after the *WSJ* story, ZeniMax filed a multibillion-dollar suit.

Since a legal action was pending, Luckey and his colleagues were immediately blocked from correcting the false narrative that ZeniMax had unleashed in the press, as well as additional claims from the legal complaint (which largely centered around the thesis that "Luckey lacked the necessary expertise and technical know-how to create a viable virtual reality headset").

Even Luckey had to admit this made for a juicy-sounding narrative. Who doesn't love a good Winklevossian tale? Who wouldn't be tempted to believe that it was actually a legendary programmer (and not a previously unknown teenager) that was responsible for creating the Rift? Sure, Luckey thought, that all *sounded* good. Except, of course, that's not what happened at all! If only people knew the whole story . . .

One reason that Luckey thought ZeniMax was probably occupying themselves with this case could have been due to *Elder Scrolls Online*, a massively multiplayer role-playing game that ZeniMax had released just one month earlier. With a budget rumored to be over $200 million, it was clear that ZeniMax had high hopes for the game. But after receiving underwhelming reviews—Mashable, for example, said, "It's bewildering to see that ZeniMax missed that mark so completely"—and paltry sales, *Elder Scrolls Online* seemed on pace to become the year's biggest flop. In fact, it now seemed almost fitting that on the same day the *WSJ* piece hit newsstands, the game director's (Matt Firor) published a lengthy blog post to address the negative reviews.

Even more costly than the making of *Elder Scrolls Online* was ZeniMax's 2009 acquisition of id Software and their ensuing investment in the *Doom* franchise. To date, this had netted one slightly above average game (*Doom 3: BFG*) and another that was stuck in development hell (*Doom 4*). If sluggish sales and squandered investments weren't enough to explain why ZeniMax was now going after Oculus, there was another factor to this situation that couldn't be ignored: Robert Altman.

Currently, Robert Altman was the chairman and CEO of Zeni-Max Media. But it was his prior career—as a financial adviser at Bank of Credit and Commerce International—that interested Luckey even more: in 1992, he had been indicted by both the Justice Department (for criminal conspiracy and concealing material facts for federal regulators) and the State of New York (on nine counts of fraud and bribery) for his suspected role in the BCCI bank scandal. Altman was later acquitted on all criminal charges, but a civil suit from the Federal Reserve led to a settlement in which he agreed not to participate in the banking industry going forward. One year later, Altman cofounded ZeniMax Media.

Luckey believed anyone with a brain could see that this guy was engaged in criminal activity.

"I'm sorry you wound up getting named personally in the lawsuit," Carmack wrote to Luckey on May 21. "Dealing with all the legal crap can suck much of the joy out of life, but it is too important to blow off . . . you are probably still thinking 'Death before injustice!', and my wife is right there with you (she still wishes we had taken the previous suit against me to trial), but the pragmatic thing is going to wind up being some kind of settlement when the cost equals the loss of productivity that we would be taxed with by continuing to fight it."

Luckey appreciated Carmack's advice . . . but it still didn't make him any less angry.

"Should I call the police?" someone asked, as the man in Oculus's lobby continued to scream. *I own Facebook! I own Facebook!*

"Let me try and talk to him first," Laird Malamed said, stepping forward.

"I'm your boss!" the man shouted, pointing at Malamed.

"Sir, can I help you?" Malamed asked.

"No, I'm YOUR boss!" he emphatically replied.

"Well, sir, actually I don't think you are. But how can I help you?"

"Well . . . you're fired!"

"Okay, you don't have the power to do that. And if you don't leave, I'm going to have to ask the Irvine Police to come. You're trespassing on our property."

"But I own Facebook. This is my property!"

"Well, sir, Facebook doesn't actually own Oculus yet."

"YES. IT. DOES."

Slowly, the man's eyes moved past Malamed and fixated on the space. He appeared ready to make a run for it. But before he could make his move, Matt Thomas—Oculus's hulking director of mechanical engineering—blocked his passage. Then, after the man continued to persist, the Irvine Police was called.

By the time 911 was dialed, the man had generally given up his shouting. Instead, he grabbed hold of a chair in the lobby and refused to let go. Even after the police arrived, the man continued to cling with all his might. After resisting their requests to let go, the man was tased right there in the lobby.

When it happened, he let out a bloodcurdling scream unlike any in the office had ever heard before. It was so loud, so pained, so proudly human. And then after that moment—albeit only for a moment—there was a slight sense that this man had just said what everyone, on some level, was feeling at the time.

CHAPTER 35

CHARGING FORWARD

June/July 2014

"WHAT DO YOU THINK OF THIS NEW GOOGLE THING?" JOE CHEN ASKED LUCKEY at the office.

"Google Cardboard? I love it . . ." he replied, before taking a big dramatic pause. "No, wait. I *loved* it. Back in 2011. When it was called FOV2GO!"

Chen finished laughing before asking what that was.

"FOV2GO?" Luckey said. "Have I really not ever told you about that? Oh, man, you should look it up. But basically Google Cardboard is in many ways a direct rip-off of FOV2GO, a project I helped work on when I was at ICT. The idea was to build an extremely cheap VR headset, not out of cardboard, but of foam core. And the first ones we built were for phones. This was when I was working on the Rift. But it was clear that phone sensors weren't remotely good enough, the displays weren't very good, and trying to run something universally on a bunch of phones is just not an optimal way to do it. I don't think it's going to be providing a good VR experience anytime soon."

By this point, Chen was scrolling through old articles about FOV-2GO. Yeah, it looked almost identical to Google Cardboard.

"If Google is claiming they came up with this on their own, then they're bullshitting because we won the best demo award at IEEE. It's a great concept, but not a great product. But hey, even if it flops, it'll still be a good way for Google to show some extra ads to people!"

Chen nodded. "It doesn't seem like a threat. But do you think it'll poison the well?"

Luckey didn't think so. His colleagues, however, were split on this. About half of Oculus thought it was a good thing—that it'd at least raise awareness about VR—and the other half thought it'd do damage to the mainstream perception of VR. And as with most cases at Oculus where the opinion was split, each half spent the next few days trying to prove why the other half were idiots. But on June 29, the playful banter was interrupted by a hot new topic that hit even closer to home.

"Did you see this?" Dycus asked Luckey, pointing out one of the many articles about Facebook's "Secret Mood Experiment," in which data scientists at Facebook manipulated what appeared in the newsfeed of 689,000 users to try and see if they could make people feel "more positive" or "more negative" through a process called "emotional contagion."

"Yeah, I saw it," Luckey said.

"The experiment was successful, by the way," Dycus said. "What does Brendan think?"

Luckey didn't know and didn't expect to find out. Because at the time, Iribe was way too busy with what had been filling most of his days: aggressively scaling up the team.

One of the deal points that Iribe was most proud of suggesting (and receiving) during his negotiation with Zuckerberg was the ability to allocate up to $700 million worth of the RSUs (restricted stock units) to those he hired before the deal officially closed. In effect, this armed Iribe what he called the "ability to supercharge VR." It also enabled him to hire those people without oversight from Facebook. So between when the deal was signed (March 25) and when it closed (July 21), Iribe licked his lips and went on the ultimate engineer-hiring spree.

One by one, lead engineers came down to Irvine. Not even for a job opportunity in many cases, but just to check out Oculus's latest demos. And then one by one, Iribe would usher these unsuspecting recruits into his office to try and bring them onto the team. Oftentimes, since these were top-tier engineers, they'd say thanks-but-no-thanks—they already had a job they liked, one that paid them well, and they weren't looking to make a move (some, simply, wanted to check out the demos).

"How much are you currently making?" Iribe would ask. And then after they replied with whatever the number turned out to be, Iribe would offer them something better—usually double or triple what they were making—and see if that changed their mind.

It almost always did. Gradually, Iribe grew the team from 60 to about 160 by the time the acquisition closed and he received this email from Zuckerberg.

FROM: Mark Zuckerberg
DATE: July 22, 2014
SUBJECT: Congrats

Congrats again on closing and I'm looking forward to working alongside you to change the world.

What you're building is amazing and I'm excited to help turbo charge it so we can deliver it to the world sooner.

This is going to be a great adventure, and I'm excited to be on it with you guys.

Even more excited than Zuckerberg was Palmer Luckey, who was grateful to be acquired by someone who appeared to believe in VR as much as he. "Virtual reality is now an unstoppable force," Luckey wrote to Zuckerberg. "I am crazy excited to work with Facebook on building the inevitable as quickly as possible."

CHAPTER 36

CANARIES IN THE COAL MINE

August 2014/October 2014

SOMETHING FELT OFF.

But Joe Chen couldn't put his finger on exactly what the issue was.

This was supposed to be a celebratory day—Oculus's official on-boarding at Facebook—and, thus far, it had indeed felt special. Never before had *every* Oculus employee been together in the same place at the same time. From those who worked at the home base in Irvine to those who worked elsewhere—either remotely or in satellite offices (like Carmack's crew or the newly acquired Team Carbon)—everyone was here, at Facebook's campus in Menlo Park, getting a firsthand look at all the resources that would now be at Oculus's disposal. And yet, for Chen, something still felt off.

"Is it just me, or are you guys picking up a weird vibe?" Chen asked Dycus and Hammerstein as they took a short shuttle ride to the building where the day's orientation sessions would be held.

Dycus, he of the famous Facebooking-evil question, stared blankly at Chen as if to say: *Oh,* now *you feel it!* Though his first impression of Facebook's campus had been positive and he was currently feeling more optimistic than he had expected. For a big corporation, Facebook certainly didn't feel very "corporate." Tons of employees were dressed casually in T-shirts (often emblazoned with the Facebook logo); and the on-campus arcade, movie theater, and quad cultivated a very collegiate aura.

"Everyone seems super upbeat," Hammerstein noted. "I mean, I know we only saw them for like five minutes! But still . . . pretty impressive considering the circumstances."

The "circumstances" were that right before they arrived on campus, Facebook—the website—had crashed and gone off-line for about thirty-five minutes. So it wouldn't have been a total shock to have seen a handful of engineers chaotically running around the place. But, no, there was none of that. Even with the outage, Facebook's campus was filled with shiny, happy people.

Chen nodded. Everything Hammerstein and Dycus had said made sense. Maybe that pit in his stomach was just nerves—just a little gurgle of uncertainty now that the acquisition was actually happening.

Whatever it was, Chen tried not to think about it. This was easy to do when he and all his Oculus colleagues received brand-new, Facebook-issued laptops.

Mac laptops, Luckey noted. He found this humorous because the single biggest theme of all the onboard presentations was "openness" and Apple, of course, famously ran closed systems on their hardware.

Obviously, that was a very different sort of openness than the kind Facebook hammered home throughout the day. From the HR rep who kicked off their onboarding session to the manager who finally dismissed them for the day, they were talking about an ethos of transparency that was critical to Facebook's mission. That's why Facebook execs didn't have individual offices; why the buildings on campus had no locks on their doors; and why employees were allowed (and even encouraged) to migrate freely between projects. Openness, as it was preached over and over, was central to Facebook achieving its underlying mission of making the world a better place.

Still though, Luckey thought: at least jailbreak that Apple shit!

"I hear ya," Chen told Luckey when he voiced his opinion, though if Chen did, it was just barely. Because by this point in the day that pit in his stomach had grown to a sharp, stiffening full-body feeling. And that's when he realized *exactly* what this sour sonofabitch feeling actually was: heartbreak.

What was there to feel heartsick about? They had just sold their company for $3 billion and, literally, everyone on the team still had a job so they'd all keep working together for years. Nothing was changing . . . except for the amount of money in their pockets! But even Chen knew that wasn't actually true. Everything was about to change, and, in fact, it had already started: Oculus's recent hiring frenzy. Competitors jumping into VR. And then the one that actually stung: how hated Oculus had become in some corners of the internet.

The days of wall-to-wall races and family dinners were over. The time of maximum impact had come to an end. Oculus's mission—to finally deliver the promise of VR—was ultimately subservient to Facebook's mission of . . . well, that was the other reason Chen's heart was breaking: all these people who worked at Facebook, all these shiny happy people, *actually* seemed to believe that they were contributing to some sort of overarching, humanity-improving mission. Honestly, these people were drunk on do-good Kool-Aid!

Everything was about to change. As he looked around him, surrounded by the relaxed presences of Luckey, of Iribe, of all these people whom he had gotten to know and love in the trenches, Joe Chen thought: regardless of what I do from here on out, I'll probably never do anything as cool as what I did at Oculus ever again.

AFTER THE ONBOARDING SESSION AT FACEBOOK, IRIBE AND ONDREJKA BEGAN more seriously discussing whether or not Oculus should move their team up to the campus at Menlo Park. And ultimately, after talking it over with their respective executive teams, it was decided that this would be best for both entities.

"Hey," Luckey said to Hammerstein. "We should move in together!"

Luckey didn't just mean the two of them; he meant their significant others (Edelmann and Howland), and also their ModRetro brethren (Dycus and Shine).

After talking it over, it was revealed that Dycus and Shine didn't want to live with anybody else; and Edelmann and Howland didn't want to live with anyone beyond their respective boyfriends. But despite four of the six initially nixing the idea, Luckey and Hammerstein remained resilient and continued to push to make it happen. So they started looking for places to rent in the Bay Area.

Shortly after searching, Luckey excitedly told Hammerstein that he thought he'd found the perfect place. "There's a junkyard for rent," Luckey said. "We can go live there and buy RVs and park them at the junkyard. It'll be cheap."

"Palmer," Hammerstein replied. "I don't want to live in a junkyard."

"But the others might want to live in a junkyard!" Luckey said.

"No, they wouldn't!" Hammerstein replied. And then polled the other four to confirm this.

"But we'll save so much money," Luckey countered.

In lieu of a reply, Hammerstein just stared at Luckey—his friend who was now worth many, many millions of dollars.

Although Luckey's junkyard play didn't work out, the others agreed to take a trip up to the Bay Area and check out a few possible places, and then five of the six—everyone but Shine, not yet sure if he was willing to move at all—agreed that it might actually be really fun if they all lived together. Especially because three of the six (Luckey, Edelmann, and Dycus) had never gone to college, the other two hadn't finished (Hammerstein and Howland) and they all kind of got excited about having a dormlike experience. So in the fall of 2014, with Chen instead of Shine, this crew of cheapskate millionaires moved into a place they'd dub "the Commune."

"WHO'S WORKING ON THE GEAR VR DEMO FOR OCULUS CONNECT?" BRUCE MC-Kenzie asked at a software stand-up meeting four days before Oculus' very first annual developer conference.

Dead silence. No one, apparently, was putting together a Gear VR demo. Though slowly, all eyes started turning towards Alex Howland.

"No, no, no, no!" Howland said. "I can't do it." She was already hustling to finish up the public release of Oculus' mobile SDK.

"But mobile release isn't for another week or two!" someone said.

"Yeah! But there's a lot of—"

"Connect is in four days!"

And so, with a sigh, Howland replied, "Okay, I'll figure it out."

For the next three days, Howland worked with the DevRel team and the Home Team to assemble a roster of viable apps and create an interface to work around them; racing against time, hurdles all the way (i.e. busted lenses, trouble flashing the Samsung phones), and able to problem-solve fast enough to have everything ready by 4:30 AM on the morning of the big show.

Meanwhile, Hammerstein had been going through a similar process for Crescent Bay—which, like Gear VR, was also being shown for the first time at Connect. He too was hustling to get all the prototypes ready and, finishing around the same time as Howland, they hopped into a rented U-Haul van and started driving down to Los Angeles.

Given that Howland and Hammerstein had nearly all of the hardware that would be demoed at Connect, the decision to travel together (and to do so in a U-Haul that proved to have very, very bad steering) was not very popular with the logistics and operations folks at Oculus. But they did it, and managed to make it safely . . .

ON OCTOBER 20—SHORTLY AFTER OCULUS CONNECT—VALVE HOSTED A LITTLE developer conference of their own. This one, however, was very much not public; instead—via a cryptic invitation—requesting the presence of a dozen or so early VR developers. There, Denny Unger (from Cloudhead Games), Alex Schwartz and Devin Reimer (from Owlchemy Labs) and a handful of other game-making pioneers all gathered in Valve's office after signing several NDAs and waited for word about what appeared to be a big announcement.

And then finally it came: Valve was partnering with HTC to build a VR headset. A competitor to the Rift. And unlike the Rift, this new headset would apparently be capable of delivering a full "room-scale" experience.

CHRIS GALLIZZI—THE VR ENTHUSIAST WHO HAD ONCE MODDED *SKYRIM* FOR the Rift—didn't yet know anything about Valve and HTC's plan to bring a headset to the market, but for the first time since being seduced by Oculus, he had concerns about what they were doing.

It had started a couple months earlier, when he received his DK2 and was surprised to find himself underwhelmed. In particular, he didn't like how Oculus's sensor system forced users into a seated-only game experience. But while he found that decision pretty understandable (this was first-gen VR, after all), he was miffed to learn that Oculus— now as a part of Facebook—had canceled the gun peripheral project that he had been working on with them for months. And as frustrating as that was, it was the *reason* they had decided to cancel Project Cannon that worried him most. Because, as it was reluctantly explained to him by Luckey, Facebook didn't want to be associated with plastic guns. Which, without context, Gallizzi reasoned was a somewhat reasonable stance. Every company has the right to decide what they will and won't allow on their platform. But what hurt about this, what made him most frustrated and sad, was that Oculus—apparently—was now a company that didn't put gamers at the forefront of that decision.

CARMACK ALWAYS PUT GAMERS AT THE FOREFRONT OF HIS DECISIONS, WHICH is part of what prompted him—in late October—to email Oculus' exec team with a handful of growing concerns.

FROM: John Carmack
SUBJECT: Gear VR

The updates this morning were a breath of fresh air, but I continue to think that it is a company weakness that the exec team fails to have meaningful strategic discussions in writing. I feel that it gives us a negative culture of hints, nudges, inference, and plausible deniability.

I think the decision to not allow Gear VR to be marketed to consumers for the foreseeable future is strategically bad for the company.

Everyone agrees that Gear VR, as it stands this moment, is not ready to be pushed at consumers—a functioning store, broader content support, and more polish in general is required. Soft launching on the Note 4 actually works out well for this, but we should be ready for consumers with the Galaxy 6 launch, and possibly earlier . . .

I was hoping that there would be a set of gates something along the line of "commerce in XXX countries" and "XXX apps in the store", but the official position is:

"No product will be pushing at consumers until Brendan is comfortable in it."

There are important consequences to this position. Brendan is not yet comfortable in Crescent Bay. Therefore, Gear VR will not be pushed to consumers until it is better than Crescent Bay.

I think I can deliver some form of full time inside out position tracking on the Note 5, but I do not expect it to be superior to Crescent Bay, and this is coming from a notoriously optimistic programmer.

Therefore, Gear VR will not be considered a consumer product for over a year.

Therefore, we will prevent Samsung from selling to the best of their abilities.

Therefore, unit sales will not be impressive.

Therefore, developers will not make money.

Therefore, we will not attract the developers we want, and we wind up screwing the developers that made a leap of faith for us.

Therefore, we fail . . .

PART 4

POLITICS

CHAPTER 37

TWELVE DAYS
IN 2015

January–December 2015

JANUARY

On January 23, at the World Economic Forum in Davos, Switzerland, dozens of the world's most prominent global leaders gathered to try something few of them had ever tried before: virtual reality. Specifically: an 8-minute, 360-degree documentary film produced by the United Nations that—via a pre-consumer "Innovator Edition" of Samsung's Gear VR headset—sought to transport viewers to refugee camp in Jordan. And following an overwhelmingly positive reaction in Davos, this film (*Clouds Over Sidra*) would be but the first of several VR films the UN would produce to screen for policy makers and big donors as part of a new approach to raise awareness and funds for key causes.[1]

FEBRUARY

On February 16, in preparation for an executive retreat at Iribe's home in Laguna Beach, John Carmack wrote up a lengthy email that discussed the state and direction of Oculus.

"Things are going OK," he said at the top. "I am fairly happy with the current directions, and I think we are on a path that can succeed." Nevertheless, there were developments and tendencies that concerned him. And over the course of 2,000+ words, Carmack warned his colleagues about numerous lingering issues—highlighted by the following things:

- "Talk of software at Oculus has been largely aspirational rather than practical. "What we want" versus "what we can deliver". I was exasperated as they talk about "Oculus Quality", as if it were a real thing instead of a vague goal. I do have concerns that at the top of the software chain of command, Nate and Brendan haven't shipped consumer software."
- "The Oculus founders came from a tool company background, which has given us an "SDK and demos" development style that I don't think best suits our goals. Oculus also plays to the press, rather than to the customers that have bought things from us . . ."
- "Everyone knows that we aren't going to run out of money and be laid off in a few months. That gives us the freedom to experiment and explore, looking for "compelling experiences", and discarding things that don't seem to be working out. In theory, that sounds ideal. In practice, it means we have a lot of people working on things that are never going to contribute any value to our customers."
- "Most people [at Oculus], given the choice, will continue to take the path that avoids being judged. Calling our products "developer kits", "innovator editions", and "beta" has been an explicit strategy along those lines . . . To avoid being judged on our software, we largely just don't ship it. For example, I am unhappy with Nate's decision to not commit to any kind of social component for the consumer launch this year. I'm going to try to do something anyway, but it means swimming against the tide."
- "It would be hard for the CEO of a sailboat company to be enthusiastic and genuine if they always got seasick whenever they went out, but Brendan is in exactly that position. My Minecraft work is a good example. By its very nature, it is terrible from a comfort position . . . Regardless, I have played more hours in it than any other VR experience except

Cinema. Brendan suggested there might be a better "Made for VR Minecraft" that was stationary and third person, like the HoloLens demo. This was frightening to hear, because it showed just how wide the gulf was between our views of what a great VR game should be."

MARCH

On March 1, HTC CEO Peter Chou surprised the audience at Mobile World Congress in Barcelona. "I've got an incredibly inspiring piece of news to share with you today," he said. "Once in a generation, we see technology that transforms the world. We, at HTC, were the first to pioneer the smartphone industry. Today, we're ready to commence the next chapter . . . with a groundbreaking partner: Valve."

Amidst applause from the audience, and the Valve logo now lighting up the screen behind Chou, he talked about how "Virtual reality will become a mainstream experience for the consumer" and then introduced HTC's first-generation consumer VR product: the HTC Vive.

In contrast to the "seated experience" that the Oculus Rift would offer, the HTC Vive promised a "room scale experience"—allowing users to move freely around an environment of up to 15 x 15 feet. In this respect, the Vive appeared to be almost like a next-gen version of the "Valve Room"; and this time, instead of all those crazy QR codes, the Vive needed only a pair of "SteamVR base stations" to track an entire room.

"But that's not all . . ." announced HTC's Executive Director of Global Marketing, after following Chou on stage. "We realize that the promise of virtual reality only *truly* becomes real when we can put product in the hand of consumers. To that end, we are thrilled to share that HTC will deliver a consumer product before this year ends."

Shortly after the Vive was introduced in Barcelona, one of Oculus' earliest hires sent an email to that summed up how he and several of his colleagues were suddenly feeling:

FROM: Brant Lewis
SUBJECT: HTC Headset
      Hard to see this as anything but a disaster for us. Needs to be seen whether they can actually manufacture the thing but on specs alone—

they have the device that I would want. And sadly—the device we set out to build 2 years ago.

- Effectively same screens and lenses. Wind out of sails there. If the[y] managed to get OLED rgb-stripe then that sucks big-time . . .
- 15' x 15' tracking volume. Uuugh. Gut punch. Nothing in our road-map counters that.
- Controllers. Need I say more? So disappointing . . .

Even if they miss their date and push into 2016, from what I've seen of CV1, the HTC device is the headset that I would prefer. It's clearly aimed at the hardcore (don't mind setting up a dedicated room) VR group. I hate to parrot the forums, but it feels somewhat true. It sucks that Oculus put so much effort on product and company expansion, that we lost our focus on product innovation and kick-ass VR.

APRIL

On April 12, Luckey had dinner at Akasaka Rikyu—a Cantonese restaurant in Tokyo—with Masahiro Sakurai, the creator of his all-time favorite game franchise: *Super Smash Bros.*

"The beauty of *Super Smash Bros* is *not* the huge spectacle," Luckey told Sakurai. "It's the balancing and rebalancing—the incredibly careful calibration that goes into the mechanics of every character. And I know you *kill* yourself to make those mechanics work. And I just want you to know how much I appreciate your talent and your sacrifice."

Sakurai bowed his head in gratitude. And shortly after a discussion about the state of the game industry, Luckey moved on to the reason he had arranged this dinner: to recruit Sakurai to Oculus. Or, at least, to convince him to make a virtual reality game for Oculus Studios.

"I am honored," Sakurai said, once again bowing his head.

With the exception of *Mushiking: The King of Beetles*—a Sega-made arcade and collectible card game[2]—every game Sakurai ever worked on had been distributed exclusively by Nintendo. So it was probably a long shot to think that he'd do work for Oculus. But there was a reason Luckey had some hope: since August 2013—when Sakurai publicly stated that "Oculus Rift VR goggles . . . [will] be a hit in the near future"[3]—he had repeatedly expressed strong interest in virtual reality; in part, because VR technology could enable him to create things that just weren't even possible in real life. As an example, Sakurai had once written about how incredible it would be to drive a transparent car; noting, "I often think to myself, 'if only this dashboard and stuff around my feet were transparent.'"[4]

Needless to say, Luckey believed that—from an artistic standpoint—Sakurai would have an interest in developing a game for Oculus. But, of course, developing games is about more than just the artistry; it's also a business. So part of the reason Luckey had wanted to meet in person was to let Sakurai know that—now with the backing of Facebook—Oculus could put together a very generous offer.

"We'll give you money," Luckey told him, "and full creative control and you can stay in Japan—or, if you prefer, we can relocate you to any place in the world you'd like. Basically: I will *personally* make sure you get *whatever* you want to make VR games."

Sakurai appeared flattered but ultimately his feelings were as follows: I want to work on VR. But if I were to work on VR, I would I want to give it my all. And it is hard for me to give my all to something when I know the audience will be small."

MAY

On May 1, Steve and Allison Spinner hosted a small fundraiser for Hillary Clinton, who weeks earlier had announced that she would be running for president in the 2016 election. In advance of this event, campaign chair John Podesta, who would be speaking there that day, was advised that "There will be four important individuals within tech in the room: Aaron Levie (CEO of Box), Padma Warrior (CTO of Cisco), Anne Wojcicki (CEO of 23 and Me and married to Sergei Brinn, co-founder [of] Google), and Palmer Luckey (founder [of] Oculus VR)."

Luckey had concerns about Clinton (based on her platform in the 2008 presidential race), but figured this would be a good opportunity to find out if any of her positions had changed. In particular, he was curious about ethanol subsidies; so he asked John Podesta if Mrs. Clinton planned to continue supporting ethanol blending mandates (because, in Luckey's opinion, "those subsidies should go towards useful tech that needs it"). Podesta thanked Luckey for the question and said that the Clinton campaign had not yet developed its position on ethanol subsidies; and that they wanted help from "smart people in Silicon Valley" in developing it.

A few weeks later, in an op-ed for the *Iowa Gazette,* Hillary Clinton reaffirmed her support for the federal ethanol mandate.[5]

JUNE

On June 22—with Facebook pondering a multibillion-dollar acquisition of Unity (in a deal codenamed "One")—Zuckerberg laid out his vision and strategy for VR/AR to Iribe, Sheryl Sandberg, Mike Schroepfer, Amin Zoufonoun and four other Facebook VPs:

"Beyond the sheer value we can deliver to humanity by accelerating and shaping the development of this technology," Zuckerberg wrote, "we have three primary business goals."

The first goal was strategic. "We are vulnerable on mobile to Google and Apple," he wrote, "because they make major mobile platforms. We would like a stronger strategic position in the next wave of computing. We can achieve this only by building both a major platform as well as key apps . . . From a timing perspective, we are better off the sooner the next platform becomes ubiquitous and the shorter the time we exist in a primarily mobile world dominated by Google and Apple. The shorter this time, the less our community is vulnerable to the actions of others. Therefore, our goal is not only to win in VR / AR, but also to accelerate its arrival."

The second goal was brand. "The weakest element of our brand is innovation," Zuckerberg admitted, "which is a vulnerable position for us as a technology company dependent on recruiting the best engineers to build the future . . . Our core social networking work is no longer new, Internet.org is extending something rather than inventing it, and AI is not yet tangible. We can do more to tell our story in each of these areas, but succeeding in VR / AR has the most innovation potential in the next 5-10 years. Of course we need to succeed in VR / AR to gain any of these brand benefits, but if we do, this will be very valuable."

The final goal was financial. "The financial goal is the most specific," he explained, "and this is where I'll discuss which aspects of the

VR / AR ecosystem we want to open up and which aspects we expect to profit from."

In Zuckerberg's thinking, the ecosystem could be broken to three major parts:

- Apps/Experiences: "Gaming is critical but is more hits driven and ephemeral, so owning the key games seems less important than simply making sure they exist on our platform. I expect everyone will use social communication and media consumption tools, and that we'll build a large business if we are successful in these spaces."
- Platform services: "These platform services [referring to things like "app distribution store, ads, payments and other social functionality"] should be cross platform. Most of the services can be offered on iOS, Android, on desktop, etc. On Android [for example] . . . if we app switch to our preloaded marketplace for purchases, we won't even have to pay Google's 30% rev share . . . This will be challenging as OS providers will try to push us out, but if we build superior services and provide things OSes need (eg Unity support), then we have a good shot at success."
- Hardware/Systems: "This category includes all of the core technology required to make VR / AR work but that has little sustainable business value independently: the headsets, controllers, vision tracking, low-level linux and graphics APIs . . . aside from brand, patent enforcement and building teams that are consistently far ahead of everyone else, this is the most difficult part of the ecosystem to build into a large business. Even when companies do succeed, no single hardware company gains ubiquity like our vision requires us to do with apps.

"Over time," Zuckerberg continued, "someone will need to tightly integrate all of the software and hardware components of this ecosystem . . . If we own Unity, we can ensure this always happens well, happens quickly and happens with our systems . . . Further, since our key services will be integrated so well and so prominently into Unity, that will put pressure on other engines like Unreal to do close integrations with our key services as well as to make sure their developers have the same access. This will help achieve our goal of spreading the important platform services—even to platforms we don't control."

Zuckerberg then elaborated at length about other benefits to owning Unity, and then ended with this: "Given the overall opportunity of strengthening our position in the next major wave of computing, I think it's a clear call to do everything we can to increase our chances. A few billion dollars is expensive, but we can afford it. We've built our business so we can build even greater things for the world, and this is one of the greatest things I can imagine us building for the future."

JULY

On July 28—at a small invitation-only event in Beverly Hills—Oculus premiered a short, animated VR film called *Henry*. Directed by Ramiro Lopez Dau and created using Epic's Unreal Engine 4, the immersive short puts users eye-to-eye with Henry, a lonely hedgehog who, on his birthday, wishes for the one thing that his spiky quills make all but impossible to receive: a hug.

Henry was produced in-house at Oculus, created by the company's recently established "Story Studios." And the following year—after 14 months of effusive praise (like *WIRED* calling it "the most important movie of 2015"[6]—*Henry* would earn Oculus their very first Emmy Award.

AUGUST

On August 25—while working on a pitch for an Oculus-based subscription model—Luckey received a request from the Comms department that reminded him how fortunate he was and nearly choked him up:

FROM: Tera Randall
SUBJECT: Make a wish request - Oculus

Hi there—we received a request from Make a Wish about a boy named Tommy (13) who's last wish is to demo Oculus and meet the founders.

It's a very quick hello at the office. Palmer, would you be up for giving him a Touch demo and talking with him for a few minutes? Timing is the next couple weeks.

Nate, I thought you could swing by after the demo and say a quick hello too if the scheduling works out.

SEPTEMBER

On September 8, Mark Zuckerberg welcomed Microsoft CEO Satya Nadella to Facebook's campus to show off Oculus, discuss the ongoing strategic alliance between their two companies and, ultimately, broker a trade between Facebook and Microsoft.

The following day, in just three sentences, Zuckerberg recapped their arrangement:

- We will ship Universal Windows Platform apps for Facebook, Messenger and Instagram.
- We will make our app install ads and audience network capabilities available for Windows developers, and we will support your other request for middleware support.
- In exchange, we're asking you to ship Minecraft for Oculus mobile and Rift.

OCTOBER

On October 13, Luckey handed a gift to Iribe.

At the time, Luckey was leading negotiations with HTC—trying to get Vive, their upcoming headset to natively support the Oculus Store—and, well, those negotiations were not going so well. Because, as Luckey had recently concluded, "HTC lacks the ability to do the deal they want to do. Ultimately, Valve is the decision-maker."

Although the relationship between HTC and Valve was undoubtedly nuanced, the net result reminded Luckey of a "partnership" from years earlier—a partnership between Apple and HP, in which HP "got the right to sell iPods."[7] From a distance, this may have looked like a win for HP: Apple's iPods were becoming incredibly popular and they—with their expertise in hardware—could manufacture

HP-branded versions of this hot product. Except that as part of the deal, HP made two major concessions: they agreed not to compete with Apple in this space for a set period of time, and they agreed to pre-load Apple's digital marketplace (iTunes) onto the computers that they sold; and, as soon became clear, that's where the real money was: on the software-side of selling hardware.

"So this is a reminder of where the real power is," Luckey said, handing Iribe the gift. It was an iPod from 2004, one with the HP logo etched on the back.

NOVEMBER

On November 10, Oculus announced on its blog that, "Starting today, you can pre-order the all new Samsung Gear VR, powered by Oculus, for only $99."[8]

The post goes on to detail a few of the improvements over the previously-released Innovator Edition (i.e. lighter, better ergonomics, redesigned touchpad) as well as compatibility with additional Samsung smartphones (i.e. the Galaxy Note 5, S6, and S6 edge).

"This is an incredible moment for virtual reality," the post ends by saying, "as people all over the world can now turn their phone into an immersive VR experience."

DECEMBER

On December 15, Paul Bettner got great news: Oculus was so pleased with the latest build of *Lucky's Tale* that they wanted to bundle that game with the Rift. This way, every single person who bought one of their headsets would—for free—be able to play the game.

This news came almost three years to the day that Bettner had first visited Oculus and it was hard not to reflect on how far he and they had both come. And yet—with the first wave of consumer VR headsets launching in three months—he knew that, really, this was only the beginning of the journey.

CHAPTER 38

AWAKEN THE SLEEPING GIANTS

January 2016

"GROWING UP," PALMER LUCKEY EXPLAINED, "I WANTED TO BE A SUPER-villain."

"Wait. What?!" Joe Chen asked. "Did you say super*villain*?"

"I did!"

"Yeah. So . . . once again: what?!"

It was January 9, the final day of that year's Consumer Electronics Show. But after a week of shaking hands, taking meetings, and doling out demos at the Convention Center downtown, Luckey and Chen were now twenty miles outside the city—waiting for the police to arrive—at a run-down Baja Fresh on the outskirts of Vegas.

A lot had changed over the past fourteen months, since the euphoria of Oculus Connect, the developers' conference organized by Facebook in late 2014. For one thing: Oculus now had a launch date for the Rift (March 28, 2016), a price ($599), and a murderer's row of competition on the horizon (from Valve, HTC, Sony, Google, and more). And for another thing: Chen had left Oculus in early 2015. Plagued by that graduation-day feeling that he had felt during the onboarding session at Facebook, he decided to join a different VR start-up, a content creation company called Vrse that had recently been founded by filmmaker Chris Milk.

Chen liked his new company, but the amount he missed Oculus could feel excruciating at times. Though, as he would remind himself, what he missed wasn't even a real company anymore; it was the memory of one that had always been juggling more questions than answers. Chen missed the seemingly unending battle that had defined early Oculus—and, just as much, he missed those who had fought beside him in that battle. After all, where else could Chen expect to find a brilliant, belligerently honest, budding mogul who—apparently—once dreamed about growing up and becoming a fucking supervillain?

"I know that sounds terrible, but I don't mean it that way!" Luckey clarified. "Like, my goal was *not* to plot nefarious schemes and try to take over the world. But when I was a kid, the supervillains were the ones who had all the coolest stuff."

By "coolest stuff," Luckey meant gadgets, gizmos, and cutting-edge research labs. The kind of stuff that one would find at Skynet or in Lex Luthor's secret lair. This, Chen thought, was actually pretty interesting. It wasn't all that long ago, Chen recalled, that technology was portrayed as an instrument of evil. Supercomputers, geolocation devices, and robotic assistants—these were all hallmarks of an evildoer. Now, these tropes were the staples of our heroes. From fictional protagonists—like Tony Stark and Rick Sanchez—to Silicon Valley icons—like Elon Musk and Steve Jobs—the "mad scientist" moniker had been replaced with words like *visionary*, *luminary*, and *disruptor*; and technology, in our cultural consciousness, had become perceived an instrument of empowerment. Sure, some people were afraid of technology (or maybe, on some level, we were all a *little* spooked), but the next generation of unknowns—and the one after that, and the one after that—just didn't send a chill down our collective spines anymore; at least nowhere close to the way that it used to. And that's because, somewhere along the way, the narrative had changed; technology became synonymous with progress, disruption became the goal, and those who had the "coolest stuff" were now the ones we called the good guys.

"Hey," Chen said. "You mind if I step out for a moment? I want to try the cops again and see if we can figure out this bullshit."

The "bullshit" was what had brought Chen (and later Luckey) to the outskirts of Vegas. What had happened was that Chen had a close friend who had grown up out here and recently inherited her childhood home. She was looking to sell it but had heard from old neighbors that the home seemed to be occupied by squatters. Since Chen was in Vegas for CES, he offered to skip the show's final day and check in on the property. Indeed, he discovered a house full of people. He contacted his friend to see what she'd like him to do, and while waiting for her reply, a woman approached Chen's car.

"What are you doing here?" the woman had asked. "Are you supposed to be here? This is private property, so you should probably move along."

It was kind of funny, Chen thought, that this was almost exactly what he had planned to ask her. But at the moment, he was unable to savor the irony. Because as the woman began to rant and gesture wildly with one of her hands, the other hand remained by her waist, guarding what appeared to be a weapon. Was it a gun or a knife? Chen couldn't tell, but when she reached for it, he had no desire to find out. So he zipped away, called up the police, and spent the next few hours waiting for them to meet him at a run-down taco joint so they could discuss the situation.

In the middle of all this, Luckey texted Chen. Did he want to grab some lunch? Chen explained to Luckey the bullshit he was dealing with, but, to Chen's surprise, Luckey didn't care; he had finished all his interviews for the day and would happily keep him company. Forty-five minutes later, Luckey showed up at Baja Fresh in a limo.

"Looks like it's still gonna be a few hours," Chen said, returning to the table. "But feel free to bail whenever. Seriously."

Luckey had no intention of bailing any time soon. He missed having Chen by his side; he missed the counsel, the friendship. "What did you think of the reaction?"

"To $599?"

"Yeah."

"I mean . . . it wasn't good!"

Three days earlier, Oculus had opened up preorders for the Rift, and when the price was revealed ($599), there were a lot of angry people, and understandably so. Not only had DK1 cost half of that ($299) and DK2 around the same ($350), but for years Luckey had preached the importance of affordability, as explicitly as saying that "if something's even $600, it doesn't matter how good it is . . . it really might as well not exist" (June 2013, *All Things D*), and as recently as September 2015, when Luckey told *Road to VR*, "We're roughly in that ballpark [$350]."

Luckey explained to Chen that Iribe's desire to keep increasing the quality of CV1 (and make it more of a "Facebook-quality" product) had resulted in a headset that was about twice the cost of what they'd originally planned. "By early to mid-2017," Luckey noted, "we should be able to get the price down to about $350–400. But none of that really matters, does it? Because I still look like an asshole."

"Yeah," Chen replied. "You kinda do."

"Thanks. I appreciate you agreeing with me!"

"Sorry, dude."

Luckey knew that a simple apology would not be enough to win back the goodwill that Oculus had lost. Nor should it be, he thought. If he were in their place, he'd be upset too. In his mind, he still *was* these people: a gamer, a dreamer, a thrifty-ass motherfucker. Which is why he was confident that, in time, he'd be able to win back their affection.

"You know," Luckey said, shaking his head, "I'm really getting concerned about this whole freedom of expression thing."

Chen laughed. "At Facebook? Or, like, in America?"

"Both. All. Everything. Did you know that our supposed allies don't have freedom of expression? Germany doesn't have freedom of speech. The UK doesn't have freedom of speech. The UK claims to have freedom of speech, but then they have a giant list of exceptions."

"Like what?"

"Let me pull it up," Luckey said, searching on his phone. "Okay, here we go: guarantee of freedom of expression . . . blah blah blah . . . *However*, there is a broad sweep of exceptions including threatening, abusive, or insulting words or behavior intending or likely to cause harassment, alarm, or distress or cause a breach of the peace . . . sending any article which is indecent or grossly offensive with the intent to cause distress, glorifying terrorism, advocating for the abolition of the monarchy, indecency, obscenity, and defamation."

"So . . . everything?"

"Yes!"

"But is this stuff actually enforced?" Chen asked.

"It can be. That's how I came across this stuff. There's a recent case in Ireland. Someone was arrested for insulting somebody on their message board!"

"That's crazy . . ."

Talk of the First Amendment soon segued into talk about the Second (which Luckey strongly believed in for constitutional reasons; Chen, too, for grew-up-in-Texas reasons), and then eventually the conversation spiraled into the upcoming presidential election.

"Am I correct to assume that Bernie Sanders is superpopular these days over at the Commune?" Chen asked.

Luckey laughed. "Yes, you would be correct."

"And you?"

"Well," Luckey began, "I think there's a lot to like about Bernie. But personally, I'm pretty jazzed about Donald J. Trump."

"Wait. What?!" Joe Chen asked, taken aback. "Did you say *Trump*?"

"Yeah," Luckey said, "I did!"

"Yeah. So . . . once again: what?!"

Luckey wasn't shocked by Chen's tongue-tied reaction. This was actually mild by Silicon Valley standards. Go tell anyone at Facebook or Google or anywhere else that you thought Trump had some fresh ideas and you'd be lucky to get a response that didn't include an expletive. For that reason, Luckey had hardly told a soul. It wasn't worth the argument. It wasn't worth explaining that *of course* he didn't think Trump was perfect; *of course* he didn't agree with Trump on every issue.

"It's crazy," Luckey said, referring to the few conversations that he'd had with liberal-leaning colleagues. "They all voted for Obama. Who, obviously, I didn't vote for . . ."

"You were Romney in 2012?" Chen asked.

"No: Gary Johnson."

"What about in 2008?"

"I was sixteen!" Luckey responded.

Chen cracked up and Luckey continued. "But my point is that just because they voted for Obama, I don't take that to mean they agree with him 100 percent. I'm not, like, oh, you voted for Obama? I didn't realize you supported civilian drone strikes!"

Luckey had guessed that Chen would be more open-minded than his colleagues; and he knew that guess had been correct when Chen followed up not with the usual tsk-tsk, but with a genuine ask of why. "Give it to me," Chen said. "Give me the case for Trump."

"First," Luckey said, picking up his phone, "I want to show you something."

Luckey opened up the Facebook app, but then remembered he had already deleted the post he wanted to show Chen. So instead he searched through his phone and was able to find what he was looking for, a Facebook post from March 16, 2011:

"So, Donald Trump says he is seriously considering running for president in 2012, since he does not think anyone else can save America. He says he will make his final decision by June. I am thinking this might be pretty awesome."

"That's from 2011?" Chen asked in disbelief. "As in a year before Oculus? That's wild! So what was it you saw back then that made you think that Trump might be a good candidate?"

"Well," Luckey explained, "I had been a fan of Trump for a long time. I loved *Art of the Deal*. I took many of the lessons to heart even when starting Oculus."

"Really?"

"Oh yeah," Luckey said, and then started rattling off a few of the things that had appealed to him about Trump. "He was a reasonable conservative who had smart beliefs on the drug war. He has a proven history of taking over failed government projects and getting them to the finish line ahead of schedule and under budget. He was outspoken against the war in Iraq . . ."

"What about his personality?"

"Obviously, I don't agree with everything Trump says. But that's true of any candidate. I ended up voting for Gary Johnson in 2012, and I'm *definitely* not on board with everything that comes out of that guy's mouth. You also have to keep in mind who Trump is talking to. He's not talking to you and me. He's talking to *millions* of people."

"Well, sure."

"He can't be everything to everyone. And he has come a long way during the primaries on being a lot more brash. I mean, nobody even expected him to last this long, let alone actually be the leading Republican candidate. He is smarter than people think."

"Maybe," Chen conceded. "But if he were *really* smart, he'd realize that being president sucks. It's a shittier gig than the one he's got!"

Luckey laughed. "That could be. Who knows, maybe I am blinded by fandom. I've been following the Donald for a long time."

After chatting about US politics, the guys eventually moved on to talking about internal politics at Oculus. "So guess who wants to lock down our system?" Luckey asked.

Chen thought for a second. Brendan was the obvious answer, but even though he wasn't totally into really open systems, he hadn't imposed his will on DK1 or DK2. So why start now? He wouldn't. Which meant . . .

"Yup," Luckey nodded. "Marky Z."

"Oh man," Chen said, shaking his head and laughing. "I'm glad I got out when I did!"

CHAPTER 39

LOCKDOWN

January/February 2016

ALONG WITH A TRIO OF OTHER EXECUTIVES—SHERYL SANDBERG, DEBORAH Crawford, and Dave Wehner—Mark Zuckerberg addressed a handful of financial analysts during the Facebook's year-end Earnings Call on January 27.

"Overall Q4 was a strong quarter and a great end to the year," Zuckerberg said. "More than 1.59 billion people now use Facebook each month . . . and when it comes to our business, we're also pleased with our continued growth."

For the year, Facebook's revenue had risen by 52 percent to reach $5.8 billion. Powered not only by those 1.59 billion monthly users on Facebook, but also by the 2.1 billion monthly users on their subsidiary platforms—with 400 million on Instagram, 800 million on Messenger, and WhatsApp ending the year at "nearly 1 billion monthly actives."

"With virtual reality," Zuckerberg said, "we've reached an important milestone. The Samsung Gear VR shipped over the holidays with our Oculus software and we're pleased with the initial reaction. This month, we also opened pre-orders for Oculus headsets . . . This Oculus launch is shaping up to be a big moment for the gaming community. But over the long term, VR has the potential to change the way that we live, work and communicate as well. The launch is an important step towards the future, and we're really looking forward to seeing how people use it."

ACROSS THE FACEBOOK CAMPUS, ON THE SECOND FLOOR OF OCULUS'S BUILD-
ing 18, a red neon sign blasted the word LOCKDOWN brightly for all to
see. And beside the word was a number, which was reduced by one with
each passing day; and today's number was "49" to remind employees
how few days remained until they officially launched the Rift.

Like Zuckerberg, everyone at Oculus was looking forward to how
people would use this product they'd devoted a piece of their life to
build. But among the executives, there was a growing concern about
how people would be *allowed* to use it—which came to a head six days
later.

"After significant consideration," Iribe wrote to Zuckerberg on February 2, "we believe staying open on PC is the right decision and will
be key to our long-term success."

By "open," Iribe meant that Rift users would be free to upload and
download software from outside the Oculus Store. For consumers, this
would be significant because they'd be able to purchase content from
other channels (like Valve's Steam or EA's Origin), as well as access
content that was prohibited from the Oculus Store (i.e., pornography).
For developers, this would be significant because it meant that they'd
be able to choose where to sell their content, as well as ensure that their
content could still be distributed even if it not approved by Oculus.
And for distributors this would be significant because they'd have the
freedom and autonomy to run their own store—and as a result, they'd
be able to compete directly with Oculus.

"What you're describing is not my instinct on how to proceed, but
I'm open-minded about discussing it . . ." Zuckerberg replied. "I'd need
to understand why this doesn't just reduce Oculus to being a peripheral
for PC and how it actually positions us better as a platform."

Over the next week, Oculus's exec team met several times to put together what they believed to be a compelling case for staying open. And
as they rushed from one meeting to the next, often passing that neon
sign in Building 18, the irony of what it said was not lost on them—as
they brainstormed ways to try and stop Facebook's CEO from locking
down their platform.

On February 12, Iribe met with Zuckerberg to explain why staying
open, or even "mostly open" was core to Oculus's strategy of funneling

users into their platform. The more open their platform was, the more it would attract consumers, developers, and distributors; and, as a result, the more Oculus's platform would grow and be the beneficiary of network effects. This was more or less the strategy that had won Microsoft the PC market and had won Google the mobile market.

History and strategy aside, there was also an ideological component to this. Iribe, Abrash, Luckey, and all of Oculus's other execs believed in an ethos of openness when it came to the PC marketplace. Zuckerberg respected their perspective but was unsympathetic to their cause. And any benefit of the doubt that the Oculus team might have hoped to receive was effectively nixed when Zuckerberg made clear that Oculus no longer seemed to have "a technological advantage" over their competition. The meeting ending shortly thereafter with Zuckerberg telling Iribe, "You're not going to wear me down on this."

Nevertheless, Zuckerberg still hoped to resolve this issue in an amenable manner. In an email sent two days later, he outlined a potential compromise: "So far we've discussed two approaches: (1) enable only apps distributed on our store to run on Rift, plus a few exceptions; or (2) enable apps distributed anywhere on Rift. A middle ground proposal is to enable apps to run on Rift that are either distributed in our store or distributed directly by developers, but not allowed versions of apps distributed in other stores."

Zuckerberg continued to outline the perceived benefits of his proposal as well as a potential pitfall ("The one thing I don't get though is how we're going to enforce that Valve obeys our terms no matter what restrictions we set"). While he gave the exec team at Oculus a lot to think about, it was something he said in closing that they couldn't stop thinking about for days.

"While I obviously value your opinions a great deal," Zuckerberg wrote, "I also view this platform strategy as the whole reason we're investing in VR so I'm not going to change my view simply because you care a lot."

If there was any uncertainty about Zuckerberg's resilience here, that was all but cleared up by an ensuing email from Brendan Marten, Facebook's head of finance for AR/VR, which stated the following things:

- "Mark sees that there are risks to closing the platform, but does not believe they are that great or irreversible, and is willing to take that chance. He believes that you can open it once closed, but going the other way is much harder."
- "Mark believes that some Oculus employees will be unhappy with closing the platform, and even that it goes against the culture and could cause unhappiness across the team, but considers that a reasonable price . . . For example, how many long-time Oculus employees (the ones most likely to strongly object to closing the platform) will walk away from their RSUs? If they don't, how many will do their jobs worse as a result? Would even highly-principled John [Carmack] walk away?"
- "Unless we present Mark with a long-term plan for Rift success that he thinks makes sense, I believe Rift is probably going to be somewhere between resource-starved and orphaned soon."
- "Either: a) we come up with an explanation for openness that Mark can believe makes sense, b) we decide to go closed, or c) there is going to be a head-on collision. And remember, even if we win that head-on collision, Mark controls budget and head count, so the Rift may find itself starved regardless."

In the midst of all this, Iribe sought counsel from Epic CEO Tim Sweeney, who promptly replied that "locking the Oculus hardware down in any way is an incredibly bad idea."

Sweeney then elaborated with the following advice: "Iterate several moves ahead. You've blocked Steam, Valve has blocked Oculus, and all the other platforms have formed battle lines. So much is happening in secret that you won't even see the enemy positions until they're deeply entrenched . . . Enough stupid crap is happening already, with Amazon refusing to sell Apple TV, Apple abortively fighting an ad-blocking war against Google, and Microsoft's broken Windows Store that doesn't sell real Windows Apps. Facebook should just keep above the fray, as a trusted and even-handed player in all ecosystems."

As Luckey pondered these words from Tim Sweeney, in came another email from Brendan Marten that triggered an idea. "My sense is [Mark] believes Google is our biggest long-term competitor . . ." Marten wrote. "Google is the actor with the most staying power financially, and it is most interested in achieving a broad VR ecosystem along the lines of what Mark wants us to achieve for Facebook. Mark needs to hear more about that in our ongoing strategic vision/discussions."

Google got Luckey thinking about Google's Android phones, and Android phones got him thinking of something that he thought might actually work.

"Are you familiar with the 'Unknown Sources' setting on Android phones?" Luckey asked Iribe during an exec meeting that week.

Iribe wasn't, so Luckey explained. "So one of the best things about Android is how easy it is to sideload apps"—by which Luckey meant running apps that came from places other than the Google Play Store. "All you have to do is go to your settings and enable 'Unknown Sources.' It's basically a way to jailbreak your phone, but with just one click instead of all that crazy iOS shit."

Luckey's pitch to Iribe was that Oculus essentially copy that trick. This way Zuckerberg could have his closed platform, and users could open it up by just clicking on one thing.

"Interesting," Iribe replied, now mentally gaming out this idea.

"And to think," Luckey joked, "we probably would have come up with this idea sooner if any of you ever dared to use a piece of hardware that wasn't made by Apple!"

But wait, one of the execs pointed out: Android's (mainstream) audience was very different than what the (early adopter and PC gamer) audience that would be buying the Rift. The implication being that—compared to the estimated 1 to 2 percent of Android users who enabled 'Unknown Sources'—Rift users would be *way* more likely to do so.

"Correct," Luckey said. "But we don't need to advertise that to Mark!"

THAT EVENING, BACK AT THE COMMUNE, LUCKEY FOCUS-TESTED THIS UN- known Sources idea with Edelmann, Dycus, Hammerstein, Howland and Shine. All five of them—each the epitome of a hardcore PC gamer—were on board, thinking this was a clever solution.

"To be honest," Howland said, "I'm a little surprised that Brendan's on the right side of history here. Maybe he should have a little talk with his buddy Nate."

Luckey understood where she was coming from, given Iribe's affection for Steve Jobs and how explicitly he had touted the benefits of "closed" platforms these past few years.

"And to be even *more* honest," Howland said, "does anyone else find it odd—nay: *unnerving*—that Zuckerberg doesn't seem to understand the market he just bought himself into?"

Accurate or not, Zuckerberg wound up signing off on the Unknown Sources ideas.

"Meeting with MZ went pretty well," Iribe messaged the exec team on February 18. "I presented an SDK licensing proposal that's similar to Android where . . . users still have the power to be open and use Steam or any other apps but devs and stores have a very hard time promoting content that requires disabling the security/safety check box . . . I think we landed in a really good place considering the alternatives."

Feeling good about this resolution, as well as the many launch titles he had been playtesting for the Rift, one thought kept running through Luckey's head—so much so, that he decided to post it to Reddit that evening: That's some good shit right there, if I do say so myself . . .

FROM AN APARTMENT IN THE NETHERLANDS, JULES BLOK FURIOUSLY CODED ON his laptop.

The code that he was writing was for a passion project—for a plug-in called Revive—that had Blok concurrently feeling anxious and excited. Excited because, if Revive worked, then Vive owners would be able to play the games that were currently only available on the Rift; and anxious because he had no idea how Oculus would react.

Maybe they would sue him. Or maybe they would ban him from their applications. Or maybe they would not care at all. At this point, it was all up in the air. But even if there might be danger ahead, Blok was determined to move forward. Because, to him, this was about something bigger than himself; this was about principles and about proving, without a doubt, that a plug-in like Revive was technologically possible.

Creating Revive, or something similar, had been on Blok's mind for about a year now. Ever since Oculus started talking about "exclusives." Because, like most PC gamers, he saw exclusives as antithetical to the PC experience. They destroyed one of the greatest things about PC gaming: that anyone could play any game on any PC. That was pure, that was noble; and, for Blok, it was also something worth fighting for. So when Oculus eschewed the open ethos of PC gaming for the walled-garden approach of consoles, Blok began to wonder if there might be something he could do. Something like Wine—an application that made Windows-based games playable on Linux platforms—but, in this case, with Oculus-based games on the HTC Vive.

To make this work, Blok determined that his best bet would be to create a compatibility layer between Oculus's SDK and OpenVR (Valve's SDK). Fairly quickly, he felt confident that he *could* pull something like this off, but he was conflicted as to whether or not he should. This uncertainty persisted until a Reddit AMA that Palmer Luckey did in January 2016.

"You might slightly misunderstand our business model," Luckey explained to a Reddit user who was concerned about exclusivity. "When we say 'Oculus Exclusive,' that means exclusive to the Oculus Store, not exclusive to the Rift. We don't make money off the Rift hardware, and don't really have an incentive to lock our software to Rift. That is why the Oculus Store is also on Samsung's Gear VR. Gear VR and the Rift are the first consumer VR devices coming out, but in the future, I expect there will be a wide range of hardware at a variety of price and quality points, much like the television and phone markets."

That information was a comfort to Blok, but what really caught his attention was when Luckey said this: "If customers buy a game from us, I don't care if they mod it to run on whatever they want. As I have said a million times (and counter to the current circlejerk), our goal is not to profit by locking people to only our hardware—if it was, why in the world would we be supporting Gear VR and talking with other headset makers?"

After reading that, Blok decided to move forward. So he waited until Oculus put out the first stable version of their SDK and then, as soon as they did, he got to work.

From the get-go, Blok knew that one of the biggest challenges would be Oculus's code signature check. This check was a standard security measure for hardware products, and it was designed to prevent hacked (or *un*checked software) from ever running. To get around this, Blok and another programmer he trusted, created an injector file, which was able to boot itself up and bypass Oculus's security system. And by late March, Blok was nearly ready to share this passion project that he had dubbed Revive.

CHAPTER 40

ENTITLEMENT CHECKS

March/April/May 2016

ON MARCH 26, DRESSED IN FLIP-FLOPS AND A RED HAWAIIAN SHIRT, PALMER Luckey flew up to Alaska and—freezing his ass off—planned to personally deliver an Oculus Rift to a customer named Ross Martin.

"This is the very first Oculus Rift," Luckey explained to someone recording. "It's signed by myself, Michael Antonov, Brendan, and Nate. So we are going to go and deliver it to Ross Martin, who was the first person to preorder. Let's go!"

In theory, this quick trip was designed to do something special for Oculus's first customer. But as Luckey spoke it was clear that this delivery was even more special for him than Martin. "Man, this is incredible," Luckey said, after passing along the Rift. "I've been working on this thing for so long and you're the first person to actually get one. It's kinda like me taking all this work, and then handing it off to you; so you have to make sure you have fun with it or something . . ."

Martin promised that he would, and after a brief conversation about virtual reality, he thanked Luckey for making this unexpected trip. "I didn't sleep at all last night. This is amazing. On behalf of the online community and everyone . . . we really appreciate all the transparency and outreach that you do."

"I try to do as much as I can," Luckey replied. He then shook

Martin's hand and ended by saying, "And now I need to get home as fast as I can, so I can keep working on the launch . . ."

UNFORTUNATELY, THE MOST PRESSING LAUNCH-RELATED ISSUE WAS THAT— contrary to everything Oculus had been saying for months—they would not actually be launching in full the following day.

Other than a few hundred units earmarked for their original Kickstarter backers, Oculus didn't have any units in the US that could go out to consumers. Worse: it was still unclear when they would be able to begin fulfilling orders. Wednesday (March 30) seemed like the most likely option.

The delay was largely due to a shortage of suitable optics for the Rift (which, in turn, was largely due to Iribe setting a quality bar for optical components that—as a member of the exec team would later describe—was "probably too high . . . We ended up rejecting lenses that were better than what HTC was shipping on all Vives.") Nevertheless, this was the situation Oculus was now in. And so, one day before what Luckey described as "the paperest of paper launches," he outlined what he saw as three potentially viable options:

1. "Ship the Kickstarter units to consumers (instead of to the original backers)."
2. "Be honest. Tell people that the initial batch, which we had hoped to get out on Monday, is now delayed till Wednesday. Reveal that all of those units will be mailed via overnight shipping for no additional charge (something we planned on doing, but have never revealed), point out that most people will be getting their units earlier than they would have had we shipped on Monday."
3. "Do nothing, be quiet, weather the storm. This is very risky, given the media attention on our launch, and it will make people far more skeptical of future problems. We need to ration our goodwill."

Looking at this list, Luckey thought the answer seemed obvious— #2: Be honest. As was often the case, he was joined in support by Nirav

Patel. But in the end, they were outvoted, and Oculus wound up chancing it with a combination of #1 and #3.

ON MARCH 28, THE OCULUS RIFT LAUNCHED TO NEAR-UNANIMOUSLY PRAISE from the press. But despite splashy headlines like "This shit is legit" (*Gizmodo*[1]) and golden pull-quotes like "this is an astonishingly well-made device" (*WIRED*[2]), and the catharsis that came from shipping the world's first modern VR headset, the day felt anticlimactic—disappointing, even—to many employees at Oculus.

To some degree, this was just human nature—the by-product of building up a dream for years and years and years. And to another degree, this felt like the hidden cost for all those moments of joy and excitement that had already been experienced at trade shows over the years. That was all part of it, as were concerns about the price, content and ceding the space to competitors. But while all these things were true, there was also something else—something at the bottom of it all—that was best summed up by the final line of Adi Robertson's review for *The Verge*.[3] "The headset you can buy today," she wrote, "is not Oculus' most ambitious vision for virtual reality—but it's a vision that Oculus has successfully delivered on."

CV1, of course, was never supposed wouldn't be Oculus' "most ambitious vision"; it was, after all, just meant to be the start. But shipping without motion controllers; and against a pair of upcoming headsets that more closely resemble Oculus' original vision (Sony with affordability; HTC with immersion) it was hard for many at Oculus—on this day of reflection—to think about what they were shipping, instead of what they were not.

"COULD YOU PLEASE STATE YOUR FULL NAME FOR THE RECORD?" AN ATTORNEY asked the co-founder of Total Recall Technologies, during a taped deposition on the morning of April 8.

"Ron Danger Igra," he replied, speaking directly to the camera.

"And could you spell your middle name, please?"

"D-A-N-G-E-R"

"Kind of like Austin 'Danger' Powers?"

"Kind of."

Ron "Danger" Igra was the entrepreneur who—two years ago, in the aftermath of the Facebook acquisition—had messaged his ex-partner, Thomas Seidl, to suggest that they could "make millions together" by "going after Oculus."

At the crux of this claim was a "Nondisclosure, Exclusivity and Payments Agreement" that Luckey had signed at Seidl's request in August 2011 (shortly after the two had met on MTBS3D). Among other things, this agreement laid out terms for Luckey to build a pair of HMDs that could playback footage from the 3-D, 360-degree "immersive camera" Seidl was building.

Given this signed agreement, plus the types of paternity issues that often arise between inventors, it would seem understandable for Seidl to explore his legal options. But back when Igra contacted him in 2014—and repeatedly in the two years since—Seidl had explicitly said that he wasn't interested in "going after" Luckey, Oculus or Facebook. Igra, however, felt differently; and in May 2015—on behalf of Total Recall Technologies (TRT), his and Seidl's all-but-defunct partnership—Igra filed a suit against Palmer Luckey alleging "breach of contract and wrongful exploitation and conversion of TRT intellectual and personal property in connection with TRT's development of affordable, immersive, virtual reality technology."

Given the peculiarities of this case—like the fact that, while working for Seidl, Luckey had never once heard the names "Ron Igra" or "Total Recall Technologies" before; or that, just three days earlier, Seidl had reiterated his desire to "veto" Igra's lawsuit—it was safe to say that the Total Recall case required far less of Luckey's attention than the ongoing litigation with ZeniMax. And eventually, eleven months later (in March 2017), Igra's case would be dismissed.[4]

At the time, of course, Luckey couldn't have known that; so he took the lawsuit seriously—seriously enough to take the uncommon step of traveling to Wailuku, Hawaii, on April 8 so that he could exercise his right, as a defendant, to face his accuser and attend their deposition.

Though as he sat there, staring down Ron "Danger" Igra—trying not to laugh when Igra's attorney claimed his presence was "intended to harass the witness"—Luckey realized that despite being three time zones away from Irvine, his mind was still all-consumed by Oculus and what critics were increasingly referring to as their "botched" launch.[5]

On April 11, podcaster Jeff Cannata tweeted, "Order processed 12 mins after Oculus went on sale. My new estimated ship date: 5/16-5/26, 2 months after 'release.' That's a botched launch."[6] Hours later, Cannata's two-month delay was put to shame when games journalist Patrick Klepek tweeted "The Oculus Rift that I pre-ordered has been bumped from shipping in April to June."[7] Unsurprisingly, frustrations about the delays were exacerbated by the relative silence from Oculus. It had now been nearly two weeks since the Rift's alleged "launch," and the only explanation that Oculus had provided was that a "component shortage impacted our quantities more than we expected . . . We apologize for the delay."[8]

Meanwhile, in the midst of that backlash, another storm was brewing. And this one hit on April 14, with Luckey quickly receiving a flood of messages from concerned developers, who were linking to something that had just been posted on Reddit: "Play Lucky's Tale and Oculus Dreamdeck on the Vive." This post—written by a guy calling himself "Cross VR"—linked back to a GitHub page where people could download a plug-in that would make those games (and other content from the Oculus Store) playable on the HTC Vive.

ON MAY 20—IN AN EXECUTIVE DECISION SPEARHEADED BY MITCHELL—OCULUS issued a software update that, among other things, neutered Revive's ability to play Oculus content on the Vive. The backlash was swift and unforgiving—with many citing Luckey's Reddit comments from just a few months earlier: "If customers buy a game from us, I don't care if they mod it to run on whatever they want."

Luckey still felt that way, but he believed this situation was different than what he had described on Reddit for two reasons: one being that, with Revive, customers were *not* "buying games from us" (since they were getting access to free content like *Lucky's Tale*) and two being that this wasn't an example of "modding" but rather an example of "piracy."

Given these circumstances, Luckey felt that if Oculus's DRM decision had been communicated better, there was good reason to think that this approach would have worked.

"But we miscalculated, and it is really screwing us," Luckey wrote to the Oculus exec team on May 25. "It is tempting to think this is a storm that will blow over like so many others—it is not. The bad media coverage, defections from usually friendly writers, trending on Facebook and Twitter, Reddit threads, and scathing videos from YouTube celebrities getting hundreds of thousands of views are only a symptom of the real problem: blowing out the trust of the people who actually want to see Oculus succeed by allowing them to see us as liars through lack of communication. That trust cannot be recovered with silence, and the longer the situation persists (especially the outright falsehoods), the deeper it will set."

"Thanks for writing this up," Iribe replied. "I'm not a fan of hardware DRM and would prefer we stick to a software only approach. What is the proposal to address this?"

The following afternoon, Luckey, Iribe, Mitchell, Binstock, Patel and Abrash got together to discuss possibilities. During this meeting—and then even more explicitly over email—Mitchell reiterated his case for the hardware DRM:

- "We didn't change anything for Rift users; the only folks impacted were those modifying Oculus software with Revive to make it run on Vive."
- "We wanted to convey a message that Vive users shouldn't expect Oculus content to work for them reliably. I think we sent that message clearly regardless of what we do on rolling this back.
- "We want to continue investing heavily in content— investment that benefits all VR developers and that the ecosystem hugely benefits from—but it needs to make sense for Oculus. Using exclusives to bootstrap our HW adoption to drive platform adoption is a critical strategy; if our fully-funded titles run on competitive hardware, it makes little sense to keep funding them."

"For me," Mitchell told the group, "this was the most important reason to make these changes. I don't want us to be in a world where we can't invest as heavily in the VR because it doesn't make sense."

"None of us want that," Iribe replied, earning nods of agreement from Luckey, Binstock, Patel and Abrash. "But given our resources, we could invest regardless of whether we think it's having a meaningful ROI."

Mitchell conceded that was a good point.

"Also," Iribe added, "as a child of the open PC ecosystem, I don't think I can publicly support or defend a hardware DRM."

Iribe's words struck a chord with Mitchell, who thought that this alone might be reason enough to roll back Oculus' update. "Especially," Mitchell said, "with there being a lot of internal unhappiness."

"We need to start communicating the background behind these decisions," Luckey said several times that afternoon. He then reiterated this again in a lengthy email on May 31 that—over the course of 10 overarching points—elaborated his thoughts on this as well as what felt like an increasingly deeper battle with Valve:

1. "Rolling back with no solid explanation wastes the PR opportunity, and leaves people thinking of Oculus as a still-evil beast that they managed to control for the moment with community outrage."

2. "We should not announce a date for rolling anything back. It should be communicated as something we will roll back once we have enough time to protect our platform and developers from piracy in other ways. People should be talking about why we are doing the 'right thing,' not when we are doing it."

3. "We need to make it explicitly clear to people that this is not an endorsement of unofficial hacks of any kind, nor is it a promise to support them . . . We want all our customers to be first-class citizens with access to all the awesome features of the Oculus platform, store, and SDK."

4. "Revive is designed to go far beyond just allowing people to play our games. It strips our games out of our platform and puts them into SteamVR, effectively turning us into a payment processor for a handful of exclusive games as far as Vive users go."

5. "As of today, we have no path to supporting Vive without integrating Steam/SteamVR/OpenVR. That is exactly the way Valve wants it. Myself and the ["Jasons"] had some good discussion on why it could make sense to wrap them in a tricky way later this year as part of a strategy to get Valve users onto our platform . . ."

6. "We really did make an honest, good-faith effort to natively support Vive at the same level of quality we support the Rift, with access to all of our platform features and content. It could be that we would have chosen to not pull the trigger and go ahead with it in the end but we never got far enough to have that choice—Valve made that impossible for obvious reasons. Valve gains nothing by letting Oculus support natively without SteamVR. They know we have a better SDK, they know we are going to have a broader/better content lineup, they know all the things I talked about above. As long as Valve stays stuck to their platform, they win. The real twist in the current situation is that as of now, they get to have the ideal business decision AND the ideal PR! We take all the hits, we play punching bag, and they reap the benefit. I think there is a way to get people looking at this in a much more honest and critical way, business vs business instead of good vs evil. We do much better in an honest comparison than the current twisted narrative."

7. "Valve gains a lot by effectively locking Vive users to SteamVR and their ecosystem. The current situation is basically the only reason they can get devs to target OpenVR/SteamVR and by extension, Steam: If the Oculus SDK supported Vive (or alternatively, if Vive did not exist), no developer would bother using SteamVR. Why would they?"

8. "We should realize that the most likely outcome of taking this stance is a stalemate. Again, Valve has no business reason to budge on this . . ."

9. "Valve is by no means invincible from a PR standpoint.

We should not treat them as an unstoppable force and immovable object. They have gotten into pickles over the last few years that were generally solved by rapid backpedaling, but their standing is not what it once was. In the specific case of VR, they have failed to deliver on many of the things promised: OpenVR is not open or independent of SteamVR, Lighthouse is not an open standard, and use of the Vive is still tied directly to Steam. On top of that, their Rift support is still jankier than Vive support, a reflection of their priorities."

10. "We can come out of this with a strong E3, a strong year, and a focus on why Oculus is great instead of why Oculus is bad. We won't ever be seen as perfect, but we can be seen as the justified winners."

CHAPTER 41

THE DEVIL IS
IN THE DETAILS

June/July 2016

"WHAT A DIFFERENCE AN INTERNET UPROAR CAN MAKE," WROTE ARS TECH-nica reporter Sam Machkovech on June 24 in a piece detailing how "After weeks of playing cat-and-mouse to block the 'Revive' work-around . . . Oculus quietly updated its hardware-specific runtime on Friday and removed all traces of that controversial DRM."[1]

"We choked!" Luckey told Mitchell, referring to Oculus' lack of messaging around this whole DRM issue. "We failed to communicate anything that would help people see things from a different perspective or see Oculus as anything but a faceless, sterile, corporate-to-the-max behemoth that was forced by the good guys to heel on a bad decision."

Days later, Luckey elaborated on this issue—on how Oculus had so badly "choked"—in an email to Mitchell and a handful of other ex-ecutives. "What is really unfortunate," Luckey wrote, "is that the lack of useful communication was not due to a lack of preparation, analy-sis, or ability. Lots of very smart people put a lot of time into solving these problems, and we had prime opportunities to use those solutions. Instead, we kept making last-second decisions to toss that work, do nothing, or let other people (including our direct competitors) define the narrative completely . . . We fell into the completely predictable outcome of reactive firefighting because we ended up not utilizing some of our strongest abilities. Oculus has more red tape and inertia than it used to, but we can and should be far more nimble and transparent than we have been . . . Maybe I am in the minority here, but I think we can do better."

For Luckey, the best way he and Oculus could do better going forward was by finally figuring out a way to get Oculus' platform onto the HTC Vive. And though this was something he had been very passionate about for a long time, the idea really started to gain traction earlier that month; at an "H2 Goal-Setting Session" where Luckey and Rubin—by virtue of using nearly all their "votes" to push the Vive issue for discussion—were able to start getting serious buy-in on the issue.

By the end of that session, even Iribe appeared convinced. So much so that he'd end the day by saying, "I think we should bring the Oculus Platform to Vive. It's The Right Thing To Do for PC VR and Oculus (as John would say)."

The following day, Rubin upped the ante by suggesting that they make this happen in time for Oculus' third annual developer conference (OC3) to be held that year in October. "In a perfect world at OC3 we would announce . . . full Vive support for the whole store," Rubin wrote; though he caveated that "The Devil is in the details of how we do this mechanically without supporting OpenVR and Steam."

That, in many ways, was what this would all come down to. And Luckey—now grinning at the opportunity—was ready to take on this challenge.

MEANWHILE: ACROSS THE COUNTRY, IN NORWICH, CONNECTICUT—ONCE known as "The Rose of New England"—a thirty-three-year-old guy named Michael Malinowski* was feeling frustrated. This had nothing to do with virtual reality, but, on the contrary, with the state of reality itself. He was irritated by what he felt had become of this country that he loved. And even more specifically—especially at times like these: awake, late at night, rocking his infant daughter back to bed—

Michael Malinowski is a pseudonym provided at request of interview subject.

Malinowski was irritated by the direction that the country was headed and the harsh realities that would await his little girl.

It hadn't always been this way. Of that, Malinowski was sure. He had grown up here—just a few miles away from where he and his wife now lived—so he *knew* how much things had changed. He had seen it all happen with his own eyes. The whole area went from being pretty well off to . . . well . . . now there was just a lot of sadness around here, every day a reminder of what was, what could have been.

The problem was that there just weren't many jobs for the "average person" anymore. The famous textile mills? Gone. The once-vibrant Thermos factory just up the river? Moved abroad. The nuclear plant that made fuel for subs? That was all but crushed by post–Cold War budget cuts during the Clinton administration. As bad as all that was, here was the honest-to-God truth: compared to most Americans—in the Rust Belt, in the heartland, in any place too small for a subway—compared to those people, Malinowski knew that he and his New London neighbors were among the lucky ones. At least they had the Foxwoods and Mohegan Sun casinos to anchor the community.

Back in the '80s and '90s, there were a ton of jobs where you could make $50,000 a year. And that was enough to build a life, raise a family, and wake up each morning still believing in the American dream. But now? Well, even if you are able to get one of those casino jobs, you're making $12 an hour. And that's a *good* job.

If the people of New London were among the "lucky ones," then Malinowski was among the luckiest of them all. After college, he had been able to find a job outside of the casinos: working in technical support for a component manufacturer. For this opportunity, Malinowski was grateful. Truly. There wasn't a day that went by without him thanking God for this fortune: the ability to support his wife, Gloria, and their two wonderful young kids. Gloria, having grown up in the Philippines, made sure that not a day went by without the family thanking their Lord and Savior. In this respect, they were a happy family—a blessed family—but still, every morning, Malinowski could not help but wonder how long all this would last.

When Malinowski had first started at the manufacturer where he worked, only a small percentage of the labor was being done overseas.

Initially, it was just customer support jobs that had been outsourced. But with each passing year, more and more of his colleagues were laid off and their jobs were shifted abroad to employees who were willing to work for a fraction of their American counterparts. And sure, the service wasn't as good—the support staff didn't speak great English, nor did they understand the products as well—but that didn't really matter. Because whatever was lost in quality was gained in enormous savings. Nowadays, almost all of it was being done overseas—scheduling, logistics, engineering, and so on—and Malinowski felt confident that he knew how this story would end: with a small skeletal crew still in the US (maybe) and everything else moved overseas.

What made Malinowski so confident of this outcome? Because that's what companies were supposed to do: benefit the shareholders as much as was legally possible! Which is why Malinowski didn't even blame his company, or any of the businesses throughout the country that were doing the exact same thing.

Malinowski turned his ire toward the politicians. *The fucking politicians! After all, they were the ones who legislated this bullshit system!* We sent them to Washington to represent our interests—not just mine, not just yours, but all of ours, for all of us Americans—and instead of improving our country they sold us out to corporate interests. They were in need of a heavy wash cycle.

The initial lineup of candidates didn't seem very promising. The front-runners, as of early 2015, were Jeb Bush on the right and Hillary Clinton on the left. Between these two, Malinowski's choice would have been Bush (although he held out hope that the Republican nomination might go to Marco Rubio, who he considered to be the best guy in the race back then). But whether it was Bush or Rubio, Malinowski worried that the only way a Republican could win was if everything went right. With the two biggest states (California and New York) going Democrat no matter what, it was just too hard to energize a country full of lethargic voters; people like his neighbors who felt "forgotten" by our country's politicians, who felt like "Republican" and "Democrat" were just two words for the same thing: "Globalist." And so a year before the election, Malinowski, like many Americans on both sides of the aisle, couldn't help but come to this conclusion: if these were the best candidates our country could come up with, then we, as a nation, really were screwed.

"Our country is in serious trouble," Donald Trump said in June 2015, announcing his plans to run for president. That statement alone—that brutal honesty—was music to Malinowski's ears. Finally, it seemed, there was a candidate willing to slice through the platitudes of politics. Not just when it came to talking about the economy, but politicians themselves: "We have losers. We have losers. We have people that don't have it. We have people that are morally corrupt. We have people that are selling this country down the drain . . . they're controlled fully by the lobbyists, by the donors, and by the special interests, fully."

Malinowski didn't immediately jump on the Trump Train, though. He appreciated Trump's tell-it-like-it-is attitude, but he didn't really take Trump seriously at first. The guy seemed more fit to be Fool than King.

But then, over the next couple of months, something funny happened: Malinowski was so entertained by Trump that he wanted more than just the sound bites that hit the news. So he started actually listening to full interviews with Trump, then watching some of Trump's campaign speeches, and you know what? Kind of like some of those old Shakespeare plays from high school, the Fool turned out to be the Wise Man. This guy actually made a lot of sense! Trump's position paper on immigration, especially, seemed to hit the nail on the head: "We are the only country in the world whose immigration system puts the needs of other nations ahead of our own."

Those were the words of a Wise Man, Malinowski thought. And even better: this Wise Man wasn't willing to be bullied by the media. When the media claimed that Trump said, "Mexicans are rapists," he pushed back and told them to actually watch the tape: where, contrary to their reports, he didn't say "they are rapists" but "their rapists" (as in Mexicans illegally immigrating to the United States were "bringing drugs," "bringing crime" and "[bringing] their rapists." And when the media still continued to run with their narrative, Trump again refused to back down. "I can't apologize for the truth," Trump explained on Fox News.

Even to a budding devotee like Malinowski, Trump was not a hero without flaws. As a man of faith, Malinowski questioned Trump's religious zeal. He also questioned Trump's morality; with the candidate having been married multiple times, failing big with a few business ventures, and allegedly discriminating against tenants in the early '70s. But, at least, Trump appeared to be happily married now, his businesses had bounced back, and there had been no reports of discriminatory behavior in the past four decades (meaning that perhaps the initial allegations were not true; or that, if they were, Trump had changed). Maybe those were all just rationalizations, Malinowski thought. Because, let's face it, Trump *was* an asshole. And four years ago, truth be told, Malinowski never would have hitched his wagon to an asshole. But after seeing the way that Romney had been treated, Malinowski had come to believe two things: (1) They're going to call our guy a "racist" "sexist" whatever, no matter who we send up there; (2) We couldn't win with the nice guy, so let's send in the asshole. At least he'll fight for us, Malinowski thought. At least he'll give hell to those America-hatin', race-baitin' liberal assholes. And let's face it, we aren't voting for saints here. We're voting for the man or woman who we think will do the best for the most number of Americans.

Feeling fired up about this election, Malinowski wanted to get involved, and the easiest way for him to do so was online. Over the years he had been a moderator on over 50 subreddits, including the following:

- **/R/ROMNEY:** devoted to supporting Mitt Romney's 2012 run for President
- **/R/WEALTH:** devoted to sharing success stories about entrepreneurs
- **/R/ALEXJONES:** devoted to exposing the "vile" lies of Infowars' Alex Jones and his "modern-day anti-intellectual cult."
- **/R/NAZIHUNTING:** devoted to sharing information about "organized racists" and targeting "racial extremism"
- **/R/STORMFRONT:** devoted to "weather-related news and information" (but really it was just a way to block the

neo-Nazi internet forum "Stormfront" from claiming the domain).

So heading into election season, Malinowski was to become a moderator for a presidential candidate he felt he could really get behind. Which eventually led him to a subreddit called "The_Donald."

Founded in July 2015, The_Donald was created to follow the news related to the then-unlikely presidential candidate. "Media hit pieces from the left and the right will be vetted," The_Donald promised in its introductory note. As a news junkie, especially one who thought that Trump was being covered unfairly (i.e., "Mexicans are rapists"), Malinowski—under the handle "jcm267"—took pride in correcting the record; and eventually, within weeks of the subreddit's founding, Malinowski was offered the position of top moderator for The_Donald.

Being "top mod" (or any mod for that matter) came with no payment. It was a volunteer position whose only perk was, as implied, that you got to moderate the conversation within the community. You set the parameters, enforced the law, and shaped the direction of the sub. For Malinowski, these responsibilities were critical, since, in the early days of Trump's campaign, some of his loudest supporters online were outright white supremacists, or part of a growing fringe of white nationalists (whose rhetoric made no claims of superiority, but rather focused on a desire to create an all-white ethnostate). Whatever the distinction between supremacist and nationalist, Malinowski wanted none of it. Which is why, to set the right tone, the first thing visitors saw upon visiting The_Donald was a stickied post at the top of the page that boasted about how "jcm267 banned a White Supremacist" and led to a link where Malinowski made clear the community's position on racism: "The bottom line is white supremacist clowns are not wanted or needed by the Trump campaign. You morons do way more harm than good." After being challenged by the now-banned supremacist, Malinowski then elaborated to say, "This is a country of immigrants from all over the world! My wife is Asian, I'm a mix of very different European nationalities. Race doesn't matter nearly as much as you White Supremacists say it does, *especially* here in the US."

Despite explicit messages like these, and many more in the months ahead, that wouldn't stop the media from calling The_Donald "a hate speech forum" (Slate) or a "Trump fan club" that "promotes eugenics" (*Washington Post*). Unsurprisingly, those who frequented The_Donald were by no means choirboys. The community's members were rowdy, argumentative, and oftentimes crass; their posts tended to range from goofy ("Ted Cruz is just a skinnier version of Kevin from *The Office*") or fraternal ("Just wanted to say how fucking awesome you guys are!") to snarky ("Hillary for Prison!") or angry ("Muslim Kills 20 Gay People in Orlando, Media Refuses to Mention His Religion"). This last example—referring to the Pulse nightclub shooting from June 2016—provides insight into why members of The_Donald (mostly conservatives) viewed their community as vital while, quite conversely, many nonmembers (particularly liberal-leaning journalists) viewed this community as "a hate speech forum." To these journalists, it was considered hateful to harp on the religion of the shooter; doing so, they might argue, was dangerous because it reinforced a negative stereotype that could put innocent people at risk. But for members of The_Donald, that concern was dwarfed by a much bigger worry: censorship—the idea that journalists would omit, withhold, or underplay relevant pieces of information (i.e., motive for mass murder) just because of how that information might be interpreted by their audience. This, to Malinowski and members of The_Donald, was the essence of political correctness—not only because it subverted facts for feelings but because minimizing key details (like the fact that the Orlando shooter had sworn allegiance to Islamic State; or the fact that he did so because of US bombings in the Middle East) seemed to make it more likely that, without addressing the root causes, similar tragedies were likely to occur again.

Most of the time, Malinowski thought, liberals didn't even notice the way in which content was censored to fit their worldview. But, with the Orlando shooting, the scope of this censorship appeared to be on full display. Especially when it came to Reddit. Because on r/news—the site's main news hub (which had nearly nine million subscribers)—any stories that mentioned the religion of the shooter were immediately taken down. As were any comments that even just mentioned the word *Muslim*. While r/news was censoring stories, Malinowski and his fellow moderators were sharing them like crazy, so much so that at one point more than half of the posts on the Top 25 of all of Reddit came from The_Donald.

For the moderators of The_Donald, shunning political correctness, while also enforcing a few of the community's primary rules (i.e., No Racism, No Anti-Semitism) could make for a fine line to walk. Malinowski, without hesitation, was up for the challenge. The fact that Trump—this no-nonsense outsider who had burst onto the scene by pointing out our politicians were all "losers" controlled by lobbyists—the fact that this guy even had a chance at all was, frankly, inspiring to him. This inspiration kept him fueled as The_Donald became so popular that it became a feeding ground for many bad actors that Malinowski (or his fellow mods) would need to ban. In one difficult-but-necessary case, Malinowski even needed to ban one of The_Donald's most prominent moderators after it was revealed that he had become a white nationalist. "Having a moderator of the largest pro-Trump sub-reddit promote people that Trump himself would disavow does nothing to help Trump," Malinowski had written in the aftermath of that removal. "It does, however, give credibility to the left wing narrative that Trump's campaign is built on racism."

That The_Donald was being covered at all by the media was a tribute to its enormous success. When Malinowski first took over in August 2015, the community had only a couple hundred members. Now, one year later, it had become Reddit's second-most-active community, boasting over 175,000 members. It had grown so powerful that even Trump himself had come by to answer questions. And yet, as popular as The_Donald had become, and as good as Malinowski had gotten at brushing off "fake news" about how evil-racist-whatever he and his fellow Trump supporters were, he couldn't shake the feeling that he ought to be doing more. There just had to be something else out there that he could be doing—to help nudge his candidate to victory, to help nudge New London back on track.

After months of racking his brain, Malinowski found what he thought might be that something: a simple-but-clever way to make a difference. But little did he know that this simple idea would have major ramifications for the fate of a company called Oculus.

"ARE YOU PRO-TRUMP?" IRIBE ASKED LUCKEY IN JUNE VIA FACEBOOK MES-senger.

"More anti-Hillary," Luckey replied. "But yes. Why?"

"Come on . . . you're brilliant."

"I don't take the choice lightly. It has a lot of thought in it."

"I'll stay out of it," Iribe said, "but Trump is bad news."

"He could be," Luckey conceded. "I just think Hillary would be worse. And there are quite a few people who agree. Including Peter Thiel. Happy to discuss in person . . ."

About twenty minutes later, Mitchell came by in person to ask Luckey if he really was planning to vote for Trump.

"Yes," Luckey told him.

"Dude," Mitchell replied. "I thought you were smarter than that."

CHAPTER 42

NIMBLE

August 2016

"THIS CAN WORK," LUCKEY TOLD JASON RUBIN DURING ONE OF SEVERAL POW-wows the two had in August. *"If* we're able to move fast, then this can definitely work."

"Yeah," Rubin agreed. "Getting this out there before they even realize what hit them will be key."

"Yup," Luckey said. "As long as we're nimble, as long as we're quick, then we've got a real shot at pulling this off."

The "this" that Luckey was referring to was a plan for Oculus to support the HTC Vive in a way that didn't sacrifice features, optimizations, and overall quality. (In other words: what Oculus had wanted to do all along). But since Oculus had been stonewalled when going through official channels, a small contingent of the company—Luckey, Rubin, and a handful of engineers—had started to put together a plan that pulled this off without any official collaboration agreements. Key to this was figuring out a way to bypass Valve.

To pull this off, Oculus would need to hook into Vive at the OS level. This way, they could control the display pipeline and directly deploy rendering techniques like asynchronous time warp and asynchronous space warp (a new feature that was going to be announced at Oculus Connect 3 in a couple months). And, ideally, this too would also be announced at OC3. It kind of *had* to be released there for this to work. Because as soon as Valve found out, they would almost definitely try to break Oculus's work-around. Which is why deploying this at OC3 was critical; it was the best way to reach the most VR devs in the shortest amount of time. And since Oculus's plan would ultimately benefit the developers—by being able to now sell their Oculus content to Vive users; and having that content remain uncompromised—all those devs would be excited by what Oculus had done, to such a degree that if Valve *did* end up breaking Oculus's work-around, there would be an enormous backlash to that decision.

"So basically," as Luckey had put it to Mitchell when this plan was first being set into motion, "we're gonna take a page out of Valve's playbook and blindside them the same way they did to us at Steam Dev Days."

With Luckey and Rubin pushing this forward, Oculus was closing in on coming up with a viable solution. Something, it seemed, that would actually be ready in time to disrupt the VR community at OC3.

MEANWHILE, AS OCULUS WAS NIMBLY SCHEMING, MICHAEL MALINOWSKI AND a few of the other moderators of Reddit's fastest-growing community—The_Donald—were discussing a potentially disruptive idea of their own: starting a political action committee (PAC) to put up catchy, semisnarky Trump billboards across America. They envisioned something along the lines of a Trump-supporting version of the Nuisance Committee (a super PAC founded by Cards Against Humanity creator Max Temkin that put up catchy, semisnarky anti-Trump billboards, billboards that said things like "If Trump is so rich how come he didn't buy this billboard?"

Not all the mods at The_Donald thought that a PAC would be a good idea. They were wary of doing anything that tried to hit up the community for funds. This was a valid concern, but Malinowski and another mod—Dustin Ward—believed that with Trump still trailing in the polls, it was important to try and implement their digital activism in the real world. Initially they had hoped to put up pro-Trump billboards just outside of the Democratic National Convention. But as Malinowski and Ward looked into how this could be done, they realized that just filing the paperwork necessary to officially start an organization like the one they had in mind would cost around $9,000, way more than either of them could afford. They'd all but given up the idea of moving forward, until August 9, when they received an unexpected message from a potential benefactor named Palmer Luckey.

THROUGHOUT AUGUST AND MOST OF SEPTEMBER, LUCKEY WAS PRIMARILY FO- cused on hacking the HTC Vive so that Oculus could natively support. Between that, preparing for the launch of Touch and speaking at trade shows around the world, the days were often exhausting. But he loved the work—oh so much did he love the work. Nevertheless, after four years of Oculus without *ever* taking a vacation, he was looking forward to a cruise that Edelmann had planned for them in late September.

Days before the trip—on Friday, September 16—the executive team decided to revise their hardware roadmap and kill the Tuzi project they had been working on for just over a year now. Luckey's thoughts on the hardware decision were mixed (he voted yes, but only if they were *certain* they could deliver Venice on time; around Q4 2018), but he thought Iribe's decision—about when to share this news with the Tuzi team—was completely unacceptable.

"Please wait to start talking to folks," Iribe messaged the exec team. By which he meant—and then reiterated by Mitchell—that they shouldn't share this news until Monday.

"To be clear," Patel replied, "are you suggesting that we not start letting folks know about any changes until Monday?"

"Yes," Mitchell said. "It's 4pm on Friday."

Patel pointed out that—with the team crunching that weekend; as they had for many months of weekends now—this meant that "People are working on things that will be cancelled."

Even so, Iribe and Mitchell wanted to wait, so that they could coordinate a plan about how to deliver the news.

"This is bullshit," Luckey told Patel. "We are literally prioritizing 'internal messaging' over our *actual* employees. That's just unacceptable."

Patel agreed. So he and Luckey started breaking the news to a handful of employees. The following week, Luckey was able to leave for his cruise with a clean conscience. But by the time he returned, everything at Oculus would be forever changed.

CHAPTER 43
INTERNET DRAMA

September 2016

ON SEPTEMBER 13, NIMBLE AMERICA FILED THE PROPER PAPERWORK TO BEcome a 501(c)(4) organization, which was a designation for nonprofits dedicated to "promoting social welfare." And in the case of Nimble America, their mission was to "Develop and advocate for legislation, regulations, and government programs to promote America first, improve legal immigration, fight corruption and stimulate the economy."

Four days later, Malinowski proudly unveiled Nimble America on The_Donald to the community's 200,000+ subscribers:

Announcing Nimble America

What we've been able to accomplish here has been amazing and much bigger than any of us and certainly much bigger than Reddit.

We believe that America has been led by poor leaders who have abandoned American principles and sold out all Americans. With the right leadership America will reverse its course towards mediocrity and globalism, becoming great again.

We've proven that shitposting is powerful and meme magic is real. So many of you have asked us, how we can bring this to real life. We wanted to do it in a way that was transparent and had purpose. Not just sell t-shirts to sell them, but to sell t-shirts to shitpost. We've worked with lawyers and RNC consultants to advise us on how to establish the proper entities to do this right, and we'll be transparent with all financial activity from Reddit. We've also worked with the Reddit admins to make sure all of our activity operates within their guidelines.

Announcing Nimble America, Inc., a social welfare 501(c)4 non-profit dedicated to shitposting in real life.

For many, the term "shitpost" might conjure up images of the worst dreck on the internet. Or, relatedly—as per places like Urban Dictionary and Know Your Meme—"shitposting" was defined as something like "worthless and inane posts on an internet messageboard" or as "a range of user misbehaviors and rhetoric on forums and message boards that are intended to derail a conversation off-topic." Over the years, however, the term had evolved. "Now people are using it to describe funny posts. Now shitposts are funny instead of shitty," one Redditor explained in response to a user who had noticed that, on many subs, people now seemed to "embrace and take pride in 'shitposting', a term that previously seemed to be more derogatory." Within the context of The_Donald specifically, shitposting referred to playful, sardonic content that—at its best, at its most "spicy"—simplified complex issues so as to cut through the noise and deliver some sort of larger, underlying truth. For example, some of the most popular shitposts on The_Donald involved Hillary Clinton's (literal and figurative) embrace of former Senator Robert C. Byrd, someone whom Clinton publicly called a "friend and mentor," yet—as these shitposts pointed out—someone who also, before becoming a politician, had been a prominent member of the Ku Klux Klan.

Now, of course, one might take issue with intellectual honesty of this shitpost (someone, for example, might point out that Byrd later called joining the KKK "the greatest mistake" he ever made) but— agree or not—the point is that crass as the term "shitpost" may sound, it's not all that different from a traditional print advertisement. Except in two ways: one was that anyone could do it (not just ad executives) and that, unlike traditional ads, shitposts were not bound by any regulatory measures. This latter point made the mission of Nimble America extremely fascinating.

As it turned out, Nimble America, a 501(c)(4), *would* be bound by political advertising standards and regulations; and since they would be dealing with billboards, the content would also need to be approved by the legal department at their chosen billboard vendor (in this case: Lamar Advertising, which operates in the US, Canada, and Puerto Rico). Bottom line: even if someone objected to the memes that appeared on The_Donald—the billboards that Nimble America would be "shitposting in real life" would need to pass legal and regulatory guidelines.

To this end, Nimble America cofounder Dustin Ward followed up Malinowski's initial post with an example of the one billboard the organization had put up already, which featured a caricaturized portrait of Hillary Clinton—a photographic image of the candidate, but with a large forehead and thick chipmunk cheeks—above text that read: TOO BIG TO JAIL.

Despite founding Nimble America with what Malinowski and Ward believed to be the best of intentions, the immediate response from members of The_Donald oscillated between annoyance (i.e., "Who asked for this?" to condemnation over what many perceived to be a cash grab (i.e., "Fuck this shit. It was only a matter of time before the gold rush came knocking.").

Although not to this degree, Malinowski and Ward had anticipated that some Redditors might react with skepticism. So, they had reasoned, the best way to prove that Nimble America *wasn't* a "scam" or "cash grab," was by highlighting that they actually had backing from a wealthy benefactor. Since their benefactor (Palmer Luckey) wished to remain anonymous, this made things a little tricky, but after giving it some thought, Malinowski and Ward came to see opportunity in the anonymity as it allowed them to play up the mystery and create a bombastic, Trump-lovin' persona (and Reddit account) they named "NimbleRichMan."

So, on September 17, to try and quell the rising skepticism online, Malinowski and Ward wrote up a call-to-arms that they planned to post from the NimbleRichMan account. But before publishing the post, they asked Luckey—who, at the moment, was in Japan for Tokyo Game Show—to review what they had written. And after he did, they published the call-to-arms.

I am a Nimble Rich Man. I support Donald J Trump, and I need your help defeating Hillary Rodham Clinton through Nimble America. AMA

America is the land of opportunity. I made the most of that opportunity. I am a member of the 0.001%. I started with nothing and worked my way to the top . . .

You and I are the same. We know Hillary Clinton is corrupt, a warmonger, a freedom-stripper. Not the good kind you see dancing in bikinis on Independence Day, the bad kind that strips freedom from citizens and grants it to donors. Hillary Rodham Clinton is not just bought and paid for. Everyone around her is, too. The elite of the country know it. They don't care. They know she is the candidate that will do what they want.

I have supported Donald's presidential ambitions for years. I encouraged him to run in the last election, but the overwhelming power, resources, and institutional advantages available to a sitting President made it almost impossible. If we allow the Clinton Dynasty to continue, the situation will be repeated.

I reached out to the leaders of this community because I am doing everything I can to help make America great again. I have already donated significant funds to Nimble America, and will continue to do so. I need your help: For the next 48 hours, I will match your donations dollar for dollar. Donate ten dollars and I will match you by flying my jet a minute less. Donate a hundred and I will match you by skipping a glass of scotch. Donate a thousand and I will match by putting off the tire change on my car. Am I bragging? Will people be offended? Yes, but those people already hate Donald. They cannot stand to see successful people who are proud of their success.

Let's generate some success of our own. Make America great again with your meme magic, centipedes of The Donald!

Since the whole point of Nimble Rich Man post was to prove that Nimble America was for real, Malinowski and Ward still needed a way to prove that they really did have the backing of a "member of the

0.001%." So for that, they turned to one of the most prominent Trump supporters on Reddit, Milo Yiannopoulos:

> **[–]yiannopoulos_m**
> Hi guys, Milo here. I have personally spoken with /u/
> NimbleRichMan and gotten to know him and his passion for
> Donald Trump and the Trump movement. As a billionaire and
> maxed out Trump supporter, he reached out to the mod team
> and was put into contact with their project to create the non-profit
> Nimble America, with the goal to "Shitpost Across America".
> I want to say this is a worthy cause that I endorse because
> they are one of the Trump faithful and will promote pro-Trump
> policies . . .
>> With love from your dangerous faggot.
>> Xoxo

As with Nimble America's original announcement earlier that day, Yiannopoulos's post was largely met with anger and annoyance, ranging from comments like "Fuck off with the money grab" to "Love you Milo but you're hardly ever here except when it helps you and nobody else." The NimbleRichMan post, however, was better received, inspiring many members of the community to speculate on his identity (and whether or not he really existed).

Several theories were thrown out—Peter Thiel, being the most popular, but Elon Musk and Carl Icahn were also in the mix. Ultimately, however, community members seemed less interested in *who* this individual was than why, if he really were a member of the 0.001 percent, he felt compelled to hide his identity. That seemed cowardly. Why wouldn't he proudly step forward as a Trump supporter?

> **[–]NimbleRichMan**
> I can only answer your question with a question: Where are all
> the wealthy, powerful, and publicly identifiable Trump supports?
>> Answer: We dare not say a word. It would destroy us. I would
> never dream of blacklisting a business for the political views of

the men who work there, but the same cannot be said for many
HRC supporters.

Any clarification wouldn't have mattered. Because by this point in the day, the damage was already done. Nimble America was perceived to be, at best, a poorly executed good idea; at worst, a total scam.

In light of this blowback, Malinowski resigned as a moderator the following day. "I do this not because we did anything unethical," he wrote, "but because the launch certainly could have been done better, much better, and as a result of a terribly executed launch the Nimble America controversy has become too much of a distraction for me to continue to be on the moderator team here." As the guy who had shepherded The_Donald from the very beginning—from the days of two hundred users to the days of two hundred thousand—this was disheartening for Malinowski. All he had wanted to do was help Team Trump. He had anticipated that there might have been a few skeptics, but not this level of outcry from his own community.

CHAPTER 44

THE DAILY BEAST

September 22–24, 2016

SOMEWHERE.

Somewhere out in the Pacific Ocean, coordinates unknown. There, from the balcony of his room on a Riviera cruise ship, Luckey looked out at the horizon: bright, blue, and brilliantly tranquil. It was perfect. Or it should have been. But right now, it didn't matter. Because nothing mattered. Except for one thing: he was about to become the most hated man in Silicon Valley.

As Luckey considered the implications, his phone buzzed with a text message:

UNKNOWN NUMBER: What the fuck.

For a moment, Luckey was able to appreciate the irony of this situation—that right now, on the precipice of all hell breaking loose, only a handful of people knew what was about to go down. And one of those people—the one texting Luckey—was a guy he barely knew, and who, despite a (much deserved) reputation for courting controversy, had, in this instance, just been trying to help a fellow Trump supporter. That guy was Milo Yiannopoulos.

MILO YIANNOPOULOS: You saw my emails, I couldn't have been clearer.

MILO YIANNOPOULOS: I said "Confirmed?" and he said "Sure thing."

MILO YIANNOPOULOS: I've never seen the Beast behave so dishonorably

The "Beast" was *The* Daily Beast and "Sure thing" was a reference to an assurance that Daily Beast reporter Gideon Resnick had given Yiannopoulos two days earlier after a brief correspondence between the two that began with this:

FROM: Gideon Resnick
TO: Milo Yiannopoulos
DATE: September 20, 2016
SUBJECT: Quick Question

Hey,

Hope you're doing well. I wanted to know if you had any knowledge of the Nimble America charity. Let me know when you get a free second. Thanks

>>>FROM: Milo Yiannopoulos
SUBJECT: Re: Quick Question

On background

All I did was verify the identity of the billionaire. I can ask him to talk to you. Nothing to do with me.

I can tell you off the record who it is if you want . . .

Even "off the record," Yiannopoulos had no intention of sharing the billionaire's identity, at least not without Luckey's permission. At this stage, he was trying to get a sense of what Resnick was really after here. Was this going to be a piece about Nimble America; or was Nimble America merely a jumping-off point to pen a "hit piece" about Yiannopoulos?

For most people, this might seem like a paranoid thing to wonder, but for Yiannopoulos it was a valid concern—the natural by-product of an over-the-top, happy-to-piss-people-off personality that was perhaps best summed up by a quote Yiannopoulos had given to the *LA Times* in late 2015: "I enjoy upsetting the right people. I love poking fun at earnest censors. I want to push the bounds of what can be said on the Internet."[1]

Given that MO, it was no surprise that journalists often wrote pieces about Yiannopoulos's behavior. In fact, that was largely the goal from Yiannopoulos's perspective. So he didn't mind it when reporters took him to task . . . as long as what they were reporting was actually accurate information. Which he believed was *not* the case with a piece that Resnick had written about him one month prior: "Breitbart Editor Milo Yiannopoulos Takes $100,000 for Charity, Gives $0."[2] Yiannopoulos believed this to be an outright lie. "Charities take a long time to set up," he would counter. "And all donations remained untouched." Regardless, that experience had led him to believe that Resnick was less interested in reporting the truth than in collecting "scalps" from political enemies (i.e., anyone who supported Donald Trump); and it led Yiannopoulos to believe that Resnick was looking to write a similar piece here, something alleging that Nimble America was a scam—that they didn't actually have a wealthy backer—and probably also alleging that any money Nimble America raised would just be going to line Yiannopoulos's pockets. So anticipating this, Yiannopoulos replied: "What is the thrust of your story? Are you coming for me again?"

>>>**FROM:** Gideon Resnick
SUBJECT: Re: Quick Question

It might not be a story, just trying to verify that detail because whoever it is didn't file paperwork for it yet.

>>>**FROM:** Milo Yiannopoulos
SUBJECT: Re: Quick Question

That's nothing to do with me. All I did was verify the identity of a billionaire I know personally.

>>>**FROM:** Gideon Resnick
SUBJECT: Re: Quick Question

Got it. Could you provide the name just so I can independently verify?
Won't say it came from you.

At this point, Yiannopoulos reached out to Nimble America's benefactor (Palmer Luckey), forwarded along this correspondence, and asked for a favor: Could I put you in touch with this reporter to get him off my ass?

Luckey, who very much meant what he had said on Reddit ("We [Trump supporters] dare not say a word. It would destroy us"), didn't love the idea of being "outed" for his political beliefs. Yiannopoulos sympathized. "A conservative in super liberal Silicon Valley? They'd *destroy* you." So seeking a solution that would satisfy all parties, Yiannopoulos asked Luckey if he would be open to speaking with Resnick provided that Resnick agreed not to reveal his identity.

"Sure," Luckey said, open to being anonymously mentioned in a potential article about Nimble America. So Yiannopoulos contacted Resnick and offered an introduction to Nimble America's wealthy backer under one condition:

FROM: Milo Yiannopoulos
TO: Gideon Resnick
SUBJECT: Re: Quick Question

You can't print it without permission from him directly. Confirmed?

>>>**FROM:** Gideon Resnick
SUBJECT: Re: Quick Question

Sure thing. We are going to attempt to verify with him directly.

And with that, an agreement had seemingly been struck. But before actually going through with this, Luckey wanted to make sure that Yiannopoulos felt that Resnick could be trusted.

After a moment of consideration, Yiannopoulos said he did feel that way. Even though he took issue with that piece from just one month earlier, he noted that Resnick had acted ethically with regard to sourcing quotes, background comments, and other attributions. Besides, Yiannopoulos ended by telling Luckey, "Protecting your sources is rule number one of journalism. I know a lot of these places play dirty, but I *truly* can't imagine that Gideon [Resnick] would go back on his word. That would be such a flagrant, blatant, violation of journalistic ethics. Nobody would commit such a cardinal sin of journalistic virtue."

Luckey agreed with this logic and gave Yiannopoulos permission to make an introduction. And minutes later, Luckey received an email from the Daily Beast reporter:

FROM: Gideon Resnick
TO: Palmer Luckey
BCC: Milo Yiannopoulos
SUBJECT: Re: Gideon—meet Palmer

Thank you, Milo, for connecting us. Moving you to BCC to continue the conversation with Palmer.

Palmer, nice to meet you! I just wanted to get some info on Nimble America. How much do you know about it? Are you involved with it? And what's the plan for it? Let me know when you get a chance. Thanks so much!

>>>FROM: Palmer Luckey
TO: Gideon Resnick
SUBJECT: Re: Gideon—meet Palmer

Hi Gideon,
Let's chat on the phone in 20 minutes?

About twenty minutes later, Luckey and Resnick spoke briefly by phone. Luckey confirmed that Yiannopoulos had no connection to Nimble America (other than to vouch that there was indeed a wealthy backer), answered a few questions (about why Nimble America had appealed to him), and explained how he had first gotten in touch with the founders of this organization (Dustin Ward and Michael Malinowski, whom Luckey would later introduce to Resnick).

Shortly after the call, Resnick followed up with a quick question:

FROM: Gideon Resnick
TO: Palmer Luckey

Thanks again for taking the time to talk. So as a quick follow-up, are you "nimblerichman" on Reddit and did you post on Saturday night about donations? According to the PAC's website, "nimble rich man" is wealthy benefactor so just checking if this is you.

Technically, Luckey hadn't actually written the post in question. But since the account *had* been created to represent him, he believed that it would have been "intellectually dishonest" not to take responsibility for anything posted under the "NimbleRichMan" name.

FROM: Palmer Luckey
TO: Gideon Resnick

The Nimble America team made the account, but yes, it represents me. As far as I know, there are no other wealthy donors to Nimble America.

>>>FROM: Gideon Resnick
TO: Palmer Luckey

So they posed as you on Reddit to solicit donations? And did you know that was going to happen?

>>>FROM: Palmer Luckey
TO: Gideon Resnick

I made the post, just not the account.

>>>**FROM:** Gideon Resnick
TO: Palmer Luckey

So you made this post asking for contributions? [link to Reddit]

Come on, Luckey thought. Obviously, I didn't *personally* write that! I would never say stuff like "Not the good kind [of freedom-strippers] you see dancing in bikinis on Independence Day" or "Donate a hundred and I will match you by skipping a glass of scotch." I mean, I don't even drink! But like I said, the account represents me, and have to take responsibility for what was posted.

FROM: Palmer Luckey
TO: Gideon Resnick

Yes, with guidance from the Nimble America guys. Like I mentioned, they would have been better off establishing a track record before pushing for this.

>>>**FROM:** Gideon Resnick
TO: Palmer Luckey

Got it. So you sent the body of that post to someone else to post on your behalf? And do you have a good contact for the person in charge of Nimble America? Sorry, not trying to be annoying.

>>>**FROM:** Palmer Luckey
TO: Gideon Resnick

I posted the body myself. Want me to email an intro to them?

Notably, at no point during this email exchange (nor during their brief phone conversation) had Luckey ever changed his mind and granted Resnick permission to print his name. Neither had the founders of Nimble America. All of which made it stunning to Luckey when Ben Collins, an editor at The Daily Beast, emailed Facebook a couple days later to say, "We're doing a story on Palmer Luckey, who told us today he helped found and donate a large sum of cash to a political organization called 'Nimble America' . . . The group was cofounded with two moderators of Reddit's r/The_Donald, which is often home to white supremacists [*sic*] memes."

MILO YIANNOPOULOS: If he pushes ahead . . . they are begging to be sued

MILO YIANNOPOULOS: It's staggering

MILO YIANNOPOULOS: Is this my fault? I panicked and asked you to do this. I know he lied and fucked you over but I feel some responsibility. I don't know what to do.

Shortly after the Daily Beast editor had contacted Facebook PR, Luckey emailed Resnick to reiterate "you do not have permission to use my identity in your piece."

"Palmer," Resnick replied, "we had an entire conversation where you discussed this with me and at no time said that we couldn't use your name . . . I would point you to journalistic ethics guides which dictate that: 'These deals must be agreed to beforehand, never after. A source can't say something then claim it was "off the record." That's too late.'"

Resnick ended his email with a link to the *NYU Journalism Handbook for Students*. Three hours later, Resnick's article about Luckey and Nimble America—co-authored with editor Ben Collins—was published on The Daily Beast with the following juicy-but-false headline:

The Facebook Billionaire Secretly Funding Trump's Meme Machine

Palmer Luckey—founder of Oculus—is funding a Trump group that circulates dirty memes about Hillary Clinton.[3]

BACK AT FACEBOOK—AS THE STORY BEGAN TO SPREAD AROUND CAMPUS— Oculus' head of communications spoke briefly with Luckey and then sent an update to a handful of execs:

FROM: Tera Randall
SUBJECT: Palmer Trump Story

The story from Palmer's conversations with the reporter about his donations to Trump just hit. Palmer is in the process of writing an internal post to share on Oculus FYI, explaining the background. I think it's important that Palmer get in front of this. Goal is to post in the next 30 minutes, and he's going to run the post by us.

I'm not commenting on this to press.

"Oculus FYI" was an internal Oculus page where company-wide messages were blasted out to employees. Typically, this meant things like holiday notices or product announcements, but—as per Randall's email—this would be the ideal forum for Luckey to dispute key details in the Daily Beast article and explain, in his own words, exactly what the hell was going on.

"You need to write something that's really authentic," Iribe advised Luckey. "Something that explains your reasoning and really feels like it's you."

Luckey agreed with Iribe wholeheartedly. In fact, after 16 months of being around people who just assumed that (like them) he was supporting Hillary Clinton, it would be kind of nice to lay out *why* that wasn't the case. With this in mind, Luckey drafted a statement that explained why he didn't plan on voting for Clinton—citing his beliefs in "protecting American markets" and "not getting into endless wars"—and why he was supporting Donald Trump.

"WERE YOU SURPRISED BY HOW THE STORY WAS REPORTED?" A FRIEND WOULD ask Luckey six months later. "Not just The Daily Beast piece, but also the hundreds of other articles that followed."

"Only a cynic could have anticipated what happened next," Luckey answered.

"What do you mean?"

"Only the ultimate cynic could have predicated how poorly the story would be reported. Even having my problems with the media that I had, I still had more faith in the media."

For about fifteen minutes after The Daily Beast story went live, Luckey clung to what still remained of his faith in the media. Wasn't there one journalist out there who would challenge the Daily Beast's narrative before it cemented—just one journalist who might stand up and say, "Can we at least *see* some of these 'dirty memes' before we ruin a man's reputation?" But any hopes of something like that were effectively extinguished by a prominent tech blogger named Anil Dash who—according to his website—had been cited in "hundreds of academic papers" and "sources ranging from the New York Times to the BBC"; and who—on the evening of September 22—tweeted the following to his 590,000 followers on Twitter:

6:15 PM @ANILDASH:

One reason every political hashtag on Twitter is filled with racist trolls? The founder of Oculus is funding them.[4]

6:18 PM @ANILDASH:

This guy, @PalmerLuckey, put some of his billion FB dollars toward explicitly funding white supremacy.[5]

In two tweets, Dash was willing to explicitly state what The Daily Beast had merely implied. And in doing so, he gave cover to other journalists who—over the next 24 hours—would push the details of this engaging narrative farther and farther and farther. At 7:04 p.m., *Boing Boing* would report that Luckey was funding a "tactical team that churns out racist, sexist, hatey anti-Hillary Clinton memes and works to make them go viral." At 8:08 p.m., *Business Insider* would report that Luckey's troll factory "may be most closely associated with the kinds of anti-Clinton Facebook memes that even strident Donald Trump supporters roll their eyes at . . ."

8:09 PM @TTLABSVR:

Hey @oculus, @PalmerLuckey's actions are unacceptable. NewtonVR will not be supporting the Oculus Touch as long as he is employed there.

8:11 AM @SAMFBIDDLE:

this guy turned out to be an immense shithead? no way

8:19 PM @LEFTWINGMILITIA:
The French Revolution 2.0 is coming & we'll see your head fall into basket of deplorables.

8:26 PM @PAPAPISHU:
Excited for the new PlayStation VR slogan "VR. . . . Without The Holocaust Memes"

8:46 PM @LUKEPLUNKETT:
Musk: let's go to Mars!
Luckey: whitesupremacistmeme.jpg

9:03 PM @GHOSTOFGHOSTDAD:
Will I catch racism if I try on an Oculus headset?

On and on it went—the tweets inspiring articles, and the articles inspiring more tweets; it was the whole machine in all its glory: a feed-back loop as outrageous as it was engaging.

For a moment, Luckey felt the fragments of previous furies—of entitlement checks, of shipping delays, of "selling out" to Facebook. But even with the familiarity of those flashbacks, Luckey could tell that this one was likely to be different due the political twist. And if he had any doubts about that, those were gone by the time he received an unexpected call from someone "in the loop" at Facebook.

"I'm not telling you this as a colleague," the caller told Luckey, "but I'm telling you this as a friend: you need to get an employment attorney."

The implication, of course, was that Luckey's job was now in jeop-ardy. Initially, he found this difficult to believe—that he might be fired for donating 10k to a barely-one-week-old non-profit organization whose mission was to put up billboards in support of the Republican nominee for President. Besides, Luckey reasoned, if his job really was in jeopardy, then surely would have received a heads up from Iribe (who wasn't just "in the loop," but was actually a member of Zuckerberg's ten-or-so-person inner-circle "M-Team.")

In addition to his relationship with Iribe, there was something else that gave Luckey confidence: the California Labor Code. Specifically, two sections of the California Labor Code—both of which had been on the books since 1937[6]:

- **SECTION 1101:** "No employer shall make, adopt, or enforce any rule, regulation, or policy . . . Controlling or directing, or tending to control or direct the political activities or affiliations of employees."
- **SECTION 1102:** "No employer shall . . . attempt to coerce or influence his employees through or by means of threat of discharge or loss of employment to adopt or follow or refrain from adopting or following any particular course or line of political action or political activity."

Nevertheless, Luckey figured it better to be safe than sorry, so he made a few calls and found an employment attorney. And it was a good thing for him that he did, because the following night—by that point shaking with rage—Luckey would compose himself for long enough to tell Edelmann that "they said that I have to resign tomorrow."

FROM THE MOMENT THE DAILY BEAST STORY BROKE, LUCKEY HAD WANTED TO issue internal and external responses as soon as possible. And as per that earlier email from the head of comms—the one about issuing a statement "in the next 30 minutes"—this had been the original plan. But that plan quickly changed after the statement that Luckey drafted was run up the chain at Facebook and deemed unacceptable to share. Specifically, as was soon made clear to Luckey, there was "no way" that he'd be permitted to post any statements that expressed support for Donald Trump.

Unsurprisingly, Luckey pushed back on this; though, eventually, doing so became (at least temporarily) moot. Because with Thursday turning to Friday, the plan was to wait until the morning and finalize Luckey's plan then.

"Are you really going to resign?" Edelmann asked.

"Fuck no," Luckey replied, momentarily refilling with the rage that had consumed him over the past couple of hours.

Anyone who truly knew Luckey knew that he'd rather die than exit from Oculus. Which meant, really, that things were about to get very messy; because, as he and Edelmann lay down for bed, Luckey was well aware that, the following morning, Facebook's plan was for him tender his resignation.

THAT PLAN, HOWEVER, WOULD SOON EVOLVE. AS DID THE COVERAGE OF LUCKEY and his donation to Nimble America. By the time he awoke on September 23, the story had metastasized to include Edelmann. "She frequently shit-talks feminist Anita Sarkeesian, mocks 'SJWs,' and tweets pictures of herself at Trump events," Gizmodo reporter Bryan Menegus wrote.[7] Articles like that were largely spurred by something Edelmann had tweeted from a Trump rally back in May: "Love the diversity of people here!"[8,9] And though Edelmann had meant those words sincerely ("I met so many great people from all different races, ages, and backgrounds!"), her comment was presumed to be sarcastic and reported as evidence of bigotry.

Luckey thought it inappropriate for his girlfriend to be brought into the discussion at all. If the media wanted to go after him, fine. But going after Edelmann—whose Twitter feed soon became so inundated with insults and threats that she disabled her account—*that was just classless.*

Luckey wanted to fight back. He wanted to defend his girlfriend, he wanted to clarify his political views, and he wanted to correct every single falsehood that had been written about him over the past twelve hours. But Facebook continued to drag its feet, though—at least—Luckey would soon be informed why: at some point since the previous evening, Facebook had decided that they would *not* be forcing him to resign. Luckey presumed that it was because their attorneys had quickly brushed up on the California Labor Code, although that seemed unlikely given what he'd be forced to do instead.

MEANWHILE, THE BAD PRESS CONTINUED TO MOUNT. AT 6:32 A.M.—WITH THE headline "How your Oculus Rift is secretly funding Donald Trump's racist meme wars"—*Ars Technica* informed its readers that "the stream of racist, sexist, and economically illiterate memes appearing in support of Donald Trump . . . is being bankrolled in part by the 24-year-old inventor of Oculus Rift." At 6:47 AM, *Motherboard* boldly asked "What Does Alt-Right Patron Palmer Luckey Believe?" And at 6:59

a.m., *Mashable* reported that Luckey had funded an organization that "aims to circulate anti-Hillary memes across the internet"; the piece also noted that "many people in the tech world were not amused"; and then linked back to a series of angry tweets (including that initial one from Anil Dash—the one about Luckey "funding" "racist trolls").

"BRENDAN, THIS IS ILLEGAL," LUCKEY TOLD IRIBE LATER THAT DAY, AFTER RE- ceiving a copy of the statement that he would be required to post—a statement that came "directly from Mark," and which Luckey would not be allowed to alter in any way:

> I am deeply sorry that my actions are negatively impacting the perception of Oculus and its partners. The recent news stories about me do not accurately represent my views.
>
> Here's more background: I contributed $10,000 to Nimble America because I thought the organization had fresh ideas on how to communicate with young voters through the use of several billboards. I am a libertarian who has publicly supported Ron Paul and Gary Johnson in the past.
>
> I am committed to the principles of fair play and equal treatment. I did not write the "NimbleRichMan" posts, nor did I delete the account. In fact, I am not even supporting Donald Trump in this election; I'm supporting Gary Johnson. Reports that I am a founder or employee of Nimble America are false. I don't have any plans to donate beyond what I have already given to Nimble America.
>
> Still, my actions were my own and do not represent Oculus. I'm sorry for the impact my actions are having on the community.

"Brendan," Luckey repeated, "this is illegal!"

Perhaps, Iribe conceded, but then—knowing what Luckey cared about most of all—he made a very good point: if you fight this, you *may* one day win a lawsuit; but even if that were to happen, you'd still be gone from Oculus.

As Luckey weighed Iribe's words, he felt the chill of what it would actually be like to be gone from Oculus—to be ejected from the organization that he had worked his entire adult life to build.

"Brendan," Luckey told Iribe, "Oculus is literally *everything* I have."

Iribe agreed, which is why he gave Luckey what sounded like brotherly advice: find Jesus and become a big, vocal Gary Johnson supporter. All things considered, this statement didn't seem like *that* big of a deal. If saying this one thing would get you off the hook for murder, you'd say that thing, wouldn't you?

Ultimately, that's what Luckey decided to do—to suppress his views and post the lie so that he could avoid losing that which he considered to be everything. For hours, Luckey and his attorneys tried to get in a few tweaks to the statement, but eventually it became extremely clear that there would be no revisions to the statement that came from Mark. Except for one thing: the line about not supporting Trump.

Why this concession? Perhaps cooler heads prevailed. Luckey would have liked to think it was because of something like that. But much more likely it was because someone convinced Zuckerberg that forcing an employee to denounce support of a politician crossed a line. Or maybe it was just because there were enough people in Silicon Valley—not many, but enough—who would know that such a denouncement was verifiably false.

Regardless, at 7:10 p.m., Luckey posted on his personal Facebook page the statement that Zuckerberg had written.[10]

CHAPTER 45

EXILE

September/October/November/December 2016

"WHY?" DYCUS ASKED LUCKEY, THE NIGHT HE AND EDELMANN RETURNED TO the Commune.

Luckey looked around the room—at Dycus, at Hammerstein, at Howland, at Shine—and saw the glares of anger, confusion, disappointment and sadness.

"What?" Luckey asked, trying his best to sound aloof. "Did something happen while I was gone?"

This broke the tension and led to chorus of reluctant oh-you laughter that gave Luckey a momentary sense of normalcy that he had never needed so badly before.

"Seriously," Dycus said. "Why would you do that? Why would you take that risk?"

"I didn't think it was a risk at the time," Luckey replied.

"I don't totally buy that," Hammerstein said.

"Our conversation was off-the-record!"

"Why are we talking about the reporter?" Shine asked. "The risk was Nimble America."

Luckey then explained how he had met the Nimble America guys, why he had decided to contribute to their organization and how ridiculously this story had been reported.

"Remember," Luckey replied, "that there was a very small amount of time and energy put into the entire lark. What I'm saying is: if it's just this tiny, stupid, little thing that you don't think is going to matter, it starts to make more sense how it was handled. If that makes sense. Basically, it's hard to look at it in a vacuum."

"Maybe," Howland said. "And look: we're obviously very sympathetic to you, and the press has treated you so unfairly—so unfairly, so unfair. But you should have thought about what you did when you did it. And probably not done it."

"Or at least provided a better explanation in your apology," Shine said.

"Yeah," Dycus agreed. "The first sentence of your apology post is one of the most insincere things I've ever read."

"Come on," Hammerstein said. "There's no way he wrote that!"

Dycus turned to Luckey and asked him if he wrote that apology he posted.

"You know I'd tell you guys anything," Luckey said. "But it's better for all of us, I think, if we don't talk about this."

"We don't need to talk about it," Hammerstein said. "Because I know you didn't write it."

"What do you mean?" Dycus asked. "Why are you so sure?"

"Just count the spaces," Hammerstein said. "Anytime that Palmer writes anything, he uses two spaces after a period before starting the next sentence. But whoever wrote that apology for Palmer only used one space."

"Multiple women have literally teared up in front of me in the last few days . . ." Facebook engineering director Srinivas Narayanan wrote to a handful of high-level execs on September 27. "[But] the Palmer issue is only one problem. There are other big systematic issues. For example, some women feel that their coworkers don't understand their challenges or worse, don't care."

"GOOD MORNING, EVERYONE," BRENDAN IRIBE SAID, WELCOMING 2,000+ DE-velopers to the San Jose McEnery Convention Center to kick off Oculus Connect 3.[1] "So a lot has happened in the community over the last 12 months. We've gone from devkits and prototypes to bringing millions of people into VR . . . And at the heart of it all, what brings VR to life is the content you create. The work that you've done has led to hundreds of VR experiences on the Oculus platform. From gaming and entertainment to education and science. The ecosystem is taking off thanks to you. Thank you!"

After a moment of applause, Iribe brought Mark Zuckerberg on-stage and passed the baton to him to deliver the meat of the morning's keynote.[2]

"So before we get started," Zuckerberg began, "I just want to say how meaningful it is that you are all here with us today. I'm looking around and I see a lot of people that we've worked with for a long time. I see a lot of people who are in virtual reality right now. And I see a lot of people who have been in the industry for a very long time. And you're all the reason why virtual reality is at the point it is today. So thank you so much and thank you for being with us today."

HA!

Watching this presentation from a laptop in his home, Luckey couldn't help but scoff at Zuckerberg's comment. He understood why Facebook wouldn't let him attend the event, but it still felt wrong not to be there, like he was a recently deceased organ donor, now in some sort of purgatory, watching someone else prance around with parts of his body.

MARK ZUCKERBERG
Our industry has made more progress in the last couple of
years than I think any of us could have really hoped for, right?
When we bought Oculus a couple years back and planted
a flag in the ground that we thought that virtual reality was
going to be the next major computing platform; at that point,
no one had ever shipped a modern consumer virtual reality
product. No one had ever seen Touch or hand presence. And
at that point, certainly, no one would have guessed that just
now, two years later, there would be more than a million
people actively using virtual reality products.

If Oculus were Luckey's organs, then Touch would have been his heart. Obviously, numerous people had helped bring Touch to life—

Nirav Patel especially—but, from the very beginning, Luckey had been the one beating the drum about the importance of hand presence. Even before Oculus's Kickstarter campaign, he had been the one prototyping hand controllers. And with the Touch release now officially slated for release on December 6, it was one of those dangling carrots that had made this hiatus bearable for Luckey, because at least he'd be back for that launch.

"NOW," ZUCKERBERG CONTINUED, "THE FIRST STEP FOR GETTING VIRTUAL RE-ality out into the world is getting the basic hardware out there. And this is happening, right? And it's happening, I think, at a faster rate than any of us had really expected. And, you know, we had a little bit of a slow start earlier this year on Rift, but now that's rolling out quickly. And we're going to get Touch in your hands by end of this year too. So we're excited about that."

Over the next fifteen minutes, Zuckerberg talked about demoing VR for world leaders, the potential of virtual education, and Facebook's incredible growth (1.7 billion users now; and presently responsible for four of the six most popular apps), before getting to the crown jewel of his presentation: standalone VR.

"So today," Zuckerberg explained, "there are two primary categories of virtual reality products: there's mobile, like Gear VR, which is great and it's affordable—you can take it anywhere with you that you want. The second category is PC VR and that's like Rift. And that is the highest quality of virtual reality experience that you can get today. It's really powerful. It is powered by a high-powered computer, which means that it's a little bit more expensive—and because you're tethered to a computer, you can't really take it with you out into the world. So we believe that there is a sweet spot between these: a standalone virtual reality product category that is high quality and that is affordable and that you can bring with you out into the world. Because it's not tethered to a PC and because it has inside-out tracking so it can track your position as you move through the world. So we're working on this now. And it's still early . . . so I don't want to get your hopes up too much!"

The audience erupted with laughter.

From backstage, Nate Mitchell cracked up as well. Zuckerberg, in his opinion, was crushing it; and, with each passing minute, giving VR more and more of the credibility it needed to go mainstream. But as much at Mitchell was loving how this was playing out, it *did* feel a little odd not to have Luckey there.

"So," Zuckerberg said, his voice echoing, "let's take a look at where we are in developing this."

On-screen, a minute-long video introducing "Santa Cruz" played for the audience.

"It is an honor to be on this journey with you," Zuckerberg said, concluding his talk.

THE WEEKS AFTER CONNECT WERE TOUGH ONES FOR LUCKEY. NEW ARTICLES about him continued to pop up daily—calling him racist, anti-Semitic, homophobic—and he was having trouble keeping track of how many "friends" had unfriended him by this point.[3,4,5,6,7] Meanwhile, he was still waiting for Facebook to give him the green light to return to work, but "a couple weeks" became "a couple more weeks" became "it'd be best to wait until after the election."

Thus, Luckey was left with a lot of time on his hands and a level of uncertainty that he hadn't felt since his days living in a trailer. To keep this anxiety at bay, he created a Google Doc called "Falsehood Collated."

The purpose of this document is to clear my head so I can maintain sanity, plan for the future, document the past, and focus on facts while they are still fresh. Below is a thread that will be reiterated throughout this document, but bears mentioning at the start:

I donated ten thousand dollars to Nimble America. Nimble America put out a billboard. It showed a caricaturized portrait of Hillary Clinton alongside the text "Too big to jail." They planned on releasing further billboards. The group (and the people who ran it) never trolled or harassed anyone, nor did they do anything racist, sexist, homophobic, or anti-Semitic at any point. There was never any kind of internet campaign to malign anyone (or, in fact, do anything beyond pushing billboards and t-shirts into the real world).

The reference to "shitposting in real life" on their website is a tongue

*in cheek reference to a term that means making internet posts with zero
information or worth, which is obviously not the actual goal of any
campaign with intent to persuade. A wide variety of media outlets and
social media influencers have ignorantly and/or maliciously accused myself
and Nimble America of trolling, harassment, astroturfing, and copying
the Clinton campaign's well-known paid anonymous internet posting
efforts through Correct the Record, unpopular with internet users on both
sides of the aisle. Nimble America handled their launch poorly and lack
organizational skills, but they are just regular guys who don't like Hillary
Clinton and do like Donald Trump, a set of opinions shared by roughly
half our country.*

*The narrative that is being created and shared by media outlets does
not align with reality, but is widely accepted by a public that cares enough
to follow it and drive clicks to it, including FB employees and former FB
employees. There are a handful of outlets making fair criticisms related to
credibility of the people involved, potential scamming, and the perception
that I should have stayed out of things entirely as a public figure, but the
vast majority of the ongoing backlash is driven by a false narrative that
goes far beyond the truth.*

*The unfortunate result is that a handful of press outlets and
influencers have managed to construct and push a narrative of several
huge lies and countless small lies that are being widely accepted by the
greater public, which lacks the interest or ability to do their own research.*

*The public cannot be blamed. The public owes me no benefit of
the doubt, and unlike the journalists involved, they have no duty to
uncover or report the truth. It takes an order of magnitude more effort to
disprove bullshit than to spread it—many of the lies being spread seem
insignificant on their own, to the point where refuting them all publicly
would be an endless task that is easy to portray in a petty light, giving
them cover to distract from their big lies and move the goalposts as far as
they can.*

*Unfortunately, hundreds of small lies stacked together and re-reported
endlessly are more powerful than most would expect. Tackling this is
easier when you can engage with people on a human level, but impossible
when elevated in the minds of many to supervillain status. Meme
supervillain . . .*

The rest of this doc—an ever-growing list—compiled examples of the most egregious articles, with brief notations about what they got wrong.

Amazingly (except, of course, not amazingly at all), there wasn't a single outlet that published any of these alleged memes. One would think that if Luckey had really been the kingpin behind a racist, sexist, anti-Semitic troll army, there would have been articles (slideshows even!) ranking the most heinous memes that Nimble America created, endorsed, or spread.

To this point, exactly one week after the Daily Beast's original "Meme Machine" piece, Scruta Games—the first developer that had threatened to cancel support for Oculus unless Luckey stepped down—published a series of interesting tweets:

- "We've failed to find any evidence backing up the Daily Beast's claim that Luckey paid for hate speech. Only a lame billboard."[8]
- "So we were misinformed about him financially backing hate speech, which was the issue we had."[9]
- "Since there is no evidence of that so far, we will tentatively resume work on Touch support."[10]

If Scruta Games, a small independent game studio, had reached this conclusion, then surely the media outlets that had reported on Luckey and Nimble America must have realized the same. And yet, whether they did or not, there were no corrections, clarifications, or retractions of the stories that had been printed. Instead, because Palmer Luckey = Racist Supervillain made for a clickable narrative, additional inaccurate stories continued to sprout. Luckey added them to his Google Document of lies.

"I just finished reading the Google Doc," Joe Chen said, speaking with Luckey by phone. "It's crazy to see it all compiled together like that."

"Yep," Luckey replied.

"It's especially crazy," Chen said, "because the *non*embellished version would been enough to get clicks. If they'd just said something like:

Palmer Luckey appears to like Donald Trump. But instead they throw in white supremacy and trolls and all this other shit."

"Never mind doxing Nicole," Luckey added. "And spreading info about her living situation. Especially things like the fact she lives with me, which is not popular with conservatives in mountain towns."

"How is she doing?"

"She is not used to the internet hate machine like I am."

"I know. I just . . . it's just hard to believe that people can get away with this stuff."

"Hang on," Luckey said, searching for something on his computer. "Okay, here it is. I have a sticky note on my desktop that has a quote from Trump: 'We have a media that is so dishonest,' Trump said. 'These are among the most dishonest people you will ever, ever meet.'"

Chen laughed.

"Yeah," Luckey said. "Even people who I assumed were friends. Or at least reasonable. No reaching out to me. No questions. Just immediate calls for my termination. Right after they block me on social media. Assuming that the media must be more honest than their friend."

"Who surprised you the most?" Chen asked.

"I won't say," Luckey answered. "I don't want to throw people under the bus like they did. Especially since there is a decent chance some of them will change their mind. Probably won't ever be good friends with anyone who mindlessly calls for my termination, but still."

"I'm sorry, man. I don't know what else to say. I just feel bad."

"Not that I expect people to suicide their careers for me, but worth noting that essentially nobody was willing to stand up for me. This reaction shows why that is the right decision. Never stand up for what you believe in when the media disagrees with you. Never stand up for someone they are hell-bent on crucifying. And don't ever dare call them out as liars."

"I'm sorry, man," Chen said. "This really sucks."

"Oh well," Luckey replied.

"One more thing; if you can't answer it, you can't answer it: When you posted that apology—the day after all this went down—why didn't you just come right out and say: Yeah, I like Trump. Here's why . . .'"

"Long story," Luckey said. "I can't really talk about that right now. But you know my creed: Do the right thing, not the popular thing."

"That's why you're you," Chen said, hoping that this might lift Luckey's spirit. "So when do you think they'll let you get back to work?"

"After the election," Luckey replied. "Hopefully things will have settled down by then."

"Yeah. That sounds about right. Unless . . . well, what if Trump wins?"

Just a few weeks from the election now, the odds of Trump winning appeared to be laughably low—so low, in fact, that in anticipation of a Clinton presidency (and in fear of further firearms regulations), Luckey spent most of election day hanging out, and shopping at, Ade's Gun Shop, Ammo Bros. and a handful of other gun shops in the Orange County area.

On election night, Luckey hopped on a plane back up to San Francisco. By the time his flight departed, it was clear that the election was going to be much closer than originally thought: Clinton was still the front-runner, but Trump appeared to be doing better than expected in the battleground states of Ohio and Wisconsin.

The trip from Orange County to San Francisco is short—about ninety minutes—but for many aboard that flight it would feel like the shape of their country had changed beneath them. Upon landing and even before Luckey's phone booted up, he noticed many passengers sobbing around him. It appeared that, holy shit, Trump was actually going to win this thing.

Moments later, Edelmann picked up Luckey at the airport—a red MAKE AMERICA GREAT AGAIN hat waiting for him in the passenger seat—and the two of them celebrated Trump's unlikely victory by grabbing drinks at a spot called Yard House, Edelmann throwing back beers, and Luckey downing delicious lemonades.

What if Trump wins? Luckey recalled Chen having asked a few weeks earlier. In theory, nothing should have played out differently. Luckey had done everything Facebook had asked of him, and it was time they let him back to work. In practice, however, things became much more complicated. If the reaction of those on Luckey's flight was indicative of the mood at Facebook, then the majority of employees at Oculus felt some combination of shocked, devastated, and/or horrified.

This cocktail of emotions seemed to crystalize in the creation of a quickly popular internal Facebook group called "Refocusing Our Mission." As per the page's introductory message—"The results of the 2016 Election show that Facebook has failed in its mission"—the conceit of this group was that Trump's victory was some sort of proof that Facebook needed to change its ways.

Although it's natural to empathize with the frustration that those joining this group must have felt, a handful of Facebook employees— people from both sides of the political spectrum—felt that something was very unnatural, creepy even, about seeing the election results as proof that Facebook had somehow failed. Because, frankly, it provided a pretty ugly answer to the question people continued to ask more and more: What *is* Facebook? Well, according to the founder of "Refocusing Our Mission" and the hundreds of employees who quickly joined and engaged, Facebook was basically some sort of social engineering tool—an invisible hand meant to guide its users toward the "correct" political beliefs.

Or to put it another way: since Zuckerberg often described Facebook as "like a utility," then this reaction was the equivalent of AT&T declaring that Mondale losing to Reagan meant it was time for them to rethink the mission of this whole phone line network thing.

Beyond the Refocusing Our Mission group, there was perhaps no better personification of Facebook's activist-driven mission than a popular, postelection internal post—published to a different internal group called "Facebook (the company) Is Broken"—by an employee in Community Operations:

> I have never felt more ashamed to be working here. This isn't a completely new feeling. I work in CO where we see the dark side of the business . . . where we censor people and claim openness, where we apply US centric policies and claim being global, where we express concern on a daily basis but never really do anything about it.
>
> But now it is different. The world is crumbling around us and we are silent . . . History will not be good to us . . . We are part of this.

I will never forgive myself for being part of this . . . I will never forgive myself for caring about things like steady income and mortgage when people are dying in my region and are being kicked out of countries.

I will never forgive myself . . .

Luckey's exile continued into late November when, finally, he was informed that the internal investigation had found no incidents of inappropriate conduct. I know! Luckey wanted to shout. You should have listened to me when I told you that two months ago! But, at this point, that was water under the bridge. All he really cared about was getting back to work.

With the ZeniMax trial coming up in January, Facebook wanted him focused on that rather than Oculus business. He was, however, allowed back into his office for a couple days in early December.

Upon his return, Luckey was greeted with a mixed reaction. Many colleagues—especially those who had been with Oculus preacquisition—were happy to see him and talked about how much the company had missed his leadership; but at the same time, there appeared to be just as many employees whose body language made it clear that they were disgusted by Luckey's mere existence. To some degree, Luckey couldn't blame them. He'd have hoped that they would have cared enough about the truth to do a little research and come to realize that much of what had been written about him and Nimble America were just plain lies. But at the same time, he could also appreciate that investigative journalism was not one of their responsibilities; and even if it were, he was confident that a sizable portion of these disgusted colleagues would still probably consider him an enemy simply because he was supporting Trump at all.

Either way, Luckey wanted a chance to win those people over. He wanted a chance to explain himself to *anyone* at the company who had a question about his character or his political views. That's why he was dying to do the Q&A that Facebook had told him he would be able to do when things calmed down. But apparently things weren't yet calm enough, because a Q&A to discuss what had happened was still not in the cards.

In the meantime, however, Luckey was now allowed to send an email to all his Oculus colleagues. And while he tried to keep it short and sweet, he wanted to make it abundantly clear that he *wasn't* the bigoted monster that had been described in all those stories; and that, for as long as it was up to him, he *was* going to be at Oculus for a long, long time.

FROM: Palmer Luckey
DATE: December 4, 2016

I am returning to the office on Monday and wanted to take a moment to say a few words to the team.

First, there's something I want to make very clear: I have never supported nor do I believe in racism, bullying, misogyny, anti-Semitism or hate speech of any kind.

A lot of people on the team have taken time to give me their feedback, and I appreciate it. This situation is difficult. I learned some tough lessons, many of them very publicly, which were hard not only on me but on teams and people associated with me. I apologize to everyone who was impacted.

External media coverage painted a picture that didn't accurately represent my values or actions. I remain committed to supporting a tremendous team that represents a diverse set of people, views, and beliefs. I know the past couple months have caused some employees to question that commitment. Please know that I will continue doing my part to make Oculus an inclusive place to work.

I am 100% committed to Oculus, and humbled to be part of what is clearly the smartest and best team in the VR industry. I plan on spending the next 50 years building and winning the future of VR, AR and whatever comes after. If anyone wants to ask questions, give further feedback, or just catch up, I would love to meet with people. Reach out and we will make it happen.

CHAPTER 46

THE HEIST, THE COMEDY AND THE FANTASY VS. THE DOCUMENTARY

January 2017

THE HEIST

"One of these pictures is cattywampus," Federal District Judge Ed Kinkeade exclaimed on the morning of January 10, pointing to a photo that ZeniMax was trying to get just moments before the jurors would be sworn in.

The photo in question—a grainy, previously undisclosed security cam image—depicted John Carmack exiting id Software during a summer night in 2013. To just about everyone in Judge Kinkeade's Dallas County courtroom—the defendants (Luckey, Iribe, and Carmack), their attorneys, and a packed house full of reporters—there was nothing unusual or spectacular about this grainy image. But ZeniMax looked forward to clearing that up, among many other things, during what would turn out to be a three-week-long trial. And shortly after Judge Kinkeade ruled in favor of admitting the cattywampus photo,

Tony Sammi—ZeniMax's lead attorney—stepped in front of the jury to begin his opening statement.

"This case has technology in it," ZeniMax's Tony Sammi began, "but really this case fundamentally breaks down into something much more simple: this case is about theft. This case is about stealing something very valuable. This case is about the Defendants stealing our technology, selling it for $3 billion—that's *billion* with a *B*—and then covering it up by destroying evidence. That's fundamentally what this case is about. It's one of the biggest technology heists ever."

Sammi paused to let that sink in with the jury and reporters covering this trial, and then launched into the journey that had led to this moment. "I have been on this case from the very beginning . . . my job in the last two and half years has been to act as sort of a lead investigator, to uncover the truth, and what have we searched through? Over a million pages of documents, terabytes of electronic data, computer source code, all to find the evidence and show it to you. So what did we find?"

Sammi then teased the jury with a sampling of what ZeniMax had uncovered:

- "a secret meeting where the Defendants hatched their plan to take our technology."
- "secret text messages between Defendants that Defendants never wanted you to see"
- "destruction of evidence, computers wiped, computers missing, USB drives wiped, computer system logs deleted, hard drives destroyed."

Before diving deeper into these allegations, Sammi decided to take a step back and give the jury some context for what they'd be hearing in the weeks ahead.

"First of all, don't let anybody tell you there's something special in the water in Silicon Valley—that 'only good technology can come from there.' All of this happened right here . . . Now, let's talk for a minute about what was going on in the world [of] virtual reality around 2011 . . . in 2011/2012 there was no commercially viable VR. And the evidence shows that what you could get that was small was cheap, but bad. It would not fool your brain. So everyone is looking for the holy grail. John Carmack said, let's solve it in software. Instead of using $100,000.00 of equipment, let's solve it in software. Make something cheap, the headset, because you have to sell it, and make the software do it. We call that software the VR engine, the VR engine. Without the VR engine, there is no VR . . . [and] what did John Carmack do after he made the VR engine? He told people about it."

One of those people was, of course, Palmer Luckey. "Carmack ran into Mr. Luckey on the internet," Sammi explained. "And Mr. Carmack said: I'm looking for parts. Will you let me buy from you some optics so I can use it for my VR engine? Now, Mr. Luckey sent Mr. Carmack something in a box just like this, a medium-sized USPS Priority Mail box. Very important for the case. What was in this box? What is the evidence going to show you? The evidence is going to show what was in this box was a screen, lenses [that cost] $11.00, a plastic housing, and a video card which is a piece of hardware that makes the screen be able to accept input . . . What wasn't in this box was: No software, no VR engine, no sensor, no reliable power supply."

After laying down the backstory, Sammi played a few clips from E3 2012—clips of John Carmack giving VR demos to reporters. "Everybody went wild. ZeniMax was the belle of the ball . . . Now, shortly after E3, the evidence shows that Mr. Iribe and Mr. Luckey connected . . . What happens? Here we go. They have a secret meeting."

On cue, Sammi pointed the jurors to a blown-up photograph that ZeniMax had been able to uncover during the discovery phase of this lawsuit. It's a photo of Luckey giving Iribe a VR demo in a dimly lit room. In the corner of this image, a timestamp confirmed ZeniMax's timeline. Even more damning: Luckey and Iribe appear to be joined by two previously unmentioned "associates."

"What happened next," Sammi continued, "is two days later, on July 6, 2012, they had papers drawn up to divide the company Oculus amongst themselves." Then about three weeks later, "Mr. Luckey and

[Iribe's longtime business partner] Mr. Antonov are sitting at Valve. What are they doing? They're showing Valve the VR engine . . . Remember the text messages I talked about that the Defendants didn't want you to see? Here they are. Here is the first one."

> **MICHAEL ANTONOV:** Can we leave it with Valve? (It means that we also have to leave SW [software] that Carmack just warned about . . .)
> **BRENDAN IRIBE:** Will they give us a video endorsement?
> **BRENDAN IRIBE:** I'd leave it if they give us an endorsement, otherwise, I'd tell them we have to get approval from Carmack/Bethesda.

"Now my time is up . . ." Sammi said after sharing a few more examples of private employee communications. "I'm going to leave you with this: Chapter one was the heist. Chapter two was the payoff. Chapter three was the coverup. You, ladies and gentlemen, get to write Chapter four."

TO HELP WRITE THAT FOURTH CHAPTER IN FAVOR OF ZENIMAX, THEIR MOST compelling firsthand witness took the stand on January 20.

"Could you please introduce yourself to the jury?" Sammi asked the witness. "You have been sitting here for the whole thing."

"Good morning. My name is Robert Altman. I am the founder, chairman, and CEO of ZeniMax Media."

"And, Mr. Altman, do you have a family? Wife, kids?"

"I do. My family is with me here in the courtroom. My wife, Lynda Carter, we'll be married this month 33 years, and my two children, James and Jessica."

"Mr. Altman," Judge Kinkeade interrupted. "They don't look like you."

"I hope my children look like my wife, sir!"

After briefly describing his path to founding ZeniMax, Altman was asked the question that everyone had to be wondering: Why did you bring this litigation?

"Well, I didn't sue out of spite, as has been suggested in this court.

And I didn't sue over sour grapes, over a lost opportunity . . . We thought we were dealing with respectable and honest people. We were dealing in good faith. And what happened was we were treated very badly by these Defendants. Our property was stolen. Our trade secrets were taken. Our copyrights were violated. Contracts were breached. We were lied to, given explanations for things that were insulting. It was all very damaging to our company. And we're asking this jury to make it right."

In contrast to that type of behavior, Altman talked about how things worked at ZeniMax—"Our motto is 'we turn square corners.' Meaning we don't take shortcuts."—and then eventually about what this case was all about: virtual reality.

"We had actually looked at VR technology for a long time," Altman explained. "Going back into the '90s. In fact, there was a photograph of the fellow who is our vice president of game development on the cover of PC Gamer Magazine wearing one of those rigs back in the '90s. So we have looked at this a long time. Bethesda Games Studios had looked at it."

Not only did ZeniMax Altman reveal that ZeniMax had a long history with VR, but shortly thereafter he let it be known that his company was the one that invented it. "The suggestion that has been made that we backed away from VR is absurd," Altman said. "We saw the opportunity of VR before the rest of the world. We're the ones who invented it. We are the ones who took it to E3 . . . But what has happened by the actions of the Defendants, we have now been forced in this position where we don't get the benefit of the breakthrough technology that we created."

Throughout the rest of his testimony, Altman expanded on this notion: stating that Palmer Luckey had "no training, no expertise, no ability to create VR"; that "the story that is being told that he invented it is just a myth"; and that, in reality, "it all came together with the special sauce that Mr. Carmack created by March of 2012. Before we ever talked to any of the Defendants."

Looking on, Luckey—clearly bothered—tried his best to keep an even face. And though he found that rather hard to do at times, something occurred to him: Robert Altman *was* right about one thing.

THE COMEDY

"One of these pictures is cattywampus," Federal District Judge Ed Kinkeade exclaimed on the morning of January 10, pointing to a photo that ZeniMax was trying to get just moments before the jurors would be sworn in.

For the third-parties in attendance—particularly the reporters and jurors—Judge Kinkeade's colorful personality went a long way towards making the proceedings somewhat bearable. Whether using words like "cattywampus" or occasionally interrupting the proceedings to ask layperson questions about the tech being described, Judge Kinkeade's personality and demeanor was a nice respite from two-plus weeks' worth of dense, often technical testimonies.

The only thing the courtroom audience seemed to enjoy more than a well-timed, well-phrased cameo from Judge Kinkeade were the moments of unintentional comedy that would occasionally come from those being questioned on the stand. Moments like these:

CARMACK BEING CARMACK

> **LAWYER:** I believe that you firmly think that nobody can own the concept, and I may agree with you there. I don't think the concept of solving for distortion is yours or mine . . . Now, *how* I choose to solve that problem, particularly in source code, that's not open to the public, right, if you're writing that code at a company, being paid for that company?
>
> **JOHN CARMACK:** No. But when I discussed at E3 using a GPU to distort an image to correct for it, that's fairly specific.
>
> **LAWYER:** Okay. Let me give you this example: I'm going to discuss with you a mode of conveyance that has four wheels, an engine and it uses gasoline and it's got pistons. And I'm going to stand here for three hours and discuss that. Can you take that and go build a car?
>
> **JOHN CARMACK:** Yeah, I probably could . . . I *did* pick up rocket science as a hobby. It wouldn't be as good as Detroit's latest, but, yeah, I think I could build an automobile.

CARMACK, AGAIN, BEING CARMACK

LAWYER: Did you know at the time that Palmer Luckey used that technology at a secret meeting in a Long Beach, California, hotel room on July 4, 2012?

JOHN CARMACK: How is it a secret meeting? I was not clear on that.

LAWYER: Yes. Let me ask that. Did you know about it at the time?

JOHN CARMACK: No, I didn't. That doesn't make it a secret.

LAWYER: Secret to you, doesn't it?

JOHN CARMACK: There's a billion meetings going on in the world that are not secret to me.

LAWYER: And did . . . didn't you think the purpose . . . are we to believe that you think that the purpose of an NDA means that when you send this technology to Palmer Luckey he can show it to anybody?

JOHN CARMACK: The whole idea of this was that he could show it to people so that he could raise awareness.

NATE BEING . . . HONEST

LAWYER: Do you think that Palmer Luckey single-handedly brought VR back from the dead?

NATE MITCHELL: No

LAWYER: Did you ever read that in an article?

NATE MITCHELL: I have

LAWYER: What do you think about that?

NATE MITCHELL: I would disagree

THE SECRET COMPANY MOTTO

LAWYER: Mr. Altman testified earlier in this trial and he told this jury our motto is "we turn square corners," and he went on to explain it. And Mr. Willits this morning came in and told the jury that's the ZeniMax motto. Have you [in your four years as a software engineer at ZeniMax] ever heard Mr. Altman use the phrase "We turn square corners"?

GLORIA KENNICKELL: No.

LAWYER: Have you ever heard those words come out of Mr. Willits' mouth?

GLORIA KENNICKELL: No.

LAWYER: Have you ever heard anyone at ZeniMax or id use that phrase?

GLORIA KENNICKELL: No. And I'm not sure what it means.

LAWYER: Okay. Is it posted anywhere?

GLORIA KENNICKELL: Not that I know of.

LAWYER: So if this was a motto, it was some kind of secret motto you didn't know about?

GLORIA KENNICKELL: Potentially, yes.

THE BICYCLE ANALOGY

LAWYER: If this jury finds that Oculus stole virtual reality technology from ZeniMax, improving upon that technology doesn't make it yours, does it?

MARK ZUCKERBERG: I don't know . . . I disagree with the premise of your question so it's kind of hard to get on top of that.

LAWYER: Alright, let's make it real simple: if you steal my bike and you paint it and put a bell on it, does that make it your bike?

MARK ZUCKERBERG: No . . . [but] I think the analogy to a bike is extremely over-simplistic here.

LAWYER: Probably.

MARK ZUCKERBERG: This would be like someone—

LAWYER: I agree with you.

MARK ZUCKERBERG:—who created a piece of, like, a bar that might go on a bike and then someone built a spaceship out of it.

THE BOX

LAWYER: In 2012, you sent a headset to Mr. Carmack, correct?

PALMER LUCKEY: Yes, I did.

LAWYER: Was it on May 10, 2012?

PALMER LUCKEY: That sounds like it's probably correct. I'm not sure if that's the exact date.

LAWYER: [*presenting a cardboard box*] And you packaged your headset into a United States Postal Service medium-sized box, flat rate Priority, correct?

PALMER LUCKEY: That's not an accurate representation of the box. USPS actually makes several different medium-sized boxes in different shapes. Some are longer, some are boxier. There are actually several different sizes . . .

LAWYER: During your deposition—under oath—you testified that you put what you sent to Mr. Carmack in a USPS medium priority box.

PALMER LUCKEY: Yes. And what I'm saying is there is more than one USPS medium Priority box. There are actually several different shapes and sizes of box that are all classed under the medium flat rate fee . . . For example, I believe that there is a long and flat wide mailer that is small. There is another that is kind of a cube that is small. So just for the sake of being accurate, I just want to point out that box is not an accurate representation, but it is also a USPS Priority flat rate medium-scale box.

LAWYER: Is the box too tall?

PALMER LUCKEY: It's actually just much too narrow. The box that I used was one of the more squared ones. The headset wouldn't have begun to fit into that particular box.

LAWYER: So your box was wider and shorter?

PALMER LUCKEY: Yeah. It was similar in size to the box that— that the prototype is currently housed in.

LAWYER: We can agree that what you sent to Mr. Carmack was, in fact, in a box, right?

PALMER LUCKEY: Yes. I had to put it into a box so that I could send it through the postal service and get it to him, because, well, that's the best way to move physical goods.

THE FOLLOWING DAY . . .

SECURITY OFFICER: All rise for the jury.

JUDGE KINKEADE: I'm so glad y'all made it back. Y'all be seated. [*noticing something in the lawyer's hands*] You have a new box . . . that's a different box.

LAWYER: In fact I do, Your Honor.

JUDGE KINKEADE: Oh, wow. Okay. Go ahead.

LAWYER: Let's start right there. Good morning, Mr. Luckey.

PALMER LUCKEY: Good morning.

LAWYER: Is this the right box?

PALMER LUCKEY: Yes. That's the same type of box that I used to ship my Rift prototype to Mr. Carmack.

LAWYER: Very good. So everything that you shipped to Mr. Carmack fit in this box, correct?

PALMER LUCKEY: That's correct.

LAWYER: Outstanding. Glad we could get that cleared up.

While moments like those served to lighten the mood, they paled in comparison to the testimony from ZeniMax's damages expert, Daniel Jackson. In fact, the testimony seemed so absurd at times, that it felt like it had been lifted from a poorly written sitcom . . .

INT. UNITED STATES DISTRICT COURTHOUSE (NORTHERN DISTRICT OF TEXAS)-DAY

MS. WILKINSON, a lawyer for Oculus, questions DANIEL JACKSON, the expert witness hired by ZeniMax to calculate damages if the jury were to rule against Oculus.

MS. WILKINSON

You do this for a full-time living, right? You're an expert witness, and you work with lawyers on litigation all the time?

DANIEL JACKSON
Well, I do do some other things, but it is a big piece of my
work, yes.
MS. WILKINSON
And you charge $595 an hour?
DANIEL JACKSON
Actually, January 1, it went to $695.
   [Ooohs and Ahhhs from the studio audience]
MS. WILKINSON
Oh, congratulations.
   [Mild laughter]

The amount of money that Jackson had made by simply being
in court over the past ten days ($50,000+) was sobering. As was the
amount that Jackson's firm had already billed ZeniMax in the lead up
to this trial ($650,000).

Although it seemed a little comical that so much money would be
spent to assess potential damages, the true comedy would come from
actual assessment itself.

DANIEL JACKSON

There are two main components that are required in order to
make a VR experience happen. You have got to have hardware
and you have got to have software. So we divide it between
hardware and software.

Logically, this made sense—a virtual reality experience requires
both hardware and software. Except, as it turned out, Jackson was not
just speaking colloquially—he meant that literally: that, when assessing
damages, a simple 50/50 split between hardware and software would
do the trick. Then using this 50/50 approach, Jackson gave ZeniMax
all the credit for the "software"; and—after breaking down "hardware"
into four categories: display, optics, ergonomics, sensor—he awarded
ZeniMax half the credit for hardware (because Carmack had added a
Hillcrest tracker ["the sensor"] and a pair of ski goggles ["ergonomics"])

DANIEL JACKSON

So . . . if 50 percent is hardware and 50 percent of the hardware is related to ZeniMax property, then 25 percent of the total is related to ZeniMax, and then all of the software, 50 percent of the software, which would give us an apportionment of 75 percent of the total value would be associated in some way with ZeniMax.

[Cue the laugh track]

DANIEL JACKSON

So if you take the $2 billion [that Facebook paid for Oculus] and you say, okay, we're going to do 75 apportionment to ZeniMax, 25 percent to Luckey and other contributors, then you would come to a one-and-half-billion-dollar measure of damages.

MS. WILKINSON

. . . You're not a technology expert, right?

DANIEL JACKSON

That is correct.

MS. WILKINSON

You don't know anything about the specific trade secrets. If I asked you to explain chromatic aberration in some kind of real technical detail, could you do it?

DANIEL JACKSON

Absolutely not.

MS. WILKINSON

Okay. So all of your conclusions that you came to in this case are based on talking to Plaintiffs' other paid experts about technical issues, right?

DANIEL JACKSON

Correct. That's not my bailiwick.

[Laughter]

MS. WILKINSON

. . . No offense, you're a CPA, but the math you're doing here is not fancy, is it? You start with $2 billion, right?

DANIEL JACKSON

Yes.

MS. WILKINSON

And you say the other expert told me that hardware and software are half, they're both equally important, right?

DANIEL JACKSON

They are equally important.

MS. WILKINSON

So what you did was you went down and said, okay, then I cut that in half and that means 1 billion is for hardware . . .

DANIEL JACKSON

Correct.

MS. WILKINSON

. . . and the other billion is for software?

DANIEL JACKSON

Correct.

MS. WILKINSON

And then you had some other expert tell you that the hardware is divided into four areas, right? Buckets?

DANIEL JACKSON

Yes.

[More laughter]

MS. WILKINSON

. . . all right. So you have no idea whether the sensor that they used, that they built themselves, that Oculus did, has anything to do with the Hillcrest sensor that Mr. Carmack used on the prototype back in June of 2012, do you?

DANIEL JACKSON

No, I do not. In fact, probably it does not, other than Hillcrest having assisted them in designing a sensor that did more what they wanted. So if you take an intellectual property contributed by someone, use it as a basis to design around it, you're still continuing the use of the intellectual property to get your product.

MS. WILKINSON

That's what you think, right?

DANIEL JACKSON

My understanding, that's correct.

MS. WILKINSON

But all that understanding isn't based on any technical knowledge you have?

DANIEL JACKSON

No.

MS. WILKINSON

And, in fact, when you say ergonomics for $250 million, exactly what is it that ZeniMax did in terms of the strap and making it comfortable that's worth $250 million?

DANIEL JACKSON

Based on the purchase price, if you allocate the hardware to the four categories, that's what it mathematically works out to.

MS. WILKINSON

Sir, doesn't that sound a little crazy to you?

DANIEL JACKSON

I'm sure a lot of people think $2 billion for a company that's never sold a real product sounds kind of crazy.

[Raucous laughter]

MS. WILKINSON

So you can't tell us what each individual trade secret is worth, right?

DANIEL JACKSON

I am *not* telling you that. That's correct.

MS. WILKINSON

Right. You don't have the ability to do that, do you?

DANIEL JACKSON

. . . I wouldn't agree with that . . . I was not asked to do it based upon the description the technical experts gave of what it took to do virtual reality.

MS. WILKINSON

Are you saying that if they had asked you to assess each trade secret and value that you could have done it?

DANIEL JACKSON

No. You said I didn't have the ability. I do valuation. I do valuation of intellectual property all the time. I have the ability. Now, I haven't been asked to do it, and I haven't done it, but I have the ability . . . I made the decision that as a bundle they should have been valued, not independently.

MS. WILKINSON

[But] you have no basis for saying that they are intermingled in all of the same value since you're not a technical expert, correct?

DANIEL JACKSON

The only basis I have is technical experts, ma'am. Not me.

MS. WILKINSON

That's all the questions I have, Your Honor.

[Laughter and applause]

Jackson also provided some additional methods to quantify the damages. But these methods were as, uh, sophisticated as his original $250-million-for-a-strap calculations. For example, instead of using a two-way split between hardware and software, another option he laid out was to do a three-way split between tracking, rendering, and display. So "If you accept Tracking and Rendering being ZeniMax contributions," Jackson explained, "Display being zero ZeniMax contribution, you would then get a two-thirds apportionment of the 2 billion, which would say 67 percent. Or $1.33 billion." Another method—an alternative to the aforementioned lump sum options—would be to assess a "reasonable ongoing royalty."

And what, exactly, might be a reasonable rate? "Twenty percent for at least 10 years," Jackson proposed. And why 20 percent? That was based on a "hypothetical negotiation" that Jackson imagined could have conceivably played out between Oculus and ZeniMax in 2012. "[This] never happened, never will," he explained, "but we've got to assume this did." Okay. And if we assume a 20 percent royalty had been negotiated, what about the time frame? Where did "at least 10 years" come from? For that there was no answer.

THE FANTASY VS. THE DOCUMENTARY

"One of these pictures is cattywampus," Federal District Judge Ed Kinkeade exclaimed on the morning of January 10, pointing to a photo that ZeniMax was trying to get just moments before the jurors would be sworn in.

The photo in question—a grainy, previously undisclosed security cam image—depicted John Carmack exiting id Software during a summer night in 2013. And though there appeared to be nothing unusual or spectacular about this grainy image, ZeniMax saw things differently: they believed that the image showed a bulge in Carmack's pocket, and that this bulge could contain some sort of trade secrets.

As Judge Kinkeade and ZeniMax's Sammi haggled over whether to admit the cattywampus photo, Luckey couldn't help but think that this image was a metaphor for ZeniMax's case. Their entire case, in his opinion, was essentially a conspiracy theory. At every turn, since they first brought this case three years ago, ZeniMax consistently took the most ordinary things, stripped them of context and reason, and then inserted outrageous alternative explanations that seemed like they had been cribbed from a spy novel. That's how this photo of Carmack exiting a building became evidence of a scheme to steal "trade secrets." Or how a Fourth of July demo for Iribe, Mitchell, and Antonov—a demo that Luckey had talked about in interviews for years—turned into a "secret meeting" of plotting thieves. In this respect, it felt almost fitting that the trial was happening just minutes away from Dealey Plaza where, long ago, President John F. Kennedy had been assassinated.

In a way, Luckey almost admired ZeniMax's creativity. But as much as Luckey loved a good conspiracy theory, he was looking forward to this one being debunked.

"I like to use a moviemaker analogy," explained Beth Wilkinson—Oculus' lead attorney—moments into her opening statement. "There is a fantasy . . . and a documentary. Luckily for you, in this case you are going to see a documentary. Because we have all the evidence. It all exists: what these people were thinking, what these people were doing. It's on tape, it's on YouTube, it's in emails."

As an example, Wilkinson pointed to something her opposing counsel had presented in his opening argument. "To make this story sell—this fantasy that we actually stole their underlying testbed (not their VR engine as they call it)—counsel showed you only a portion of some texts . . . these are the actual messages. And they only showed you these first six lines. Okay? And they made it sound like if we can get an endorsement, we will leave the testbed . . . [but a few lines later]

Mr. Iribe says, 'Just tell them we need to get approval from Carmack to leave the software, so we'd rather come back and just do the demo when Gabe is here.' They got all these messages. They wanted you to think that he actually was willing and did leave it . . . That's why you need to see all the evidence. That's why you need to see the documentary and not listen to the fantasy that Plaintiffs are trying to sell you."

Situations like that text message example popped up several times throughout the trial. The most amusing of which occurred during the cross-examination of Robert Altman—in response to his comments about ZeniMax's history with VR. "We had actually looked at VR technology for a long time," Altman had stated. "Going back into the '90s. In fact, there was a photograph of the fellow who is our vice president of game development on the cover of PC Gamer Magazine wearing one of those rigs back in the '90s."

Well, as it turned out, that vice president of game development—Todd Vaughn—had been a deputy editor at *PC Gamer*. And yet, even with this new information brought to light, Robert Altman stood firm—saying, "I stand by my testimony."

As Luckey looked on, listening to the testimony of this man who claimed that Oculus' origin story was but a "myth"; and that he had "no training, no expertise, no ability to create VR"; he realized that Robert Altman *was* right about one thing: the magic that made Oculus—the "secret sauce" as Altman called—*did* begin before that 2012 E3. But to capture the full story, Altman needed to go back in time even further than *Doom 3: BFG* and the infamous *Rage* testbed.

The right place to begin was 2009—back to a time when Luckey really *did* lack any VR expertise. To describe how he got from utter ignorance to the prototype he shipped Carmack, it all sounds so simple; as if the breakthroughs that made the Rift special were merely yes-or-no options. Should this headset have a wide field of view? Yes. Well okay, then! Should the headset be lightweight? Yes, yes, yes. And what about cost? Should it be high-cost or low-cost? Well, people do tend to prefer cheaper things . . . so let's go with that option!

These reflections about the early days paired nicely with something Zuckerberg said during the trial that really resonated with Luckey: "This team . . ." Zuckerberg said at one point during the trial, "they really care about virtual reality. This is going to be, like, their life's work. When you look back 10, 20 years from now, I think that's going to be the thing that they are most proud of in their lives is they built this experience and contributed to this."

To Luckey, he believed these words with every fiber of his being. And he couldn't wait for this trial to end and finally get back to work.

CHAPTER 47

THE VERDICT

January/February 2017

"SEND A MESSAGE," ANTHONY SAMMI IMPLORED THE JURY, WRAPPING UP HIS closing argument. "It's not right. Facebook is a $350 billion company. It is an elephant in the room. It is a 900-pound gorilla. It doesn't care. Make it care. You can't do this. There is right and there is wrong. Please make it right."

Judge Kinkeade nodded to Sammi and then turned to the jury. "Ladies and gentlemen," he began, "you are the sole and exclusive judges of the facts. You should determine these facts without any bias, prejudice, sympathy, fear, or favor, and this determination should be made from a fair consideration of all the evidence that you have seen and heard in this trial."

Over the next few days, Judge Kinkeade instructed, the jury would need to reach a unanimous decision on fifty-eight different questions. Questions like:

- Did ZeniMax prove, by a preponderance of the evidence, that any Defendant misappropriated the trade secrets claimed by ZeniMax and id Software?
- Did Palmer Luckey fail to comply with the nondisclosure agreement?
- Did any of the following Defendants contributorily infringe upon any of ZeniMax or id Software's copyrights?

If the jury answered yes to any of the questions like these, then it would be up to them to come up with a monetary amount to compensate ZeniMax for the damages.

"Y'all have already worked awful hard," Judge Kinkeade finished by saying, before sending the jurors off to deliberate. "I know you're tired. I can look at you and tell. Your eyes are all red and everything, kind of like mine. But . . . be prepared to come back Monday and work hard. I'll be back then rested and ready, and so will you. I will see you back then."

"I SHOULD BE BACK NEXT WEEK!" LUCKEY EXCITEDLY SAID, SPEAKING WITH Edelmann from back in his hotel room. "Depends on how long the jury takes, obviously. But the point is I'll be back soon. It'll be good to get back to normal."

Edelmann couldn't tell if by "normal," Luckey meant like how things were before the trial or if they were going to finally let him go back to work. But at the moment, he sounded exhausted, so she didn't want to press the issue. Besides, she was dying to get some details about how things had gone in Dallas. "If you're allowed to talk about it," she said. "Or is it still, you know, supersecret and whatever?"

"I can speak about most things now," Luckey told Edelmann. "We rested our case. It's in deliberations now, so I should be in the clear. And let's see . . . where should I begin? Well, first of all, the judge is awesome. He's a very strong-willed judge. He's a Texan. He's a really cool guy. He tells lots of fun stories and anecdotes the whole time. I want to hang out with him with all this is said and done."

"That's amazing!"

"And he does something called 'sidebar discussion.' Basically the side of his desk is the sidebar and—most judges don't do this, they're very hands off when it comes to the attorneys—but almost every day he would invite our lead counsel and their lead counsel up to the sidebar for twenty or thirty minutes to talk with him about everything going on. And he's just brutally honest with them. Oh! And one time . . . so the attorney for ZeniMax is this guy Tony Sammi, and he'd sometimes snap his fingers at us when we were testifying and so during one of the sidebar discussions, the judge told him: "You know how many people have snapped their fingers at a witness in my court, Mr. Sammi? None! It's *never* happened. If you do it again, I'm going to shoot your hand off."

Edelmann cracked up.

"And then he did it again!"

"No way. What did the judge do?"

"He got *super*pissed."

Edelmann continued laughing.

"Oh! You're gonna love this," Luckey said, moving on to the sitcom-worthy performance of ZeniMax's damages expert. "the things he said, they're just laughable and unbelievable. I mean, the way came up with the value is so absurd. Like one way, he literally just split the acquisition price into four equal parts. He basically goes"—Luckey then used an over-the-top dopey voice to try and recapture the moment, "Well, Palmer Luckey said in his Kickstarter video that there are three things that make the Rift unique: low-latency head tracking, wide field of view, and an immersive stereoscopic 3-D display. And we made the software. And we did the head tracking and put a strap on the proto-type Palmer Luckey sent so . . . $1.5 billion, please."

"No way."

After recalling some of the greatest hits, Edelmann asked Luckey if he thought that they would win.

"You never know how juries are going to react," Luckey said. "Because they don't have all the background, you know? And a lot of this stuff is supertechnical. So they probably can't always tell if the outrage is righteous or not. So it's really hard to predict a jury verdict, but I think we're gonna come out ahead. I think we're going to win."

Luckey's optimism, however, would soon be dampened by a phone call from Mitchell.

"Bad news," Mitchell said. "I'm calling to let you know that we're restructuring and you're not going to have an executive assistant anymore."

"Um, okay," Luckey replied, his chest suddenly feeling heavy. "Are you still going to have an assistant?"

"Yeah."

Heavier, heavier.

". . . but I'm gonna have to share with somebody," Mitchell clarified.

"Wait," Luckey said. "What other people aren't going to have assistants that had an assistant before?"

"Uhhhh, I don't know."

That's bullshit, Luckey thought. But whatever. As long as I'm not being fi—

"Also," Mitchell continued, "I've got some more bad news: you're going to have to clean out your office."

"What?!" Luckey scoffed, so abruptly angry that the heaviness was now gone. "What the fuck, Nate?"

"It wasn't my decision! They told me they need more meeting rooms and they need to turn your office into a meeting room."

"That doesn't make any sense," Luckey replied. "When we were acquired, one of the only things that I asked for was that I be able to keep my office. It's part of my deal, Nate, that I get to have an office."

"Well, I don't know anything about that," Mitchell said. "You'd have to talk to Brendan about that."

"Nate . . ." Luckey began, trying to convince himself the timing was just a coincidence—that this would happen within *hours* of Facebook resting their case . . .

"I'm really sorry about all this."

"Nate," Luckey tried again. "Can you just tell me . . . am I about to get fucked?"

"No! Oh man, I'm sorry if I gave you that impression. It's all just restructuring stuff."

DAYS LATER, ON FEBRUARY 1, JUDGE KINKEADE ASKED THE JURY IF THEY HAD reached a verdict.

"Yes, Your Honor," the jury foreperson replied.

"All right. I'm going to go directly to the questions," Judge Kinkeade said, referring to the fifty-eight questions the jury had been tasked with answering. Did ZeniMax prove, by a preponderance of the evidence, that any Defendant misappropriated the trade secrets claimed by ZeniMax and id Software? "[reading] Answer 'yes' or 'no' for each Defendant."

The answer, the jury found, was no. They did not believe that

Oculus, Facebook, Luckey, Iribe, or Carmack were guilty of misappropriating trade secrets. This was huge. This was the big one. As expected there was quiet but palpable sense of relief from Luckey, Iribe, and Carmack.

"So we jump over to Question Number 7," Judge Kinkeade continued. "Did any of the following Defendants directly infringe upon any of ZeniMax or id Software's copyrights?"

Yes, the jury found. And did any of the named Defendants vicariously infringe upon any of ZeniMax or id Software's copyrights?

"Yes," Judge Kinkeade announced. Palmer Luckey and Brendan Iribe were both guilty of this, the jury believed. And, they believed, the amount of damages that ZeniMax and id Software suffered as a result of this infringement was $50 million.

Indignant, incredulous, Luckey's mind filled with fury. But that would just be the tip of the iceberg.

"Did Palmer Luckey fail to comply with the nondisclosure agreement?"

The answer to this, the jury found, was yes. And what sum of money, if paid now in cash, would fairly and reasonably compensate ZeniMax and id Software for that? $200 million.

Notably, the jury believed that due to ZeniMax's lengthy delay in filing charges, Luckey himself was not personally liable. He was absolved by the "doctrine of laches." Nevertheless, Oculus as a whole was still deemed liable, which meant that by midway through the questions, ZeniMax had already been awarded $250 million.

Ultimately, ZeniMax would be awarded a total of $500 million, with the additional sum coming from charges related to "false designation." This charge—of false allegation—had been previously defined by Judge Kinkeade to mean "any person who makes commercial use of any word, term, name, or symbol, or combination thereof that is likely to cause confusion as to that person's affiliation, connection, or association with another person, or that misrepresents in advertising the nature, characteristics, quality, or geographic origin of that person's goods or services." The jury believed that Oculus, Luckey, and Iribe were guilty of misrepresentation (liable for $50 million, $50 million, and $150 million, respectively).

To those on Oculus's side of the aisle, these figures were baseless and outrageous. "Particularly," Ms. Wilkinson explained, "Plaintiffs did not present any damages evidence for either of those claims." Even Carmack, who himself was not found liable for any damages, was appalled by the verdict, ZeniMax's tactics, and their allegations of "non-literal copying." Needless to say, all the defendants—Oculus and Facebook; Luckey, Iribe, and Carmack—left Dallas with severe misgivings.

Ultimately, Facebook would succeed in appealing the damages. On June 27, 2018—nearly eighteen months after the jury verdict—Judge Kinkeade would reduce the damages by half, to $250 million total, specifically eliminating the damages that went directly against Brendan Iribe and Palmer Luckey. In response to the 2018 ruling, Facebook vice president and deputy general counsel Paul Grewal would say, "a positive step toward a fair resolution, and we will be appealing the remaining claims."[1]

CHAPTER 48

THE SEEMINGLY IMPOSSIBLE CHALLENGE

February/March 2017

THE WEEK AFTER THE TRIAL—RIGHT BEFORE HE WAS SCHEDULED TO RETURN to Oculus—Luckey took Edelmann on vacation. To Disney World in Orlando, Florida, where he hoped to make up for their last attempted getaway and finally give Edelmann an uninterrupted vacation. That, however, would not be the case. Because shortly after sitting down to watch a Disney-themed musical performance at the America Garden Theatre in Epcot Center, Luckey got the call he'd been dreading for days.

"I need to take this," he said, excusing himself.

The call was from his lawyer, and it was to let him know that Facebook was planning to fire him and *not* pay any of the money he was scheduled to receive in the coming years. Which, due to how the acquisition had been structured (in a very backloaded way) was the majority of the money he'd receive for selling his company. Luckey was in disbelief. There had to be something that he could do, there *had* to be a way he could fight back. He told his lawyer that he cared much less about getting that money than he did about remaining with his company.

"Honestly," Luckey said, "I'd work there for free. I just want to go back to my company."

But by the end of the call, Luckey realized there was pretty much a zero chance of that happening. The only "fight" that remained was to try and get a *portion* of the money he was still due—which seemed possible, if Luckey would be willing to leave the VR industry and sign a non-disparagement agreement (so that, in perpetuity, he'd been unable to say anything bad about Oculus, Facebook or the employees at either entity).

It took a few moments for the reality of what he'd just been told to sink in, and then—in sharp contrast to the smiling strangers and upbeat Disney music—Luckey felt like he was going to puke. But after realizing that would accomplish nothing, he found a nearby bench and started jotting down a bunch of down stream-of-consciousness notes on his phone:

Could have settled, Luckey typed, writing up his thoughts and feelings. *Could have settled, knowing FB did not share my interests or want to keep me around as they claimed—no point in defending your abuser . . . instead, was team player and company man . . .*

Luckey wanted to scream, but stopped himself; as with the puking, there was no point.

Told to pack up office on the 31st? Did not even wait for a verdict to push me out the door. I am not stupid . . . Incentives totally misaligned . . . blew my ability to negotiate or defend myself . . .

This is the second time they are pulling this, can't just let them drag it out further. It is just not reasonable to expect me to be able to work effectively when you formally try to get me fired every few months with no cause . . . .

Shit or get off the pot. Fool me once, shame on you, fool me twice, shame on me . . . and yet . . . and yet . . . I just want to stay at my company and keep working on VR . . .

"Is everything okay?" Edelmann asked, reconnecting with Luckey after the show.

No-nope-never-again, he wanted to say. And soon he would tell her all of that and more. But looking at her—so kind, so pretty, so perfect—he didn't want to ruin her vacation. So he smiled, told her everything was okay, and asked her questions about the show he had missed.

FORTY FEET TALL AND FIFTY FEET WIDE: A PHOTO OF MARK ZUCKERBERG, HALF smiling, brightened a dimly lit auditorium. Beneath this image, dressed sharply in a black leather jacket and looking out from onstage, Caitlin Kalinowski addressed hundreds of attendees at the fourth annual Lesbians Who Tech Summit in San Francisco on February 24.

CAITLIN KALINOWSKI

Years ago, Mark Zuckerberg began inviting all Facebook employees to a weekly Q&A. Each Friday afternoon, Mark puts himself in the hot seat and gives all 16,000 employees of Facebook really honest answers to their toughest questions. When a CEO of a company this size signals that kind of transparency and trust, it makes me feel safe. Not just because I can surface questions at a Q&A and have them answered, but also because I feel like I can go directly to my leadership in my company if I have a question or even want to challenge what they're saying.

Now hitting a rhythm, and gaining momentum from an increasingly engaged audience, Kalinowski began to gently pace the stage.

CAITLIN KALINOWSKI

The world has changed since the election. It's been really unsettling. Knowing you can trust the company you work for—not only to hear your voice, but also to answer the tough questions you have and to ask tough questions back—is critical. Last week, in a letter to the public, it was Mark who surfaced the question that was on everybody's minds. He asked: Are we building the world we all want?

With each poignant point, the momentum in the auditorium grew.

CAITLIN KALINOWSKI

For me, and so many others, the answer is easily no.

And it grew and it grew—the symbiosis between speaker and listener becoming palpable—until . . .

CAITLIN KALINOWSKI

But we have to remind ourselves that while the administration does not reflect the consciousness of the majority of the country, it does represent a powerful minority with a really large platform . . .

. . . slowly and suddenly that momentum stopped growing . . .

CAITLIN KALINOWSKI

What's also shifted—and I don't know if you feel it too—is that the workplace and certainly the tech sector are becoming politicized to an extent I've never witnessed. As we've seen in the headlines, people are judging their leaders based not on their business acumen or talent, often, but instead on how they voted.

. . . and then it stopped completely . . .

CAITLIN KALINOWSKI

I worry that fear is driving the agendas right now instead of trust. And it's beginning to permeate our workplaces in ways that could compromise the integrity of what we're all doing.

It had now been several weeks since the ZeniMax trial ended, and still Palmer Luckey was at home. His office—scratch that, his *former* office—sat untouched. Certainly not a conference room.

CAITLIN KALINOWSKI

In an environment like this one, how do we lead? How do we continue to build the world that we all want? My first piece of advice on this is don't unfriend anyone.

Elsewhere, lawyers negotiated fiercely. Lawyers for Facebook, lawyers for Luckey; each wanting such different things.

CAITLIN KALINOWSKI

That may sound a little flippant, coming from a Facebook employee, but you really can lose valuable context and information when you shut people out.

For Facebook, they just wanted him gone. And to never pay him a single dime of what they still owed him for selling them his company. For Luckey, what he wanted was to go back to work. It wasn't about the money, he said, and they laughed at his obvious bluff. But then he offered to let them keep the money. Because, really, this was about the job, about going back to his company. Except there was a problem: it wasn't his company; and there was no longer any job.

CAITLIN KALINOWSKI

Research has shown, time and time again, that positive experiences with someone we fear, or dislike, or who say things that we fear, or who make us uncomfortable, can result in changes of opinion. I think that's why gay marriage is legal, in part; and that's why it was one of the most rapid civil rights wins in history. Everybody knows someone who is gay, almost everybody now. We all probably know somebody who's Republican too.

In the auditorium, the momentum was still gone. But there was always time for laughter; and so there was that bond for a moment.

CAITLIN KALINOWSKI

That wasn't supposed to be a funny line!

That only made it funnier. Even Kalinowski couldn't help but chuckle.

CAITLIN KALINOWSKI

There might be somebody in here. I think that's part of what I'm trying to say, to go off script for a second. It's an environment right now that feels so charged . . . that I'm really worried we're not actually listening to each other at all. I think we owe it to ourselves to sit with this discomfort a little bit more than we have been.

Meanwhile, as negotiations continued, Luckey tried to get in touch with Iribe. But he could not be reached.

CAITLIN KALINOWSKI

None of these great breakthroughs of communication come from taking the easy road. So my second piece that I want to talk about is Designing for Everybody. One of the greatest challenges of my career has been launching the Oculus Rift. We had to build one device that fits the 5th percentile female (in size) to the 95th percentile male. It's a huge difference. And we could only ship one device. It had to fit beautifully because the second you think "oh, my headset doesn't quite fit right," you're no longer at the edge of the skyscraper, or being chased by the dinosaur—you're thinking about your headset.

Luckey was able to reach Mitchell. Another one of his Oculus "brothers." But all Mitchell could say was that there was "no role" for Luckey at Oculus.

CAITLIN KALINOWSKI

To build the Rift, over 100 engineers, project managers, and designers spent over two years designing 300 parts . . . all designed seamlessly to fit one piece of hardware . . . in service of transporting you to another world.

No role? NO ROLE? Luckey had literally just spoken to a department head, who had told him that he'd love to have Luckey on his team.

CAITLIN KALINOWSKI

You *cannot* make product like that without an open design process and very strong feedback loops. Different people have to try the product. We have to get a lot of different opinions. Silos don't work. They kill the project.

Finally, Iribe returned one of Luckey's messages. He didn't have time to meet—because he had a flight later in the day—but he could do a call. Luckey agreed but came to campus anyway in the hopes of getting in a room with Iribe.

CAITLIN KALINOWSKI

The same rules apply to community. As people rally around the cleansing of their boards and teams in support of having environments of people who think like them—as our filters get stronger and more rigid—our work can become compromised. We have to think differently to design brilliantly.

Iribe was not in his office, but Luckey caught him leaving campus. What the fuck is going on? And Iribe said it had nothing to do with his politics. Bullshit! Not bullshit, Iribe explained, there was just no team that had a place for Luckey.

CAITLIN KALINOWSKI

So the third thing I want to say tonight: in my experience, if you're successful, be as successful as humanly possible. But if you're successful, you can make a big difference. Last year, I got up here and talked about how to turbo-charge your effectiveness. How to navigate office politics. I would never have imagined that this year I'd be up here talking about how to navigate actual politics. But my message to you is the same: get insanely focused on what you're good at. What strengths you. And do more of it. The bigger your success, the larger your platform. I want to own my mostly liberal gay views and listen to people who don't agree with them. I want to own the fact that I've been fighting for gay rights, reproductive rights, and equal opportunities for women and queers in tech, but acknowledge I haven't been good enough at fighting against socioeconomic disparity or racism. I want to keep designing the technology of the future. And I want the most talented people building it alongside me regardless of their politics. The more successful I become, the more I play to my strengths. The stronger my positioning is.

Luckey pointed out that if that were true—that no team had a place for him—then that would mean Iribe had no place for him on his team.

CAITLIN KALINOWSKI
With that comes responsibility. At Facebook, we value contributions from everybody, regardless of their gender or any other status.

That would mean that either Iribe was lying, or he did not want Luckey on his team.

CAITLIN KALINOWSKI
We do not tolerate harassment or discrimination. There's some practices and mindsets that don't belong in any office. Many of us here in this room are familiar with the types of lines being drawn over our identity, our sexuality, but never so strongly our politics . . .

It has to be one or the other. Either you're lying to me to cover for Facebook; or you're betraying me personally. Which one is it?

CAITLIN KALINOWSKI
We work in the best industry in the world. We are paid to reinvent, to design the new reality. That challenge may feel as if it's gotten harder over the last few months. People are feeling over-constrained, and under-inspired like we're facing the impossible. But isn't that every engineer's dream? The seemingly

impossible challenge. Design has always been messy and now it's going to get messier. It's going to continue to get messier.

So what did he say? Which answer did he pick?

CAITLIN KALINOWSKI
My first rule of prototyping has always been to begin with the gnarliest part. It's time to suit up, get back on our bikes, and head for the mud.

He said, "Gotta go. Time for my flight."

"SO I'VE GIVEN THIS A LOT OF THOUGHT," HAMMERSTEIN SAID, "AND I THINK the best way to describe Brendan is this: he seems like the kind of person that dogs wouldn't like."

Luckey cracked up, as he and Hammerstein covertly moved through the Facebook campus during an evening in February. By this point, he knew his days were numbered—and if he wasn't going to get to leave with the money he was owed, then he was making sure that at least he'd be leaving with the trophies that he had earned.

With this in mind, Luckey and Hammerstein entered Building 18 (Hammerstein's pass still worked) and the two of them made their way to the common space bar where years of Oculus relics adorned a smattering of drinks.

"Do you want to take them all?" Hammerstein asked.

"No," Luckey replied. "Just the ones with my name on them."

Which, as it turned out, was pretty much all of them anyway.

CHAPTER 49

EMPLOYEE NUMBER ONE

March 2017

WEEKS BEFORE HIS FINAL DAY, WHEN THE WRITING WAS ALREADY ON THE wall, Palmer Luckey called up his old friend Joe Chen.

"Joe!" Luckey boomed when Chen answered the phone. "Are you still working at Vrse?"

"Uh, yeah."

"Nope. Not anymore. You're gonna come help me start a new company."

"I am?"

"Yup!"

LATELY, CHRIS DYCUS HAD BEEN THINKING ABOUT A SIMILAR CONVERSATION he'd had with Luckey years earlier. The one where Luckey told him he should skip college and "come work with me." Given everything that had happened after that—the incredible ride that followed—that Skype conversation had always been one of Dycus's favorite memories.

Now, though, that memory was corrupted. It wasn't just the firing that bothered Dycus the most, but the fact that, for the past six months, no one at Oculus or Facebook had provided a single explanation about what was going on with Luckey, or why he had been fired. From the moment that Daily Beast article went live, there had been nothing but obfuscation, doublespeak, and question dodging from the management team at Oculus and Facebook. What kind of bullshit was that? Where was that famous Facebook "transparency" that we've all been hearing about for years? And then amazingly, even after all that, Oculus's founder gets canned and the only "explanation" provided is this little boilerplate bon voyage message from Brendan:

> I wanted to share an update on Palmer. We've been talking with him for the past few months about what's next, and the outcome of those discussions is that this will be Palmer's last week at Oculus.
>
> I've known Palmer for nearly five years and building Oculus with him, Nate and Michael has been one of the best experiences of my life. I'm proud to call him a co-founder. Palmer did so much more than create this company—he helped start a revolution and inspired the world to believe in VR.
>
> I want to thank Palmer for all he's done for Oculus and VR. I know I speak for many when I say he'll be dearly missed. His last official day is Friday, and he's happy to meet with folks 1:1. Please feel free to reach out.
>
> We're working with Palmer to share this news publicly at 1 p.m. today. I appreciate you keeping this confidential until then.
>
> Palmer, you're always going to be a friend of Oculus. Thank you.

That was the final straw. Feeling angry and insulted, Dycus wrote an impassioned post to everyone at Oculus:

> The entire situation surrounding Palmer's embarkment from the company has been the most poorly-handled, least-transparent internal affair that has ever transpired in the history of Oculus. You can believe me on that, because now that Palmer's gone, I am the longest-standing Oculus team member; I have been around for everything.
>
> This is NOT a post about Palmer's political associations, perceived or actual. This is a rant of my opinions on how Oculus handled the whole situation internally, in regards to communication with its own employees.
>
> We all know how in September last year, a story was written about Palmer that created an enormous PR nightmare both internally and externally to the company. Palmer soon after "went on vacation" for an

extended period of time. He came back for about one week before the Zenimax trial, then "went on vacation" again until he was fired.

Here is all the internal communication I could find regarding Palmer during that time:

October 14 Q&A: Info given—"Palmer is still out of the office. When he gets back, he'll be prepping for the Zenimax case. We're still figuring out what he'll do long-term."

November 17 Town Hall: Info given—"Palmer has asked to take more time off. When he gets back, he'll be prepping for the Zenimax case. According to HR, Palmer hasn't broken any big policies. When he comes back, he'll be working on the case."

December 4 Palmer issues an internal apology: Info given—"Media coverage of the event paints an incorrect picture of me and I don't believe in racism, bullying, hate speech, etc. I plan on working in VR/AR for the next 50 years."

February 10 Q&A: Info given—"Palmer won't be a public face of the company anymore. He's still working on legal cases. We're still figuring out what he'll do long-term."

March 30 Palmer has been fired: Info given—"Palmer has been let go. He'll be missed dearly."

That amounts, to me, to a whole lot of nothing said in the entire 6 months surrounding the issue. There was zero transparency here. Brendan's post telling us about Palmer's departure literally only 30 minutes before the news went live to the world does NOT count as transparency. There was no explanation for why Palmer was "on vacation" for months (why couldn't he be in the office while his fate was being decided?). I do not believe that Palmer willingly stayed away from Oculus and all his coworkers and friends, and I doubt preparing for trials required every single day of his time in those months. In addition, did the push to fire him come from Facebook or from Oculus? A combination? Did Oculus want him to stay around? Nobody knows.

In addition, Brendan's post went up one day before Palmer's official last day. Palmer was not present on campus on his last day. I've known Palmer for many, many years, and I guarantee you that was not his choice. He would have wanted to come say bye to the people he'd been working with for years.

Many teams have had going-away parties for leaving members—I've
been to a few just for the hardware team when interns leave. We had
going-away parties for interns but not for the FOUNDER of the entire
company? Are you kidding me? If Oculus values their founder so little, it
doesn't inspire a whole lot of faith in me that I'm valued very highly as
an employee.

This whole thing has been incredibly un-Oculus. Either our
transparency has greatly regressed, or the (lack of) transparency shown
here wasn't entirely Oculus' fault, in which case, I don't have a lot of
faith in the transparency of our parent company either.

For many at Oculus, Dycus's post perfectly voiced the frustration,
confusion, and uncertainty that they had been feeling for months.
Most of these people were sympathetic to Luckey; they felt that Luckey
had been treated unfairly, and that with Oculus in flux, his leadership
was desperately needed. But even among those who didn't care much
for Luckey—those who felt that Oculus had outgrown him, or that
his politics made him a cancer to the team—they, too, heard echoes
of themselves in Dycus's post. Because they, too, just wanted answers.

But that would never happen. No grand elaboration followed. Instead, just another short message from Iribe—a response to Dycus and
all those who felt the same—and then, just like that, the topic was put
to rest. Never to be addressed by the leadership at Oculus or Facebook
again.

The sentiment from this team is completely understood. Our policy is
to not discuss the specifics of any employee's departure. The reasons
employees are no longer at the company are personal, so we have
this policy out of respect for each person's privacy. Although this can
sometimes feel at odds with our core value of being open, it's important
that we apply this policy consistently, to everyone. That said, any
departure is hard for those of us left to carry on the mission, let alone
the departure of a co-founder who was a main inspiration to Oculus since
inception.

Anyone who is interested is very welcome to reach out to Palmer
to share how much he meant to them. I know he enjoys hearing from

everyone and is very proud of what we've created together and where we aim to go next.

Six weeks later, Dycus published another lengthy post. This time, to say good-bye:

I am sad to inform you all that this Friday, May 19th, will be my last day here at Oculus/Facebook. I've been presented with a new job opportunity that I just can't pass up. Unfortunately, I can't say what it is, other than it's a start-up in SoCal that is currently in stealth mode. Hopefully I'll be able to share more in the future!

I'll be spending the rest of the week documenting all my code, PCBs, CAD, etc. and wrapping up/passing off my current projects. If you've ever had me make anything for you and you think you might need another of that thing in the future, let me know so I can be sure to provide you with the files!

Why am I leaving Oculus? I would have gladly stayed at Oculus for at least a couple more years, but this new job came up and it really sounds like something I want to do. I miss the startup days at Oculus—some of my favorite memories of this company are from that time. Don't get me wrong, I absolutely still love Oculus and I enjoy the work I do here, but I want to go back to that fast-paced, everybody-does-everything sort of work that startups provide.

I've met and worked with tons of fantastic people over my years here; you all are awesome! I'd love to stay in touch, so please send a friend request to my main Facebook account, Christopher Dycus. This account (Chris Dycus) was created solely for work stuff when Oculus was bought by Facebook, and while I may check it occasionally, it will go largely unused. I will be moving out of the area back down to beautiful, sunny SoCal in the coming weeks, but I will be back to visit pretty often.

I hope to see you all again in the future and I wish Oculus the best of luck. It's been awesome working with everybody here. We brought VR back from the dead and created a whole new industry and gaming platform! I have lots of faith in Oculus and its people, and can't wait to see future products come out. I'll definitely be watching closely.

Like I said, my last day is this Friday, the 19th. I'm planning to stick around as long as I can, so I'll probably be available to chat if you'd like.

Though I am a little disappointed I won't make it to my 5 year Oculus anniversary—only 2.5 months away! Oh well. Somebody else will have to get that title first instead!

Chris Dycus, employee number 1, out.

The "stealth mode" start-up that Dycus was referring to here was Anduril: a new defense tech company founded by Trae Stephens, Matt Grimm, Brian Schimpf and—and course—Joe Chen and Palmer Luckey.

CHAPTER 50

HE'S BACK

April/May 2017

"I DON'T WANT TO BE ANOTHER EDUARDO," LUCKEY TOLD CHEN, REFERRING TO the famously ousted—and, in Silicon Valley, largely considered "forgotten"—Facebook cofounder Eduardo Saverin.

For a second—sitting in Luckey's home as they discussed the future of Anduril—Chen was slightly taken aback. This was such an uncharacteristically vulnerable thing for Luckey to say, but it kind of made sense that Luckey would bring this up now as the two of them were up late at night reviewing a pitch deck for their new company. When they went in to meet with these VCs, Luckey didn't want to be perceived as "another Eduardo."

"Dude," Chen said, shaking his head. "Never gonna happen. For one thing, Eduardo was just a money guy and you're, well, you're *you*; and two, just seeing how quickly things have come together over the last month . . . I mean, I didn't exactly know what I was getting in to, but it didn't take me long to realize: holy shit, it's happening . . . again!"

Luckey nodded. To prevent the conversation from getting too cozy, too vulnerable, he sped into talking about some new drone technology he'd heard about, the pros and cons of LIDAR technology, and "totally unrelated, but pretty awesome," Luckey told Chen about a trip that he and Nicole were going to take in May. An actual vacation, finally. To go attend an anime celebration in Tokushima, Japan, called "Machi Asobi."

"SO FACEBOOK FIRED YOU BECAUSE OF YOUR POLITICAL VIEWS?"

"I can't comment on that," Luckey replied to a twentysomething Japanese VR enthusiast. He was one of several—most in colorful cosplay—who had gathered around Luckey to ask him questions (or for his autograph) as they all waited for a panel at Machi Asobi to begin.

"Because NDA?" another one of the enthusiasts asked.

"I'm sorry," Luckey replied. "I can't say."

"That is bullshit!" a girl in the huddle said. Luckey, stiffly, did not react in any revealing manner—no sigh, no smile, no shrug—as those around him all vigorously nodded in agreement with the girl who had seemingly concluded that what had gone down at Facebook was bullshit. "*They* are bullshit."

Shortly after this, Yui Araki—a budding VR developer who helped manage the business of Japan's most well-known virtual reality evangelist, GOROman—came over to take Luckey and Edelmann to their seats for the panel they were attending. Araki, who had been friends with Luckey for a few years by this point, had spent the past few days serving as a sort of translator/chaperone for him and Edelmann while they were visiting for Machi Asobi, basically just making sure that Luckey and Edelmann caught all the nerdiest shit (and had the means to translate their geeking out if need be). As Araki pulled Luckey away from the huddle, she overheard several fans say to Luckey a few phrases that required no translation.

"You a great man!"

"You are the reason VR happened!"

"You will be very successful again!"

Hearing these comments, Araki was pleased. It reminded her of how "wrong" it had felt when Palmer was not at the previous year's Oculus Connect; and about how, at that event, she had felt exploited for her race and gender in a weird sort of American way that she noticed happening more and more. "I attended Women of VR," she wrote to the event-planners, "and was asked 'Can I take a photo?' by a man who is in black T shirt with no logo nor badge. He said nothing about what he was going to use my photo [for] or what organization he belong to . . . the next day, my photo was on the stage [at the Keynote]. I was very surprised. 'Using the photo of who you think is minority on the Keynote without your explanation and their permission' is the diversity you have? I don't think so . . . I think true diversity is to think every developer and consumer all over the world equally important and to respect every VR creator from indie, student to AAA title developer in any countries equally. I know you have the true diversity. Just show your diversity in Oculus Connect."

She ended her note by sharing displeasure about Luckey's absence. "I wanted to see Palmer in the Keynote. I wanted to see his vision about the VR in the future . . . When I met him for the first time, he was so happy to play my first VR game and gave me feedback honestly and said thanks so much to me for developing VR content because hardware is just a paperweight without content. He played my VR game so seriously, actually he is the best score holder! I was just a nonprogrammer-beginner-Asian-woman who started learning Unity three months ago. I think that's true diversity."

The panel that Araki, Luckey and Edelmann were attending that evening was a roundtable of content creators, featuring their mutual friend GOROman. Taking their seats in the audience, they were eager to see GOROman talk about his journey from creating *Mikulus* for DK1 to his current work, to the future of virtual reality. Seated beside GOROman—an unexpected panelist—was Seiji Mizushima, who had directed an anime film about virtual reality that Luckey had loved called *Expelled from Paradise*. What a nice surprise, seeing Mizushima up there, but then again everything over the past few days had been.

Flying out, Luckey hadn't really known what to expect. It had been almost nine months since he last attended any kind of expo, convention, or trade show. If you took away the ZeniMax trial, it had basically been nine months since he was in any sort of situation where he was likely to be recognized. And so, coming to Japan, Luckey wasn't sure what type of reception would await him.

Fortunately, Machi Asobi turned out to be everything he had hoped for. It was almost like he'd gone back in time to right before the Daily Beast started spreading lies. The people here, his fellow anime lovers, they couldn't have been nicer. They didn't care about his political opinions; nor did they look down on him like his exit from Oculus was an embarrassing fall from grace. All anyone who recognized him wanted to talk about was anime, virtual reality, or his "bold" choice of attire.

That bold choice—to cosplay as Quiet from *Metal Gear Solid V: The Phantom Pain*—had Luckey dressed in thick boots, torn nylon stockings and a skimpy black bikini top. And he was not alone, joined by Edelmann in matching attire.

"I am willing to admit," Luckey had told her, "that you look *slightly* better than I do."

Edelmann smiled and as fans snapped photos of her and Luckey—both then, and throughout the trip—she was reminded of something she had said to a friend six months prior: "I think I would have done just fine with an ordinary life: the kids, the mortgage, the white picket fence—I would have been fine with that. But because I'm able to do cool shit, I feel like I have to. For all the people who can't."

Luckey and Edelmann's matching outfits were apparently such a hit that it was covered by tech journalists back in the US. Though Luckey appreciated that these reporters found his love of cosplay to be newsworthy, he didn't much appreciate that just about everyone had still-uncorrected stories about him and Nimble America; and that, of course, those inaccurate stories got linked back to in these new ones. Fucking journalists.

Actually, that wasn't fair. There were still a handful of good journalists out there, Luckey believed, one of whom was a reporter for MoguraVR—a niche Japanese outlet—and with whom Luckey decided to do his first interview since leaving Oculus.

When speaking with MoguraVR, it was quickly apparent that Luckey could not discuss his exit from Facebook or go into much detail about the projects he had worked on while he was still there. But it was also clear—to the reporter, to his readers, and even to Edelmann—that Luckey truly loved taking a step back to talk about technology:

Palmer, you were influenced from the sci-fi novel Snow Crash if I recall right. Many people are thinking that as technology progresses we are getting closer to a sci-fi world. If you had to create an anime right now what would it be about? What kind of future do you imagine?

The truth is I actually have several ideas for a sci-fi anime. It's something I think about in my spare time but I don't know if I ever will realize them. One of them is about a future in which automation is common-place. All cars are self-driving and all the work is done by computers. Humans no longer know how to drive a car and have become unable to think with their own heads . . . The only exception is a group that is rebelling against this society. They can assemble, repair, and drive their own cars. They do not use any handy technology like computers and are thinking with their own heads, they are living freely without being swallowed up by the system.

Is this a dystopian future like the ones you see in The Matrix and The Terminator?

What I want to depict is a good future. It is not a dystopia where the machines are controlling everything. But you could say that it is a bit like a dystopia. I want to show how technology that is good for society can at the same time make society lose its greatest strength: individuality.

During that interview, Edelmann had seen a spark on Luckey's face that she hadn't seen for a long time. And she missed that; because she knew that *he* missed that. The good news was that every day, Luckey seemed to be feeling more and more like himself. Getting back to work was a big part of it; the long days spent building something out of nothing; the late nights spent plotting with old friends like Chen and Dycus. It was kind of like the early days of Oculus, again.

From where does your passion for VR come nowadays?

On an almost weekly basis I find content that makes me want to say: "VR is the future, I have to make the future of VR happen sooner." What I fear the most at the moment is that the speed at which VR is spreading at is not fast enough, and that people are going to lose their excitement for VR at this rate. [So] what motivates me right now is "How can we make VR more attractive for everyone," "How can we speed up the popularization of VR?" and "How can we help VR developers to succeed

as a business?" At the moment it is difficult for VR developers to sup-
port themselves just with developing VR content.

Getting back to work helped. Creating, modding, scrapping; those was all key to growing back what had been lost. Coming out to Machi Asobi was also key. The people were incredible, the costumes were insane, and it also didn't hurt that Luckey would be flying home with an extra suitcase to bring back all the trinkets and memorabilia he had acquired, an overindulgence that Luckey jokingly explained to Edelmann was just what happens when you carry around thousands of dollars in cash (which was something he did when traveling abroad). In addition to work and taking this trip, there was one other big thing, too: the internet. For as long as he could remember, the internet had been his home away from home. And after a long hiatus, Luckey had returned to the good, bad, and ugly of the internet one month earlier with a short, simple, six-letter message posted on Reddit: *im back.*

"Come up here, Palmer!" GOROman said, waving for his friend to join the panel.

For a moment, Luckey wasn't sure if he should go. Would that intrude on the panel? Will I mess up their flow since I don't speak Japanese? But GOROman continued to wave him forward, and soon the audience began cheering for Palmer Luckey to join them onstage.

Okay. Here we go. As Luckey began moving toward the stage he decided that he wasn't going to let the crowd down. So he started running—sprinting!—right on down the aisle. And then instead of taking the stairs and walking onstage—you know, like a normal person—he catapulted himself, belly flopping onto the stage and sliding underneath the panel table . . .

It has now been five years since I founded Oculus. It went by really
fast . . .

Luckey then popped up on the other side—to a round of applause, of course—then he took a seat beside the other panelists and weaved himself into the conversation.

Eventually, they started talking about the potential of using VR to create anime. In fact, this idea had been so appealing to Seiji Mizushima, the director Luckey admired, that Mizushima wanted to actually build a suite of animation tools.

"Palmer, can you use your superpowers to help make that happen?" a translator asked, relaying this question on behalf of the moderator.

"Well," Luckey replied. "I only have one superpower . . ."

He dramatically paused, so the translator knew not to wait. Then, as soon as the translator finished, fully delivering the setup to his punch line, he reached into his pocket. And with a defiant grin that told the world that he was really back, would always be back, could never be defeated, Palmer Luckey—the showman, the tinkerer, the kid who had founded Oculus—pulled out a fistful of cash from his pocket, somewhere in the neighborhood of a few thousand dollars, and then showered the *Expelled from Paradise* director with some of the money that he had made in paradise.

. . . I can't even believe that it really has been five years. Every day of my life was fulfilled. Looking back at it—at all that time—it really went by in a flash.

ACKNOWLEDGMENTS

THE PROCESS OF WRITING THIS BOOK WAS A TALE UNTO ITSELF. AND UNFORTU-nately, it wasn't always a very pleasant one. For that reason, I am particularly indebted to three people: Alex Glass (my agent), who always fought for me and the quality of what often seemed like a never-ending book; Julian Rosenberg (my manager), whose enthusiasm gave me the confidence to keep following my vision; and Matthew Daddona (my editor), who was brought onboard late in the game and then delivered a John Franco–like performance for the ages.

In addition to that trio, I'm also especially indebted to Dey Street and HarperCollins. Given how the scope of this book evolved, it would have been very easy to just say thanks-but-no-thanks. But you stuck with me every step of the way—always in service of putting out the best book possible—and then topped it all with incredible support from publicity and marketing, as well as a world-class effort from your tireless production team (to whom I owe a big apology for all those last-minute additions and tweaks).

Speaking of apologies: no one deserves a bigger one than my wife, Katie. I'm sorry—so genuinely sorry—for all the times I said the book was "almost done" and then turned out to be very wrong. I'm grateful for your patience, your love, and your friendship; and in so many ways, and for so many reasons, this book wouldn't exist without you.

Similarly, I'm grateful for the love and just-a-phone-call-away support of my parents, my brother, and my grandmother—each of whom bucked the trend this time around and *actually* read the chapters I shared with them in advance!—as well as the love, support, and family-style banter of AL, UC, JC, and HC; Uncle Brad, Aunt Erica, Tyler, and Amelie; Itchy Sander (aka "Super Kitty") and everyone in the DeLarber family who came through for Katie and me when we really, truly needed you most. Unfortunately, my favorite member of the DeLarber family (sorry, Uncle Tim) is no longer with us; but even though she won't ever get the chance to read this book, her relentless kindness forever impacted me; no doubt impacting the words in this book, as well as anything else I may write in my career. Thank you, Judy. Thank you so much.

I'd also like to thank all the friends and fellow writers who read material, provided feedback, or talked out story issues with me throughout the process of writing this book. Specifically: Gordon Bellamy, Josh Benedek, Nick Bilton, Jay Carlisle, Aaron Couch, Noah Davis, Grant DeSimone, Joe Durbin, David Ewalt, Billy Gallagher, Heather Gilchrist, Rick Gershman, Ash Heritage, Tom Kalinske, Dan Kim, Josh Kleinman, Ellen Beth Knapp, Jon Lajoie, JJ Maguire, Matty McFeely, Dave McGrath, Brian Nathanson, Josh Pincus, Cheryl Quiroz, Michael Quiroz (bonus points for first connecting me with Oculus!), Adam Raff, Nick Robalik, Alex Rubens, Jeff Ryan, Paul Scheer, Peter Sciretta, Eric Silver, Jordan Sternlieb, Jonah Tulis, Dave Waldman, and Brianna Wu.

Last, I want to thank those who lived this story. Not just for giving me great material to work with; and not just for giving me so much of your time and trust to help me tell this story; but—no matter the fate of virtual reality—I want to thank you all for the passion, patience, and sacrifice it took to try and resurrect a technology that was supposed to be long since dead.

NOTES

PROLOGUE

1. "Facebook to Acquire Oculus." Facebook, March 25, 2014.
2. "Oculus Joins Facebook." Oculus, March 25, 2014.

CHAPTER 1

1. Robertson, Adi and Michael Zelenko "VOICES FROM A VIRTUAL PAST: An Oral History of a Technology Whose Time Has Come Again." *The Verge*, August 25, 2014.
2. Heilig, Morton. "El Cine del Future." *Espacios,* January-June 1955.
3. Heilig, Morton. Sensorama Simulator. U.S. Patent 3,050,870 filed January 10, 1961, and issued August 28, 1962.
4. Sutherland, Ivan. "The Ultimate Display." Proceedings of the IFIP (International Federation for Information Processing) Congress, 1965.
5. Sutherland, Ivan. "A Head-Mounted Three Dimensional Display." Proceedings of the 1968 Fall Joint Computer Conference, 1968.
6. Mokey, Nick. "We Have Virtual Reality. What's Next Is Straight Out of 'The Matrix.'" *Digital Trends*, December 19, 2016.
7. Zachary, G. Pascal. "Artificial Reality: Computer Simulations May One Day Provide Surreal Experiences." *The Wall Street Journal,* January 23, 1990.
8. Rayl, A.J.S. "THE NEW, IMPROVED REALITY: Will the Ultimate Connection Between Humans and Computers Become the Ultimate Escape?" *The Los Angeles Times*, July 21, 1991.
9. Meyer, Kenny. "Thompson CSF to Purchase VPL Patents." *CyberEdge Journal* (Issue 19, Volume 4), January/February 1994.
10. Strassel, Kimberly A. "Success Has Been Virtual for Many VR Companies." *The Wall Street Journal,* March 13, 1997
11. Robertson, Adi. "VIRTUAL REALITY PANIC: Have We Learned Anything from the VR freakout That Happened over 20 Years Ago?" *The Verge*, June 20, 2014.
12. Rizzo, Albert, Mario Grimani, Arno Hartholt, Andrew Leeds, and Matt Liewer. "Virtual Reality Exposure Therapy for Combat-Related Posttraumatic Stress Disorder." *Computer* (Volume 47, Issue 7), July 22, 2014.

13. Aguinis, Herman, Christine A. Henle, and James C. Beaty, Jr. "Virtual Reality Technology: A New Tool for Personnel Selection." *International Journal of Selection and Assessment* (Volume 9), March/June 2001

14. Bolas, Mark, David M. Krum, J. Logan Olson, and Evan A. Suma. "A Design for a Smartphone-Based Head Mounted Display." 2011 IEEE Virtual Reality Conference, March 19–23, 2011.

15. Hamilton, Ian. "Oculus Team Building Immersive Gaming Goggles." *The Orange County Register*, December 2, 2012.

16. Luckey, Palmer. "Truly Immersive (AKA 'Holy Crap This Is Real') VR simulation" *MTBS3D*, September 25, 2011

17. Luckey, Palmer. "Oculus "Rift": An Open-Source HMD for Kickstarter." *MTBS3D*, April 15, 2012.

18. Luckey, Palmer. "First Post on Oculus!" *Oculusvr.com*, April 14, 2012.

CHAPTER 2

1. Onyett, Charles. "RAGE REVIEW: A Visually Stunning, Relatively Safe Shooter from id Software." *IGN*, October 3, 2011

2. Takahashi, Dean. "Review: id Software Grows Up with Next-Generation Shooting Game Rage." *Venture Beat*, October 3, 2011.

3. Schiesel, Seth. "Wasteland of Mutants and Thugs." *The New York Times*, October 11, 2011.

4. Kushner, David. *Masters of Doom: How Two Guys Created an Empire and Transformed Pop Culture*. New York: Random House, 2003.

5. Abrash, Michael. "Valve: How I Got Here, What It's Like, and What I'm Doing." *Ramblings in Valve Time*, April 13, 2012.

6. Carmack, John. "Parasites." *Slashdot* (Comments Section), June 2, 2005.

7. "Products: Personal 3D Viewer." *Sony.com*, September 23, 2011.

8. Tone. "LEEP on the Cheap." *VR-TIFACTS*, March 23, 2011.

9. Luckey, Palmer. "Oculus "Rift": An open-source HMD for Kickstarter." *MTBS3D*, April 15, 2012.

CHAPTER 3

1. Carmack, John. "A Day with an Oculus Rift." *MTBS3D*, May 17, 2012.

2. JamesB. "Why John Carmack's Rocket-Powered Goggles Won E3." *PC Games*, June 12, 2012.

CHAPTER 4

1. Pedriana, Paul. "High Performance Game Programming in C++." Presentation at 1998 Computer Game Developer's Conference, May 1998.

2. Dobson, Jason. "Product: Emergent, Crytek Sign Scaleform GFx SDK." *Gamasutra*, March 28, 2006.

3. Hollister, Sean. "Inside Gaikai: How to Make Cloud Gaming as Easy as Watching YouTube." *The Verge*, June 4, 2012.

4. Parrish, Kevin, "Gaikai, WikiPad Creating "World's First" Gaming Tablet: Gaikai Will Provide the Streaming Games for the 'World's First' Gaming Tablet, WikiPad." *Tom's Guide*, May 4, 2012.

5. Welsh, Oli. "John Carmack and the Virtual Reality Dream: VR Is Back—and It Works! We Chat to id's Mad Professor and Try His Amazing Homebrew Headset." *Eurogamer*, June 7, 2012.
6. Frum, Larry. "My 5 Favorite Highlights from E3." *CNN*, June 7, 2012.
7. Smith, Graham. "John Carmack Is Making a Virtual Reality Headset, $500 Kits Available Soon, Video Interview Inside." *PC Gamer*, June 6, 2012.

CHAPTER 5

1. Abrash, Michael. "Valve: How I Got Here, What It's Like, and What I'm Doing." *Ramblings in Valve Time*, April 13, 2012.

CHAPTER 6

1. Campbell, Colin. "So How Much Does It Cost to Develop for PlayStation 4?" *Polygon*, July 24, 2013.
2. Chiappini, Dan. "Sony slashes PS3 development kit cost." *GameSpot*, November 20, 2007.

CHAPTER 7

1. Carmack, John. "Re: Doom 3 in Rift on the Verge." *MTBS3D*, May 31, 2012.

CHAPTER 8

1. Parekh, Rupal. "Cause-Marketing Panel Calls on Ad Industry to Help with Darfur." *Ad Age*, September 25, 2007.

CHAPTER 9

1. Cacho, Gieson. "'Guitar' Hero: Danville Resident Was Instrumental in Designing Controller." *East Bay Times*, October 15, 2008.
2. Lewis, Peter. "Broadband Wonderland: Nearly Everyone in South Korea Has Internet Access That Puts Americans' to Shame." *Fortune*, September 20, 2004.
3. Yun, Kyounglim, Heejin Lee, and So-Hye Lim. "The Growth of Broadband Internet Connections in South Korea: Contributing Factors." The Asia/Pacific Research Center, Stanford University, September 2002.
4. Wei, Will. "What It's Like Inside a 'PC Bang' in South Korea." *Tech Insider*, October 18, 2015.
5. Tack, Daniel. "Studying Player Commitment to MMORPGs." *Forbes*, November, 2012.
6. Kosner, Anthony Wing. "OUYA $99 Open Source Gaming Console Blows Up on Kickstarter by Saying 'Hack Me, Please!'" *Forbes*, July 11, 2012.
7. Orland, Kyle. "$99 Ouya Wants to Bust Down Console Gaming's Walled Gardens." *Ars Technica*, July 10, 2012.
8. Totilo, Stephen. "$99 Ouya Console Hits $950,000 Kickstarter Goal After Just Eight Hours." *Kotaku Australia*, July 11, 2012.
9. Carmichael, Stephanie. "Ouya Surpasses Its Funding Goal on Kickstarter by $7.6M." *Venture Beat*, August 8, 2012.
10. Satterthwaite, Mark. "United to Create—Over the Edge Entertainment." *Inside Mac Games*, October 25, 2004.

11. Metz, Cade. "Unity—iPhone Code Swap Approved by Jobs (for Now)." *The Register*, September 3, 2010.
12. Meeks, Caitlyn. "Congratulations to Asset Store Top Seller, Brady Wright!" *Unity Blog*, April 27, 2011.
13. Kamio, "Interview: Super-seller Michael Lyashenko of Tasharen Entertainment." *Unity Blog*, April 7, 2012.

CHAPTER 10

1. Valve. "Handbook for New Employees." Valve Press, 2012.
2. Peterson, Andrea. "Gabe Newell on Valve's Intimate Relationship with Its Customers." *The Washington Post*, January 6, 2014.
3. Fahey, Mike. "Gabe Newell Just Made My Little Pony Fans Extremely Happy." *Kotaku*, April 5, 2012.
4. Plunkett, Luke. "Gabe Newell's Knife Collection Is Terrifyingly Impressive." *Kotaku*, February 8, 2012.
5. Kuchera, Ben. "Gabe Newell Reportedly Slams the PS3 and Asks for a Recall. WTF?" *Ars Technica*, January 16, 2007.
6. Hafer, T.J. "Gabe Newell: 'I think Windows 8 is a catastrophe for everyone in the PC space.'" *PC Gamer*, July 25, 2012.
7. Favre, Jean Pierre, Brigitte M. Jolles, Olivier Siegrist, and Kamiar Aminian "Quaternion-Based Fusion of Gyroscopes and Accelerometers to Improve 3D Angle Measurements." *Electronic Letters* (Volume 42, Issue 11), May 2006
8. Abrash, Michael. "Valve: How I Got Here, What It's Like, and What I'm Doing." *Ramblings in Valve Time*, April 13, 2012.

CHAPTER 11

1. Plunkett, Luke. "A Closer Look at the 'Woodstock' of Video Games." *Kotaku*, September 27, 2012.
2. Crecente, Brian. "The Traveler's Guide to Quakecon." *Kotaku*, August 2, 2011.
3. Pinsof, Allistair. "This Is (Was) QuakeCon." *Destructoid*. August 10, 2011.
4. Luckey, Palmer, and Nate Mitchell. "Oculus Rift: Step into the Game." Kickstarter, August 1, 2012.
5. Sheridan, Conor. "VR Headset Oculus Rift Kickstarter Successful on First Day." *GameSpot*, August 1, 2012.
6. Orland, Kyle. "Oculus Rift Head-Mounted Display Finds Funding from Developers." *Ars Technica*, August 1, 2012.
7. "Oculus Rift Virtual Reality Headset Gets Kickstarter Cash." *BBC*, August 1, 2012.

CHAPTER 12

1. Boychuk, Ben. "Review: Chess with Friends for iPhone." *Macworld*, December 17, 2008.
2. Hodapp, Eli. "'Words with Friends'—Asynchronous Online Scrabble." *Touch Arcade*, October 14, 2011.
3. Jackson, Nicholas. "Wait, Zynga Paid How Much to Acquire Words with Friends?" *The Atlantic*, July 7, 2011.

4. Cutler, Kim-Mai. "Zynga Paid $53.3 Million in Cash and Stock for Newtoy, IPO Filing Shows." *AdWeek*, July 1, 2011.

CHAPTER 13

1. Gaudiosi, John. "Meteor Entertainment Exec Mark Long Explains the Virtual Reality of Oculus Rift and HAWKEN." *Forbes*, August 27, 2012.

CHAPTER 14

1. Patel, Nirav. "The Adjacent Reality Open Source Wireless Head/Hand Tracker." *MTBS3D*, July 19, 2012.
2. Patel, Nirav. "Snow Globe and the Adjacent Reality Tracker." *Eclecticc*, May 7, 2012.
3. Wingfield, Nick. "Virtual Reality Companies Look to Science Fiction for Their Next Play." *The New York Times*, February 16, 2016.
4. Terdiman, Daniel. "Leap Motion: 3D Hands-Free Motion Control, Unbound." *CNET*, May 20, 2012.
5. Pierce, David. "A Look Inside Leap Motion, the 3D Gesture Control That's Like Kinect on Steroids." *The Verge*, July 26, 2012.
6. Ngak, Chenda. "Leap Motion 3D Motion-Control Device Changes How We Interact with Computers." *CBS News*, May 21, 2012.
7. LaValle, Steve. *Planning Algorithms*. Cambridge: Cambridge University Press, 2006.
8. Luckey, Palmer. "I Am Palmer Luckey, Designer of the Oculus Rift—AMA!" Reddit, August 28, 2012.

CHAPTER 16

1. Poon, Timothy and Brian Crecente. "Oculus Lawsuit Testimony Details ZeniMax's Negotiations with VR Company." *Polygon*, January 19, 2017.
2. Biron, L. "Quality, Not Quantity, at I/ITSEC 2012." *DefenseNews*, December 2012.

CHAPTER 17

1. Uhrman, Julie. "Growing Up Geek: Julie Uhrman." *Engadget*, October 17, 2013.
2. Rogers, Bruce. "Julie Uhrman's OUYA Out to Disrupt $70 Billion Gaming Business." *Forbes*, December 3, 2013.
3. Klepek, Patrick. "Ouya Nabs Games from Airtight, Minority Media." *Giant Bomb*, February 28, 2013.
4. Robertson, Adi. "Ouya Announces 'Soul Fjord,' an Exclusive Disco-Infused Dungeon Crawler from 'Portal's' Kim Swift." *The Verge*, April 30, 2013.
5. Totilo, Stephen. "Ouya Will Be Getting Tim Schafer's Big Kickstarter Game. Other Consoles Won't, Ouya Boss Says." *Kotaku*, February 6, 2013.
6. Corriea, Alexa Ray. "TowerFall Creator Talks Story Modes, Inspiration, Pricing and 'Yomi.'" *Polygon*, July 19, 2013.
7. McElroy, Griffin. "'Human Element' Prequel Announced as First-Party Exclusive for Ouya." *Polygon*, July 19, 2012.

CHAPTER 18

1. Lang, Ben. "Former Activision Executive Laird Malamed Joins Oculus Inc." *Road to VR*, January 24, 2013.
2. Baker, Liana B. "Activision's Brainy Toys Take Over." *Reuters*, June 5, 2011.
3. Stevens, Jon. "The Rise and Fall of Guitar Hero." *Nintendojo*, September 16, 2014.
4. Ako, Jasmin. "Gaming Initiative Helps Students Win Funds for School and Charities." *Daily Trojan*, April 18, 2012.
5. Peterson, Steve. "Former Activision SVP joins Oculus VR." *GamesIndustry.Biz.* January 23, 2013.
6. Ingraham, Nathan. "Oculus Rift: Deep Inside the Immersive, Disorienting Virtual Reality Gaming Experience." *The Verge*, January 7, 2013.
7. Topolsky, Josh. "Exclusive: Valve Said to Be Working On 'Steam Box' Gaming Console with Partners, Could Announce at GDC." *The Verge*, March 2, 2012.

CHAPTER 29

1. Threewolfmtn. "So no way to confirm this, but my friend works in the same building as Oculus, and he ran into Mark Zuckerberg taking the elevator to Oculus' floor." Reddit. January 29, 2014.

CHAPTER 30

1. Albergotti, Reed, Douglas MacMillan and Evelyn M. Rusli. "Facebook to Pay $19 Billion for WhatsApp." *The Wall Street Journal*, February 19, 2014.

CHAPTER 34

1. Hoffman, Liz and Reed Albergotti. "Oculus, Facebook Face Challenge to Rights Over 'Rift'" *The Wall Street Journal*, May 1, 2014.

CHAPTER 37

1. Anderson, Mark. "Can tearjerker virtual reality movies tempt donors to give more aid?" *The Guardian*, December 31, 2015
2. Gantaya, Anoop. "Sakurai Works on Mushi King." *IGN*, December 15, 2005.
3. My Nintendo News Admin. "Super Smash Bros Creator Really Likes Oculus Rift." *My Nintendo News*, August 29, 2013.
4. Nakamura, Toshi. "Smash Bros. Creator Wants to Drive a Transparent Car." *Kotaku* (Australia). May 16, 2014.
5. Henry, Devin. "Clinton: Support and improve ethanol mandate." *The Hill*, May 28, 2015.
6. Watercutter, Angela. "The Most Important Movie of 2015 is a VR Cartoon about a Hedgehog." *WIRED*, July 28, 2015.
7. Levy, Steve. "How Steve Jobs Fleeced Carly Fiorina." *WIRED*, October 1, 2015.
8. Oculus. "Samsung Gear VR now available for pre-orders at $99." Oculus Blog. November 10, 2015.

CHAPTER 40

1. Aguilar, Mario. "Oculus Rift Review: This Shit Is Legit." *Gizmodo*, March 28, 2016.

2. Rubin, Peter. "Review: Oculus Rift." *WIRED*, March 28, 2016.

3. Robertson, Adi. "Oculus Rift Review." *The Verge*, March 28, 2016.

4. Halfacree, Gareth. "Total Recall's suit against Oculus VR, Luckey dismissed." *Bit-tech*, March 13, 2017.

5. King, Hope. "Oculus Rift totally messed up its launch." *CNN*, April 12, 2016.

6. Cannata, Jeff (@jeffcannata). "Order processed 12 mins after Oculus went on sale. My new estimated ship date: 5/16-5/26 . . ." Twitter. April 11, 2016.

7. Klepek, Patrick. "The Oculus Rift that I pre-ordered has been bumped from shipping in April to June." Twitter. April 11, 2016.

8. Kuchera, Ben. "Rift shipments delayed for months in botched launch (update)." *Polygon*, April 12, 2016.

CHAPTER 41

1. Machkovech, Sam. "Oculus reverses course, dumps its VR headset-checking DRM." *Ars Technicai,* June 24, 2016.

CHAPTER 44

1. Ng, David. "Gamergate Advocate Milo Yiannopoulos Blames Feminists for SXSW Debacle." *The Los Angeles Times*, October 29, 2015.

2. Collins, Ben, and Gideon Resnick. "Breitbart Editor Milo Yiannopoulos Takes $100,000 for Charity, Gives $0." *The Daily Beast*, August 19, 2016.

3. Collins, Ben, and Gideon Resnick. "Palmer Luckey: The Facebook Near-Billionaire Secretly Funding Trump's Meme Machine." *The Daily Beast*, September 22, 2016.

4. Dash, Anil (@anildash). "One reason every political hashtag on Twitter is filled with racist trolls? The founder of Oculus is funding them." Twitter. September 22, 2016, 6:15 PM PST.

5. Dash, Anil (@anildash). "This guy, @PalmerLuckey, put some of his billion FB dollars toward explicitly funding white supremacy. Peter Thiel approved the acquisition." Twitter. September 22, 2016, 6:18 PM PST.

6. Pyle, Hunter. "Can I Be Fired for My Political Beliefs or Activities in California?" Hunter Pyle Law. August 30, 2018.

7. Menegus, Bryan. "Palmer Luckey, Millionaire Founder of Oculus Rift, Loves Donald Trump and Dates a Gamergater." *Gizmodo*, September 23, 2016.

8. Cutler, Kim-Mai (@kimmaicutler). "@PalmerLuckey's girlfriend." Twitter. September 22, 2016, 8:11 PM PST.

9. Cava, Marco della, and Brett Molina. "Facebook Millionaire Luckey Aligns Himself with Alt-Right, But Only If You Squint" *USA TODAY*, September 26, 2016.

10. Zuckerberg, Mark. "Palmer Freeman Luckey: I am deeply sorry . . ." Facebook. September 23, 2016, 6:44 PM PST.

CHAPTER 45

1. Iribe, Brendan, Opening Keynote at "Oculus Connect 3" Conference, Oculus, October 5, 2017.

2. Zuckerberg, Mark. Opening Keynote at "Oculus Connect 3" Conference, Oculus, October 5, 2017.

3. Mendelsohn, Tom. "Oculus Rift Inventor Palmer Luckey Is Funding Trump's Racist Meme Machine: How Your Oculus Rift Is Secretly Funding Donald Trump's Racist Meme Wars." *Ars Technica*, September 23, 2016.

4. Wong, Julia Carrie. "Who Is Palmer Luckey, and Why Is He Funding Pro-Trump Trolls?" *The Guardian*, September 23, 2016.

5. Dewey, Caitlin. "The Three Types of Political Astroturfing You'll See in 2016." *The Washington Post*, September 26, 2016.

6. Wagstaff, Keith. "Oculus Co-founder Palmer Luckey Is Apparently Funding Anti-Hillary Memes." *Mashable*, September 23, 2016.

7. Pearlstein, Joanna. "Palmer Luckey Is the Worst, but He Doesn't Reflect Silicon Valley." *WIRED*, September 23, 2016.

8. Scruta Games (@ScrutaGames). "We've failed to find any evidence backing up the Daily Beast's claim that Luckey paid for hate speech. Only a lame billboard." Twitter, September 29, 2016, 1:20 PM PST.

9. Scruta Games (@ScrutaGames). "So we were misinformed about him financially backing hate speech, which was the issue we had." Twitter, September 29, 2016, 1:23 PM PST.

10. Scruta Games (@ScrutaGames). "Since there is no evidence of that so far, we will tentatively resume work on Touch support." Twitter, September 29, 2016, 1:23 PM PST.

CHAPTER 47

1. Korosec, Tom. "Facebook Payout in Oculus Copyright Spat Cut to $250 Million." *Bloomberg*, June 27, 2018.

ABOUT THE AUTHOR

Blake J. Harris is the bestselling author of *Console Wars: Sega, Nintendo, and the Battle that Defined a Generation*, which is currently being adapted for television by Legendary Entertainment, producers Seth Rogen and Evan Goldberg, and Scott Rudin. Harris has written for *ESPN,* IGN*, Fast Company,* /Film, and The AV Club, and appears regularly on Paul Scheer's *How Did This Get Made?* podcast (on which he interviews the biggest names responsible for the worst movies ever made). He lives in New York City with his wonderful wife and their stinky cat, Itchy.